The Natural History of the Present

Front cover drawing by Alee Corbalis

Back cover drawing by John Batki

ISBN: 1-4564-8863-5
ISBN-13: 9781456488635

The Natural History of the Present

Nick Woodin

2011

TO DAUI

Table of Contents

Introduction

Some people trace the state of our warming, polluted world to capitalism, with its ability to motivate human effort towards ever more wealthy material lives. Long before capitalism, agriculture greatly increased our effect on the natural world, largely through allowing the human population to increase by many times. The first humid breath of global warming most likely came from the clearance of forests for agriculture. However, fossil evidence suggests human hunters with spears caused a major wave of extinction of large animals 11,500 years ago in the Americas; and that humans using spears, throwing sticks and fire caused one 30-40,000 years before that in Australia. The fundamental problem with human influence on the natural world is not capitalism (state or corporate) but the human evolutionary imperative to multiply and thus to occupy every possible environment. Like any plant or animal, we remake that environment to make it more suitable for us. Digging sticks, fiber carrying bags, stone blades, fire, crops, the development of machines, the use of fossil fuels, the organizing and motivating abilities of the state, increased our abilities. How far we can go isn't clear. There may indeed be limits: we need space to lie down, to move around, and in which to grow food (even if the food is grown in stainless steel fermenters on nutrients derived from sewage). But we could build steel and glass cities above ground (the linear cities of Paolo Soleri), faced with solar cells, their size limited only by the materials, or excavate caverns many hundreds of yards or kilometers underground, like those of ancient Cappadocia or modern Norway (so employing geothermal heat), or build marine cities, powered by wind, or by the difference in temperatures between ocean currents, on floating platforms tethered to the bottom of the sea. None of this interests me here. What interests me are the possibilities of a modern life in a working natural world, surrounded by and supported by working ecosystems. Ideally, half or less of any landscape would be used by people for the production of commodities of economic value. Some equilibrium would exist between the tribe of humans, the tribe of mice, the tribe of herons, the tribe of salmon. People would accept restraints. We would eat salmon but we wouldn't eat all of them; and we wouldn't destroy their habitat.

While modern life is defined by capitalism, with its myth of an economic meritocracy, the practice of pure capitalism is a myth. It exists in no society. Life under it for most of the population would be nasty, brutish and short, as those who accumulate capital use it to protect their gains, and most of the population remains poor. As a writer has remarked, markets are a great way to organize economic activity but need adult supervision. The move from an agricultural to an industrial economy, which exacerbated the economic and social failures of capitalism in practice, led to the idea of the social welfare state, with free public education, universal health care, old age and unemployment insurance, some redistribution of wealth from the rich to the poor. Modern western societies vary in how far they go toward establishing such a state. Among those societies that go the farthest, such as the democracies of northern Europe, we find the richest, the healthiest, the best educated, and those with the most upwardly mobile populations.

Capitalism can also be modified to save the natural environment. A focus on the value of clean water helps. So do two legal doctrines: the Law of Nuisance, under which an individual may not use his property to cause legally recognized difficulties for his neighbor (consider the rain of metals from the atmosphere, or the warming planet, as well as plumes of industrial discharges in waterways); and the Public Trust Doctrine, which puts the use of certain landscapes, such as riverbanks and shorelines, along with their related benefits (such as swimmable waters and fisheries) under the protection of the state, which must maintain them for the common good. In practice, both these doctrines are difficult and expensive for individuals to use, since specific harms must be traced to specific causes. Some writers argue that the notions behind the Public Trust Doctrine form the basis for protecting ecosystem functions in general.

My motives in writing this book are unabashedly sentimental. I am attached to the green world in which, where I live, the ground freezes up in early November (better have your shovel in the shed), and the wood warblers return in a flood in May. My attachment is a misperception on my part, since this green world is recent (at least in its details), certainly no more than 6000 years old. It is also much less abundant than the one I remember from 50 years ago, when as a fifteen year old boy the robin chorus woke me up at dawn. The world for the last 10,000 years has been one of shifting baselines: my world is less abundant than the one my father saw in the 1920s, when there were still chestnuts in the eastern forest, and ducks filled the coastal bays; and his less abundant than my grandfather's in the 1880s, when deer were extinct in Vermont and Rhode Island, but shorebirds were still shot for market on the New England beaches. (An amateur ornithologist, as a young man my maternal grandfather amassed a collection of bird skins.) Without the heat-trapping gases of modern civilization, my world would be replaced by several hundred feet of ice within another few thousand years. It will now probably be replaced with a warmer and greener world. ('Probably' because the results of modifying climate—"poking the climate beast" in Wallace Broecker's phrase—are unpredictable.)

This book continues the debate between the cornucopians and the Malthusians. Malthus' original concern was with population (which increases geometrically) overtaking food supplies (which increase slowly, if at all—the great production of young animals, most of which don't survive to adulthood, is one of the arguments for evolution). But the use of fossil fuels buried Malthusian fears in a new world in which world population increased six times and economic output tens of times. More people became a lot more prosperous, if a substantial part of the population remained as poor as before. The cornucopians think the more people the merrier (they may admit some limits here); the more people, the more good ideas, the better things will be in the human world. The Malthusians look at the natural world—the declining richness of metal ores; the shrinking amount of fertile farmland per person; the tropical forests going up in smoke; the birds, butterflies and fish moving north; the rising, acidifying seas—all the result of more, and more prosperous, humans; and think we're lost. Over the last two centuries the human world has become more and more a virtual one supported by immense amounts of non-renewable energy. Is it sustainable? Who knows? After coal and oil, there's methane from undersea clathrates, terrestrial and oceanic uranium, deuterium (also from the sea), helium-3 from the moon. Renewable (solar) and geothermal energies are probably not sufficient at today's levels of population and energy use, but may be so with a smaller, less energy demanding society. What will happen? We'll see.

Part I
The Problem with Economics

Chapter 1
Capitalist Economics and Our World

The usefulness of market economics comes from its reduction of human activity to matters of efficiency and profit. Specialization has a long history in modern life: children are educated in school, not the home, reducing the influence of parents; work and life are separate activities; one does not grow one's food or sew one's clothes—farmers grow food and garment workers sew clothes. Entertainers provide entertainment. Specialization of economic and social roles goes along with economic development. Social roles, such as schooling and childcare, become economic professions, performed by teachers and nannies. Thus new jobs are created by parents working outside the home. One could argue that economic specialization makes economic development possible. Distinctions among work, entertainment, and schooling are obscure in peasant, or gathering and hunting cultures, where all or none of one's activities might be considered work, and few, if any, yielded a profit; but together they constituted a life: that whole life was what children learned. While economic development has made possible the continuing elaboration of modern life, economic development also requires that elaboration. Without increasing consumption, how can capital grow?

In early modern times in Europe, walking was replaced by riding in carriages, then carriages were replaced by trains, which were replaced by automobiles and airplanes. A long time ago people drank from streams, then from hand dug wells or public fountains, then had piped household water, then drinkable piped water; now people choose from a thousand kinds of carbonated drinks and bottled waters. All this costs the individual more: soda or apple juice costs more than tap water, or a sip from the local brook, which is anyway no longer drinkable because of the side effects of economic development. In New York City, maintaining a car costs 8 to 10 times more than taking public transportation, and despite the state of traffic in the city, where in 1960 cars moved at half the speed of horse cars in 1907, about half the households have one. (The number is close to 100% in the suburbs.) The additional costs per person mean that each person supports a higher level of economic activity, and that the cost of necessities has fallen relative to total income. Early twenty-first century Americans spend 8% of their income on food, while the figure for eighteenth century France was 90%. Much improvement has been recent: in turn-of-the-century America food, clothing and shelter accounted for 80% of a family's budget; in 2000 one-third. So more income is available for other things. Spending income on other things employs the people who provide them; as their incomes rise, they also spend more, and the general level of economic activity rises further—the capitalist gift of growth.

Economic growth has some counter-intuitive consequences. If you earn more than, say $6 an hour, it is probably not worth your while to grow your own food: carrots and cabbage are too cheap. Walking across the United States, a journey of 200-300 days, is much more expensive than flying. A one way ticket from New York to San Francisco costs $150. Food and drink for 200 days costs several times that. (Even biking, which takes 30 days, costs considerably more.) Partly, the direct costs of walking are so high because one can no longer drink the water one comes across, hunt or gather one's food, camp out beside the road, or sew new shoes from animal skins. (While one could probably live on road kill and curl up for the night in a culvert, clean drinking water would be a problem.) Capitalist development has made flying cheap, walking expensive and travel more polluting; walking after all generates no extra carbon dioxide, and flying somewhat more than driving (3-6 times more than taking the train; though not if one counts the construction and maintenance of stations and railbed). A Sioux or a Tewa would have set out on foot on a trading journey of some hundreds of miles to realize what he or she considered a profit (perhaps a change of scene was also a benefit); or a medieval Frenchman walk from Arras in northern France to Saint James of Campostella in Spain for the good of his soul. Along the road the pilgrim would have been given food and lodging by others for the good of their souls. To walk the length of France at that time took 40 days. It was somewhat faster by horse: 20-22 days from Flanders to Navarre in the fourteenth century. For capitalist traders the time such journeys took represented a cost. Some profited by it; the distances allowed bankers in Florence and Siena, such as the Medicis, to use the different rates of exchange in local currencies as a substitute for charging interest. Charging interest constituted usury, a sin. Periodic improvements in transport would give a tremendous boost to trade.

While capitalist development divides up the social and material world to create more scope for profit, scientific investigation divides up the natural world in order to understand it. Science cannot yet deal with whole views of the world, though some scientists are trying. So our world is made up of many pieces that don't fit together. We have many worlds, not one whole. Such disconnectedness is a rather recent development. Most pre-industrial conceptions of the world were wholes. The Sumerians, one of the earliest irrigation societies, lived in the valley of the Tigris and Euphrates rivers, in the flat, hot countryside of what is now Iraq. All agriculturists, but especially irrigators, have to be able to predict the seasons, and the Sumerians, like all irrigators, were astronomers. They mathematically interpreted (explained, in modern terms) the cyclical movements of the planets and stars. These predictable heavenly cycles were used to predict the cycles of earthly life—the time of the river floods, the time to plant, the time to clean the canals; the time to build new ones. The cycles of the heavens determined the cycles of human society, and the cyclic life of the society that of the man or woman in it, one life within another, microcosm within macrocosm. Like the stars, the people were fixed in their professions and social classes, the son of a farmer became a farmer, the son of a potter a potter; the society was one whole in its place under the night sky. While such ancient worlds are wholes, they have their limitations. Upward mobility is rare. Doubters are banished to other worlds. The Catholic Church came close to banishing Galileo to the world of the dead for teaching that the earth circled the sun. (It didn't help that he had a bad attitude toward the Church.) Tribal groups banish those whose behavior isn't socially acceptable. Banishment in this case usually means death from starvation or loneliness. Many have pointed out that the word for non-Greeks in Greek is barbarian. This is true in many languages, where the word for man is often the same as that for a member of the tribal or linguistic group. Fundamentalist Christians and Muslims banish nonbelievers to hell. In early modern Europe (1500s-1700s) the Church taught that God had ordained a station in life for everyone, and that you fulfilled your destiny by performing the duties of your station and were rewarded with heaven at the end. In such a world ambition was frowned upon, education frivolous, and the wealthy regarded as blessed. So wholeness has its price.

While the idea of an unfettered market economy is a modern sort of whole outlook, examples of the destructiveness of market economics are many. One is the abundance of abandoned downtowns throughout the United States. City downtowns, built for streetcars and foot traffic, were unadaptable to automobile transportation. Automobiles require parking space, which didn't exist downtown. Land was cheaper outside the city, so shops and investment moved to the suburbs. Another is the abundance of leaking mines. About 40% of rivers in the western United States are polluted by mine runoff. Reclamation of surface coal mines comes to 4% of production costs in a free market, and is therefore affordable, while reclamation of hard rock mines (most of those in the West) is not economically possible at current prices for metals. In either case clean water requires government intervention, since from a market point of view, reclamation costs are costs without returns. If rivers are to be clean, unpolluted water must be given a value Then there is the continuing failure of the Atlantic right whale population to recover, despite the end of hunting. The cause is probably high levels of anthropogenic nitrogen in coastal waters; these, together with overfishing, have shifted the plankton soup the whales feed on to mostly plants, rather than a mix of plants and animals, and put the whales on a vegetarian diet, which doesn't allow them to put on sufficient weight to breed. (They are also killed in collisions with ships and get entangled in fishing gear.)

There are the silty rivers of the American Middle West, once clear and full of game fish, waterfowl and turtles, now used for barge transport, hydroelectric power, sewage disposal, disposal of industrial effluent, and (paradoxically) water supply. One thing that reduces fish populations is the estrogen in sewage effluent. Estrogen interferes with their reproduction and development. Added to an Ontario lake in concentrations similar to those found near sewage outlets, estrogen caused a 2000% drop in the minnow population in two years. Holding sewage flow in tanks for two weeks rather than four days and shifting between tanks with aerobic and anaerobic conditions allows bacteria to remove more nitrogen and seems to break down chemicals like estrogen but is more expensive, not worth the expense if clean water, wildlife habitat, riverine wetlands and fish are valued at zero. Under the same agricultural region lie pools of ground water contaminated with fertilizer and pesticide, some with pools of perchloroethylene collecting below, and pools of motor oil and gasoline sitting on top. Such hydrocarbons don't mix well with water, though they have an affinity for the fats and oils in living things. The chloroethylenes are solvents, degreasers, dry cleaning fluids. Much ground and surface water and most processed foods are contaminated with them. Perchloroethylene interferes with the action of hormones, attaches to chromosomes, cripples immune systems, and overstimulates the activity of some enzymes. About half the population of the United States gets its drinking water from ground water. Ground water eventually becomes river water, and thus everybody else's drinking water, and the water in which fish and amphibians live and waterbirds swim. Bacteria will eventually break down most of these chemicals but not before they have accumulated in our bodies, and those of the other beings with which we share the planet (the process takes too long in nature).

There is acid precipitation, a result of the sulfur and nitrogen oxides produced by the combustion of fossil fuels, which has resulted in the dying trees, declining birds, acidified lakes and calcium leached soils of the higher elevations of the eastern United States, Central Europe and Scandinavia. About 70% of soils in the eastern United States are calcium poor. As the rain leaches more calcium out and the soils further acidify, aluminum ions are released from their oxidized, and harmless, state. Aluminum is the most abundant positive ion in the soil and once it is available, trees take it up instead of calcium, and transfer it to their crowns in an attempt to neutralize the acidic water condensing on their needles. The acidic water in the rain degrades chlorophyll, and if it forms trichloroacetic acid, acts as a defoliant. But aluminum is toxic to plants and animals and interferes with the trees' metabolism. Leaves lower in calcium from the calcium leached soils are less nutritious for grazing insects and the insects make less nutritious meals for birds, who must accumulate calcium for their eggshells. Nesting songbirds lay fewer eggs in acidified northern European woodlands. With less calcium available, soil animals such as earthworms, millipedes, pillbugs and snails, food for thrushes and other birds of the forest floor, decline. Such effects may explain the strong declines of nesting songbirds on the high plateaus of the eastern United States. The first snowmelt in the spring carries much of the winter's acid precipitation (the accumulated snow and rain) into streams, and mobilize the accumulated acids from the soil, killing young trout and essentially sterilizing the water.

Research increasingly implicates the human environment in the growing epidemics of asthma, autism, adult onset diabetes and attention deficit disorder. The increase in diabetes is probably caused by changes in diet, specifically by the rise in the use of corn syrup, cheaper than cane or beet sugar, to sweeten ever larger bottles of soft drinks, while the incidence of the other three disorders (asthma, autism, crazy kids) is likely influenced by the chemical soup in which we live. We are surrounded by man-made hydrocarbons of unknown toxicity. Virtually all the atrazine an herbicide used on cornfields in the Great Lakes Basin is still in the Great Lakes, as is most of the DDT that periodically wells up from their depths. Atrazine is detectable in middle western brooks and in spring rains over the northeastern United States. In very small doses (one thirtieth of the Environmental Protection Agency's standard for drinking water) atrazine makes male northern leopard frogs hermaphroditic. Atrazine is also an immune suppressant. So populations of northern leopard frogs in agricultural environments are deformed (deformed by parasites their immune systems can no longer handle) animals of uncertain sex; and their populations have plummeted. Atrazine also interferes with shell formation in mussels. Many chlorinated hydrocarbons are toxic at two levels, one high, and one very low, where their concentration approximates that of hormones in the blood. (The higher concentrations are apparently ignored by the body's endocrine system.) Similarly, very low levels of 2,4-D (one seventh of the Environmental Protection Agency's recommendation for drinking water) has the greatest effect in reducing fertility in mice.

There are inexplicable clusters of childhood leukemia and breast cancer. Some cancer clusters may really be inexplicable (so-called statistical artifacts common to small sample populations), and some increases in disease a matter of better diagnosis. Some may be the result of atmospheric circulation having concentrated radioactive fallout in certain places during the testing of nuclear weapons 40 years ago. (One argument for this is an apparent rise in the incidence of breast cancer in American women who were adolescents living downwind from 1957-1963). Two rather interesting explanations have been put forward to explain clusters of childhood leukemia. One suggests such clusters may be related to the ease of movement of modern people. Genetic studies indicate that many people in rural areas of Europe are descendants of people who have lived there for hundreds or thousands of years. If a nuclear powerplant or nuclear fuel reprocessing facility is built in such a place (remote places are favored and nuclear facilities are always suspect in terms of cancer related diseases), people move into the area for the jobs, from hundreds or thousands of miles away. Their newly born children are exposed to leukemia-causing viruses endemic to the area, for which their mothers—not from the area—have transmitted to them no immunity. In this case connections with radioactive hazards are incidental. This scenario is currently discredited but my point is that the hazards of development are not obvious.

A second explanation concerns clusters of leukemia in children who live near industrialized river estuaries in Britain. Such children have unusual levels of alpha radiation in their tissues and two to three times the normal rate of childhood leukemia. Large river estuaries like the Severn, surrounded by fossil fuel burning industry, by a large car and truck traffic, and much space and water heating, end up with a considerable load of hydrocarbons in the water. They fall into the water from the air, or wash in off the land. Hydrophobic, the hydrocarbons tend to float on the surface and are concentrated by tidal action at the interface of fresh and salt water. The chemicals are mobilized in wind driven spray, which is carried inland and inhaled by people, settles out on their clothes and migrates into their houses. The hydrocarbons contain uranium and radon as contaminants. (Radioactive materials are natural contaminants of oil and coal.) These materials or their radioactive daughters (radioactively unstable lead) are concentrated in bone and fat. The release of alpha particles during further radioactive breakdown irradiates developing blood cells in bone marrow, and is thought to cause the leukemias. Children, whose cells divide more rapidly, are more sensitive to ionizing radiation.

Economics is concerned with prices, not with intrinsic values, so things of little intrinsic use—diamonds, gold, Cabbage Patch dolls, Pet

Rocks—may be priced highly by the market. A cup of coffee costs the same as several apples, a vacation as much as a car. Production of goods of little intrinsic value, like that of necessities, also generates toxic materials and global warming gases. In general, scarcity sets prices. Social goods (just laws, free public education) become valuable only when they threaten market stability, and ecological ones (like clean water) only when they become scarce. Abundant resources (including labor in some markets, and timber, land and water in many developed societies until recently) are cheap. Disposal of the waste products of production into air or water is free.

The social problems of capitalist societies are more likely to be ameliorated than the biological ones. Social problems of capitalist, or market, societies include dying cities, whose industries have moved on; the working homeless, who don't earn enough for shelter; the non-working homeless, often mentally ill, who live on the street; the great spread of incomes in many capitalist or capitalist-socialist countries, which reduces growth and social mobility; the unaffordability of medical care. Biological problems include the upward curve of man-made carbon dioxide in the atmosphere, which is changing the climate, and the rising level of nitrogen in coastal ecosystems, which degrades them back to more primitive ecological structures: more algae and jellyfish, fewer oysters and striped bass. Capitalist systems are open but not inherently democratic: without state direction, capitalism inevitably produces inequality in incomes that is likely to undermine democratic systems of government.

To keep up with the present rate of global warming, plants and animals would have to move poleward at 30 feet a day. But as far as global warming goes, the colder, less populated regions have experienced it most. High latitudes respond more quickly to temperature changes because of the feedback effects of ice and snow. New snow reflects 80% of incoming sunlight, while water absorbs 90%, bare earth somewhat less, and thus unmelted snow leads to more snow and cooler temperatures, while less snow leads to bare ground, more absorption of heat, and warmer ones. In the modern Arctic, spring comes 10 days earlier, summers are warmer, sea ice is less abundant and thinner, permafrost is thawing, and the pulse of summer productivity on land is greater, than 20 years ago. Red squirrels in the Yukon give birth 18 days earlier than 10 years ago. This change may be partly genetic, that is, evolved change favoring earlier breeders, and not behavioral. In 1998 Arctic terns were laying eggs 18 days earlier than in 1929. Robins and barn swallows visit Banks Island in the Canadian Arctic, and the island has thunderstorms: birds and weather for which the Inuit have no words. Sea ice is melting. Annual fluctuations in the extent of Arctic sea ice correspond almost exactly with the length of the melting season, which has been increasing by 5 days per decade. The Arctic without sea ice will be warmer and wetter. Since the 1970s Alaskan mean temperatures have risen 5° Fahrenheit (F.) in summer, 10° F. in winter. Legal travel days for heavy equipment on the tundra of the Arctic Slope (requiring six inches of snow and ground frozen to a foot) have fallen from 200 days a year to 100. Bark beetles, now able to produce two generations a summer rather than one, and whose females lay many more eggs in warmer temperatures, have killed 95% of the spruce trees on Alaska's Kenai Peninsula. For a time, this was the largest insect infestation in North America. Now pine and spruce trees are dying throughout the forests of the west. Forests covering 150 million acres of the western United States and Canada have died from warming-related beetle infestations over the last ten years. As they burn or decay, the trees release more carbon dioxide to the atmosphere, creating a positive feedback. The situation is not only hard on seals and polar bears, which live on sea ice. Gray jays at the southern edge of their range in the Canadian province of Ontario are declining. Fall temperatures in this region have risen 5° F. in the last 30 years. The jays breed at the end of winter and in the autumn lay up a supply of perishable food (up to 50 pounds a bird: berries, mushrooms, insects, small mammals) for the winter and for the breeding season. In the warmer temperatures the stored foods rot, leaving the jays starving and in poor condition for breeding.

Being animals of the subtropics that carry our climate with us in clothes or buildings, we find life on the perennial polar ice threatening. But for many animals the pack ice means life. It lets mammals like seals and polar bears live over an open ocean rich in fish and squid. Algae live throughout the ice but are concentrated in the bottom half inch or so, which, saturated with seawater, is soft. The algae remain dormant during the lightless winters but begin to grow with the return of sunlight. Eventually the mass of algae in the ice exceed that in the water, and provide a concentrated and accessible food resource for krill, the basis of many food webs in both Arctic and Antarctic oceans, and also for nematodes (non-swimmers, but very abundant in some locations), ciliates, bacteria, rotifers, copepods. So the ice helps support the abundant avian and mammalian life (whales, penguins, bears, seals, arctic foxes, ravens, scavenging gulls) of Arctic and Antarctic waters. The colored algae, by absorbing the sun's heat, may speed the melting of pack ice in the spring. Melting ice seeds the water with algae, which reproduce in a bloom (marking the place of the recent ice); crustaceans and zooplankton feed on the algae, and fish on all of them.

Loss of the ice is likely to reduce the biotic abundance of the polar regions. The presence or absence of sea ice—by changing currents and the transport of cold water out of the polar oceans—also influences weather patterns thousands of miles away. The spring thawing of sea ice in the Antarctic influences rainfall on the soybean fields in southern Brazil, the anchovy harvest off Peru, Sahel winds. The disappearance of 20% of the pack ice about Antarctica since 1950 corresponds with a reduction of krill of 40% per decade since 1976 (when the disappearing ice was noticed), and a decline of emperor penguins of 50%, of Adelie penguins 70%, in the last 30 years. Ways of coping with ice vary: spectacled eiders winter in small openings in the Bering Sea, kept open by the warmth of their bodies (aggregations can reach tens of thousands of birds). They remain amidst the ice in order to feed on the rich invertebrate fauna of the sea bottom.

But the effects of warming are worldwide. In general, the North Temperate zone has earlier springs, later falls, and more intense rains than 40 or 50 years ago. Summer temperatures in the southern parts of the United States are now 4° F. higher than optimal for seed formation in grains. Increased storminess causes more frequent high tides in the Mediterranean. More frequent very high tides constitute much of Venice's flooding problem, along with a locally rising sea level; while the absolute level of the Mediterranean is falling because of less water inflow from its rivers. There is heavier wave action in the North Atlantic and more frequent and more violent storms in the tropics. Since 1980, 18 new species of warm water fish have been caught off the coast of Cornwall, England. Fish can swim anywhere in the ocean, and thus are a good indicator of changes in sea temperature; no new species had been caught off Cornwall between 1940 and 1980. Sea level rise has increased recently from about 1.8 mm per year to about 3 mm per year (about 10 inches per century), partly from thermal expansion of the warmer water, partly from the melting of glaciers and from increased continental drainage. Man made additions to continental drainage include land drainage for agriculture and construction and the pumping of ancient ground waters not renewed by rain. Melting of the Greenland ice cap would raise sea levels by about 23 feet, of the West Antarctic ice sheet by 16-20 feet; or 39-43 feet total. A 33 foot rise in sea level would inundate land with a population of one billion people and much of the world's most fertile farmland. Such events are becoming more likely. Ice is an excellent insulator and for a long time it was thought glaciers would take thousands of years to melt, since for half the effect of a rise in surface temperature to penetrate 3000 feet down into an ice sheet takes 7000 years, but summer melt water streams on the ice's surface drain down through cracks to the bottom of the ice sheet, melting ice as they go. The melt waters pool at the bottom, lubricate the mass, find an outlet, and help the glacier slide more rapidly towards the sea. Once floating in the sea, ice melts. A rise in sea level of 1-3 feet in the next 30-40 years no longer looks farfetched. (In 1991 the Corps of Engineers estimated that a rise of three feet would cost the United States $270-$475 billion; a 2009 estimate of global costs from a rise of half a meter—20 inches—came to $28 trillion.) Three million years ago, when the carbon dioxide level in the atmosphere was last at 350 ppm, the temperature was 2-3° C. higher than now and sea level was 75-80 feet higher. Getting there may take centuries; or it may not. Since the oceans are not perfectly connected, some places may see more sea level rise earlier. Melt water from the Greenland glacier will tend to stay in the North Atlantic for some time. For the first 50-100 years of melting, the sea level rise along Greenland and the east coast of North America would be 30 times as great as that in the Pacific, that along the European coast 6 times as great. And the strong circumferential currents about Antarctica may prevent Antarctic melt waters from reaching the rest of the world for centuries.

———

Ecological disasters eventually become human disasters. Inuit women who eat a traditional diet of fish, whale blubber, seal meat, birds, berries and caribou, produce breast milk with 10 times the load of chlorinated hydrocarbons (PCBs, DDT, and their metabolites and relatives) than women who live 1000 miles to the south. These materials are implicated in causing cancers and birth defects and suppressing immune systems. Heavy metals, such as mercury, in the milk are also high. Mercury causes impaired neurological development in children. Sources of the chemicals are probably global, but some of the chemicals in the Inuit near Hudson's Bay have been traced to industrial installations in the state of Alabama. Are these installations liable?

Perhaps Inuit women should no longer nurse their babies. Breast milk helps ward off diabetes and cancer in later life; its fatty acids boost brain growth and its antibodies prime the immune system, so breast-fed babies have fewer infections. Breast-feeding protects the mother from breast and ovarian cancer and reduces her level of stress. But the Inuit are not alone: 10-15% of American women of childbearing age have more mercury in their blood than the Environmental Protection Agency considers safe. The mercury is transferred through the placenta to the fetus, where it affects the development of the brain, kidneys and liver. Approximately 600,000 children a year in the United States are born with unsafe mercury levels. A quarter of all North American women have levels of toxic compounds in their breast milk that would make it unfit for human consumption if it were cow's milk or apple juice. These include anti-bacterial agents in cleaning compounds, pesticides, chemicals from detergents, from artificial musks in perfumes, from inks, paints, cosmetics and plastics. Most mothers don't know this. If such mothers nurse their babies, they purge their bodies of these materials, transmitting 20% of their body burden of fat-soluble chemicals to the (much smaller) child in six months. In a sense, nursing protects female mammals from chemicals in the environment. While the levels of chlorinated hydrocarbons in male seals grow throughout their lives, those in females level off once they reach reproductive age and start transferring the chemicals to their young. Such transfers (which do not require nursing) are already a serious problem in some fish. Small amounts of PCBs, transferred by a female eel to her eggs, are fatal to developing eel embryos; which probably explains the recent crash in European eel populations.

The chemical industry, like the coal-fired electric power industry, is a cornerstone of the modern world. Many industrial chemicals and metals accumulate in human fatty tissue. These include pesticides, metals from burning coal (such as mercury, cadmium and lead), fire retardants, additives to detergents and plastics, and musk fragrances in perfumes. They accumulate through diet or through exposure to them in air and water. One breathes in waterborn chemicals with the steam of the shower, or absorbs them directly through the skin. The fat of a middle-aged American male contains 177 detectable organochlorines. A small fraction of the population, perhaps 5%, accumulate metals more efficiently and so are more at risk. (Such people, for instance, are probably at risk from mercury amalgam fillings in their teeth.) In general, once exposure to bio-accumulating chemicals ceases, a person sheds half of his load of fat-stored material every seven years. A large fraction comes out in feces and thus remains in the environment, though bacteria will eventually break down most of the chemicals and immobilize the metals. In the Inuit (as with people in more temperate regions) their diet continues to concentrate these materials, and the metabolites of PCBs in their body fat increase 10 times from age 18 to age 66. Sooner or later, this will be expressed in disease. PCBs affect sexual development, among other things. Concentrations of PCBs comparable to those in human breast milk in industrialized countries turn the eggs of red-eared slider turtles from male to female. Polar bears in the Norwegian Arctic, where air pollution from Europe concentrates, develop both male and female genitals. Seals and dolphins in the North Atlantic basin, at the top of fishy food chains that accumulate the rain of chemicals falling on the North Atlantic (and washed into it through rivers) are dying of diseases their depressed immune systems can no longer handle. Several persistent organic pollutants, such as DDT, dioxins and their breakdown products, suppress the immune system. The population of beluga whales in the St. Lawrence estuary has fallen from 30,000 animals to about 30. Originally reduced by hunting, the population hasn't recovered. The remaining belugas are full of tumors and virtually infertile. Their bodies have among the highest recorded levels of toxic chemicals found in living organisms (a distinction shared with the salmon and seal eating killer whales off the northwest coast of the United States). Most of these chemicals are industrial and agricultural materials that have washed into the Great Lakes. The chemicals in the water are taken up by planktonic algae. The algae are eaten by zooplankton, which are eaten by small fish, and eventually concentrated in the larger fish and eels the whales eat. Two toxins found in very high concentrations in the beluga, the insecticides chlordane and toxaphene, have no history of use in the Great Lakes/St. Lawrence Basin. However, the St. Lawrence Basin occupies 500,000 square miles, so its waters concentrate airborne chemicals from a wide area. Chlordane and toxaphene probably arrived on the wind from the American South, where they were once used extensively.

One way or another, the environment is implicated in 80% of cancers. Some effects come through choice, as with smoking and diet; and some not, or at least not so obviously. An example of the nonobvious comes from epigenetic effects on genes. Epigenetic effects involve environmentally mediated changes in gene function that don't involve changes in the DNA sequence. Genes can be turned on or off through such things as a mother's diet, or a grandfather's smoking. Such changes are heritable and may skip a generation or continue for several. Epigenetic effects are thought to explain why fathers who started smoking before puberty have prepubertal sons who are heavier than normal; and why women whose grandmothers were short of food between the ages of 9 and 12 live longer. (Other such transgenerational effects include the effects of DES, a synthetic estrogen, taken by pregnant women on the reproductive organs of their daughters.)

Industrially produced chlorinated hydrocarbons are implicated in many cancers. Now widespread in the environment, chlorinated hydrocarbons are produced naturally by forest fires, volcanoes and some marine algae (traces of halogenated hydrocarbons have been found in the oil of whales killed before such compounds were manufactured), but most of them are produced by people. Some are harmful, some not. (Overall, the natural compounds seem to be less harmful.) Some may be involved in epigenetic effects. Many of the several thousand varieties of chlorinated hydrocarbons in the Great Lakes come from the chlorine bleaching of paper. Other chlorinated hydrocarbons include the dry cleaning chemicals and solvents dichloroethylene, trichloroethylene and perchloroethylene, which are found in about a third of American surface waters and also in about a third of American drinking water and in most American processed foods. Polynuclear aromatic hydrocarbons (PAHs) are found in the haze of tire dust and unburned gasoline that hangs over freeways in Los Angeles; PAHs are ubiquitous in lake sediments in North America and Europe; PAHs in the air downwind of coal burning power plants cause genetic mutations in mice and birth defects in birds (thus the crossed beaks and other abnormalities in the herring gulls of the lower Great Lakes). The nonylphenols used in detergents and in the manufacture of plastics act as hormone mimics; their effect on the developing brains of larval fish has resulted in the fish inhabiting some British rivers being overwhelmingly female; such so-called estrogenic chemicals are also implicated in breast cancers. A woman's risk of certain breast cancers is related to her lifetime exposure to estrogen. So the action of estrogen mimics may be direct; or through lowering the age of menarche and thus increasing the number of menstrual cycles a woman has during her lifetime. (The average age of first menstruation in American women has fallen from about 17 in 1900 to less than 12 now. Most of this change is probably due to diet.) Nonylphenols leach out of things like plastic toys into human skin and out of food containers into food. They seem to affect larval fish by stimulating an enzyme that converts testosterone to estrogen. (Atrazine, a common herbicide, also an estrogen mimic, may work the same way.) The unpronounceable phthalates are used as plasticizers; phthalates are one of the most abundant industrial chemicals in the environment; some are estrogenic, some carcinogenic.

Then there are the heavy metals. As an anthropologist has pointed out, high lead levels, along with increased fire frequencies, eutrophicated lakes, disturbed soils, and early successional forest vegetation, are signs of human occupation. Increasing metal contamination of northern hemisphere lake sediments during the twentieth century has been found wherever studies have been done. Cadmium, used in automobile greases, washes from roadways into rivers; cadmium is used also as a stabilizer in plastics, from which people can absorb it directly. Cadmium is not necessary for life but is implicated in some cancers. Mercury, a nerve toxin, is used as a biocide in paper making (it kills bacteria that would disfigure the paper). It is still sent, though much less than formerly, with the wastewater from paper manufacture into waterways, where bacteria convert it to methyl mercury, in which form it enters the food chain and accumulates in fish. About 90% of mercury can be profitably removed from wastewater streams, the rest not. The major environmental sources of mercury are the burning of fossil fuels, especially coal, for electricity generation; of gasoline and diesel for transportation; metal smelting; and garbage incineration (the last because of unrecycled mercury batteries in trash). Mercury is used in the production of chlorine (one of the bases of the modern chemical industry). It also leaches naturally from rocks and soils. Since the 1970s mercury has been increasing in the atmosphere at 2% a year, so in a few years its concentration will have doubled. Mercury contamination of forest soils is thought to be the reason all species of forest thrushes in North America, except the hermit, are declining. Inactive metallic mercury falling from the air is changed to methyl mercury by bacteria in damp forest soils; biologically active methyl mercury ends up in the invertebrates the thrushes eat.

———————

Industrial chemicals are distributed worldwide by the same processes that distribute the sun's heat from the tropics to the temperate regions and poles. The earth's atmosphere insulates the earth and raises its average temperature from -2° F. (earth's temperature as a radiative black body, determined by its albedo and distance from the sun) to 59° F. (its temperature thanks to the heat absorbing gases in its atmosphere). For most of earth's existence, its temperature has remained within the range of liquid water. Air heated by the sun rises at the equator, along with much water vapor (a line of clouds marks its rise along the Intertropical Convergence Zone) and sinks as compressionally heated dry air at about 30° North and South. Thus the deserts that circle the earth at those latitudes. Cooled by radiation to space, air sinks at the poles. The tropical and polar circulations, driven by the relative strength and weakness of sunlight in these places, drive an indirect circulation over the North and South Temperate Zones. The spin of the earth and the placement of oceans and continents drive other, smaller convection cells. Heavy metals, radioactive materials, chlorinated and aromatic hydrocarbons, nitrates and sulfates enter the atmosphere from the stacks of power plants, incinerators, metal smelters, car tailpipes. Pesticides and herbicides evaporate from the fields and forests on which they have been sprayed. Warm, rising air lifted by the sun and given a twist by the earth's spin carries them poleward, until the air begins to cool and sink, or the materials condense out onto droplets of rain or fog, and are carried downward and deposited on the ground, or in the ocean or rivers. If they settle on water they are absorbed into the thin bioactive skin of the sunlit surface with its dense populations of single celled organisms (plants, animals, viruses, bacteria).

The materials are then incorporated into the food chain, usually by being taken up in cellular fats. As these materials pass up the food chain, their concentrations are biomagnified from 10,000 to 1,000,000 times in living tissue; that is, their concentration in living tissue increases geometrically as one organism eats another; the longer the food chain, the greater the concentration. (So lake trout from lakes with short food chains are safer to eat, and small fish are safer to eat than large ones.) The chemicals are also concentrated by wind patterns, by ocean currents (such as the spinning tropical gyres), in estuaries where fresh and salt water meet, by fish migrations (sockeye salmon, which die after spawning, release the PCBs in their bodies into their natal Alaskan lakes), by seabird colonies. The ponds below arctic seabird colonies, fertile oases rich in nitrogen and phosphorus from the birds, and thus with plant and animal life, are also contaminated with the chlorinated hydrocarbons and metals the birds accumulate through eating fish, which they take from far at sea. So the distribution of heavy metals and persistant organic chemicals worldwide becomes surprisingly egalitarian but varies with the terminal location (wet and cold are worse) and the length of the food chain (short is better). Mercury from power plants burning coal ends up in Minnesota walleyes far from any power plant, in concentrations making the fish unsafe to eat, and in the arctic char, whitefish, lake trout and pike of Canada's Northwest Territories, also far from any power plants. The fish-eating loons of the northeastern United States and Canada, downwind of the continent's industrial chemistry, are much more contaminated by mercury than those in the center of the continent, or those on the West Coast, an advantage that will disappear as China develops; and, thanks to the thoroughness of atmospheric mixing, DDT sprayed on cotton fields in Brazil ends up in lake trout in Lake Michigan. The United States currently exports several tons per day (9 in 1994) of pesticides whose use is banned here; but the atmosphere returns them to us.

The air in the troposphere, the lowest level of the atmosphere, mixes in each hemisphere (North and South) on a time scale of a few months. It takes about a year for pollutants to cross the Intertropical Convergence Zone and mix through the atmosphere as a whole, during which time many of them settle out. Thus the egalitarian distribution of manmade chemicals, but with fewer of them in the south, which has less land, less industry and fewer people. Material carried by the atmosphere moves inexorably north (south in the Southern Hemisphere), with an eastward component given it by the earth's rotation; material that settles on land and is not washed by rain into waterways is re-volatilized by sunlight and continues its journey poleward, perhaps going through several more such settlings and re-volatilizations, during each of which it may be concentrated by rain or snowmelt in waterways, until it settles out on the lichens of the barren lands, and is concentrated in the flesh of the herbivores that graze them (lemmings, muskox, caribou), or in the northern oceans, lakes or rivers, and ends up in their fish, birds, whales, seals, and finally in the Inuit. The material may be buried for a time in the Arctic ice.

Such scenarios were mostly unimaginable by the scientific community in 1948, when use of DDT and other synthetic chemicals was soaring. Or were they unimaginable? The dangers of tetra-ethyl lead were well understood in the 1920s, when—thanks to pressure from General Motors and the Standard Oil Company, who had built a factory to produce it— it was introduced into gasoline to raise the octane level. Adding lead to gasoline was cheaper than refining gasoline further. When lead was finally banned from American gasoline in 1996, seven million tons of lead had been deposited into the atmosphere and along American roadsides. Seventy million children had been exposed to high blood levels of lead. As for the effects of coal burning on climate, the possibility of global warming from the burning of fossil fuels had been suggested at the end of the nineteenth century, by a Swedish physical chemist who buried the memory of an unhappy marriage in several months of calculations. (After leaving him, his young and very pretty wife wrote him letters describing her joy at being free of him, and refused to let him see his son.) But the effect on climate of raising the concentration of carbon dioxide in the atmosphere was thought to be slow and perhaps advantageous. Synthetic chemical production in the United States rose from one billion pounds in 1945 to 400 billion pounds in the 1980s and is higher now. Worldwide production is of course many, many times this.

Market economics always functions against a cultural background. This becomes very clear when one looks at different market oriented societies. In Norway and Japan, countries with standards of living equal to or greater than that of the United States, the spread of incomes between the lowest and highest paid employees of a company rarely exceeds 10 times. That is, if the minimum wage is $15,000 per year, the head of the firm would earn no more than $150,000. In Sweden the ratio is 13 to 1, in France 15 to 1, in Britain 24 to 1. In the United States in 1980 the average CEO earned 40 times the wage of the average manufacturing employee. Now that CEO earns 475 times the employee's income. Under current law, much of the CEO's income is sheltered from taxes. The point is not that the American worker is worth that much less, economically speaking, than an American chief executive officer (such facts are probaby incalculable), but that the public ethos allows chief executive officers to ask that much more, and to pay that much less in monies collected for the public good. In the United States getting rich is considered a right. But how rich? A so-called welfare state that guarantees elderly people a livable income, provides for universal medical care, gives people paid vacations, maternity leave, afford-able childcare and long-term unemployment insurance, also buffers itself against economic disaster, by providing a reliable, recession proof flow of income and purchasing power. Redistributing wealth from the wealthy to those who will spend it guarantees a certain level of consumer spending and thus helps moderate market failure. Skewed income distributions, by limiting consumption, tend to inhibit economic development. Great con-centrations of wealth are thought inimical to democratic rule (they are also associated with higher rates of anxiety, illiteracy, teen pregnancy, homicide, mental illness and seem to limit life expectancy and educational achieve-ment for everybody in the society). Unequal income is a problem in parts of Africa, Asia and Latin America and is becoming one in the United States. A common way to counter the development of extremely skewed distribu-tions of wealth is through inheritance taxes (progressive income taxes also help). Andrew Carnegie, a self-made nineteenth-century Scottish-American industrialist, supported the notion of inheritance taxes. Carnegie believed in hierarchical societies, but ones in which the individual's social position depended on his economic merit, not on inherited power or wealth. In-heritance taxes, by recycling wealth, keep the social order fluid. They also mobilize the movement of money.

Capitalist societies exhibit long-term cycles of growth and decline. For 20% of the history of the United States, the gross domestic product contracted. Markets, constantly evolving, can create massive economic instability. So certain sectors of all market economies are supported by the state. In the United States the list is quite long and includes fossil fuel energy, war materials, real estate, agriculture, and the production of most virgin materials, such as timber and metal ores. Food, fuel and housing are among the few necessary productions of a modern economy. As John Kenneth Galbraith has pointed out, the most long-standing, pure market, that in agricultural commodities, must be regulated to prevent the natural cycle of boom and bust from destroying agricultural producers, with the result of food shortages and high food prices. The cycle is simple. As prices for crops rise, farmers increase production, usually taking out loans to do this. As supplies of farm crops increase, prices fall, making the loans difficult to pay back. Farm profit margins are low, perhaps 3%, so small changes in prices can have a big effect on farm profits. As farm income falls, profits drop at local businesses, such as banks, hardware stores, farm equipment dealers, coffee shops, doctors, veterinarians, real estate agents, insurance agencies. If bank loans are unrepayable, farms get auctioned off, real estate values fall, and services supported by property taxes, such as schools and roads, are hurt. After a few years of low prices, farm production drops, shortages develop, prices for crops then rise, the surviving farmers buy up the vacant land and production rises again. Corporations with capital, such as insurance companies, also buy up farms, which leads to land being farmed by tenants rather than owners. One solution to this situation is for the government to buy up surplus crops and store them until the price rises. Another is to support crop prices at the cost of production, no matter how much is produced. This is current U.S. policy. It is not the best social or environmental solution. Its original purpose was to keep food prices low. By doing this, it has turned into a support system for corporate farmers.

Consistently low, but supported, agricultural prices in the United States, along with greater and greater mechanization and chemical use, and the rising cost of land, have created more and more farm consolidation: larger and larger farms, more corporate farms. Because of the cost of land and of installing drainage, the American Corn Belt has always had a high percentage of its land in absentee owners (about half of cultivated land at the turn of the twentieth century); such owners, along with farmers who rent land, are interested primarily in what they can make off the land in the short run; they are not interested in the future of the landscape. Such large farm operations, private or corporate, produce most of the country's food and receive most of the government support payments. Typically, 20-30% of the income of large farms consists of government payments; if the farm is irrigated, the level of government support is over 90% (mostly in the irrigation infrastructure and the subsidized cost of water). Agricultural subsidies in 2000 in the United States amounted to half of farm income. Large farms also cause most of the environmental problems of agriculture. A side effect of farm consolidation in the agricultural landscape has been a fall in the rural population, and thus a decline in local economies and in public and private services (schools, stores, medical clinics), and also a widening in the income gap between the relatively well-off corporate farmers and the relatively poor laborers. A poorer population results in a further decline in tax-supported local services, just as the need for such services increases.

Most developed societies support the price of agricultural commodities and regulate production. In France, agricultural policy is seen both as supporting the food supply and as supporting the late medieval French agricultural landscape: the appearance and the taste of the land (in wine, honey, salt, butter) some Frenchmen call the soul of France. The French developed methods of replanting hedgerows in Normandy, for instance. Hedgerows are useful as windbreaks, and important as habitat for birds, small mammals, pollinators and other invertebrates. The hedgerows had been removed to make the fields larger and easier to work with modern equipment. Restoring them restores the medieval landscape. Similar small-scale agricultural operations are supported all over France. These include sheepherders, wheat farmers, small wine makers. Shepherds still follow herds of sheep grazing along the sides of country roads, or on the grassy wastelands near new housing projects. Such government subsidies represent social as much as economic choices. If the United States made similar choices, instead of the ones it now does, it would support biologically appropriate agriculture, distribute income more thoroughly, substitute photovoltaic and wind power for coal and establish universal health care. Declining cities would be rebuilt with huge parks. The natural landscape would be allowed to recover. One could call the American economy still colonial, in that (in large measure) we trade agricultural and forest products for manufactured goods and oil. The rate of erosion from agricultural land in the United States is 10 to 15 times the background geological rate. Cutover timberland also erodes. The soil and nutrients lost from the land end up in rivers and the sea, shortening the life of dams, making clean water more expensive and eliminating native fisheries. Supporting better agricultural practices, treating abandoned or polluted lands as a resource, redistributing income and developing small scale renewable energy would lead to an economic boom. As a candidate for election in 1996 suggested, a one time tax of 15% on the net worth of all Americans worth $30 million or more would raise enough money to pay off the national debt. (It would probably take more now.) The money now paid in interest would be freed up to be used elsewhere; and the rich would make their money back in three years.

Economics also functions against a biological background. Plants change levels of atmospheric gases, modify nutrient cycles, engage in chemical warfare, promote (or suppress) wildfires, create shade, alter the temperature and humidity near the ground. Bacteria change the planet through photosynthesis, decomposition, respiration, nutrient cycling and fixation, by initiating processes that allow other organisms to colonize new environments. Grazers and browsers mold the vegetative structure of their habitat. The living earth is a biological organism. Fossil fuels, limestone and phosphate deposits, soils, sedimentary iron deposits, the amount of carbon dioxide in the atmosphere, the presence of oxygen—all are signs of life. Photosynthetic cyanobacteria began to oxygenate the atmosphere about 2.7 billion years ago. The oxygen precipitated iron out of the anoxic seas; the iron oxide was uplifted over hundreds of millions of years; and weathered of impurities to form concentrated iron ore, or hematite. The oxygen was poisonous to much of existing life. Such life, adapted to a world without oxygen, retreated underground, and new, oxygen dependent life arose. The modern world (modern ecosystems), with flowering plants, pollinators and mammals, began during the Jurassic Period 150 million years ago, when the level of oxygen in the atmosphere was close to 12%. Oxygen levels started rising in the middle of the Cretaceous Period about 100 million years ago during a burst of plant growth (never reached again), and reached the modern level of 21% perhaps 50 million years ago (some claim longer). There it remained. Stabilization of the oxygen level implies a balance between biospheric photosynthesis, which produces oxygen and carbohydrates out of carbon dioxide, sunlight and water, and biospheric respiration, which uses oxygen to turn carbohydrates into carbon dioxide, water and heat (the heat dissipates to space). Lower levels of oxygen would make it hard for large mammals, which have less efficient lungs than dinosaurs and birds, to survive. (But at levels above 25% oxygen most forests would burn; at 30% oxygen very wet vegetation would burn.) Above ground life is now dependent on oxygen and dips in the level of atmospheric oxygen correspond to major episodes of extinction.

As for carbon dioxide, its level in the atmosphere until recently had been falling. Some writers think the stage was set for our cooler glacial world by the evolution, 390 million years ago, of root systems in plants. Roots secrete carbon dioxide into the soil, where it reacts with soil minerals, breaking them down and making them available to plants, and locking much of the carbon up in the process. The death of roots also adds carbon to the soil. Soil storage of carbon may have resulted in a 45% drop in atmospheric carbon dioxide and a slow planetary cooling (reinforced by the absorption of more carbon dioxide by the cooling ocean; the soil and the ocean hold most of earth's recyclable carbon; much carbon dioxide was also locked up as coal and oil during the Carboniferous 300 million years ago).

This is the ineluctable background of modern life. But economics focuses relentlessly on the human world. Economics assumes that the natural world can be ordered according to market needs and within market time scales. It assumes that biological goods, such as forests, fish and beachfront property, can be consumed as fast as the market allows, at negligible economic or human cost: wild fish, salamanders, semipalmated plovers, and other organisms of little apparent economic value, pay the cost. (And beaches, no longer able to migrate because of the buildings on them, vanish and are replaced by sea walls.) Once a given resource has been exploited, the profits, if any, are invested in other enterprises. Most people don't lose: wild salmon runs are replaced by farmed salmon; the profits from logging redwoods are invested in setting up web pages. Individuals such as fisherman and loggers may lose, but that is part of the creative destruction of capitalism; they can get other jobs; and the economy as a whole gains. As natural wealth disappears, the creation of wealth depends more and more on human manipulation. The exploitation of natural resources may remain high, or even rise, their lower value per acre (as in second growth timber), or lower unit value (the declining percent of metal in an ore) compensated for by more efficient machines, shorter rotation times, and greater use of fossil fuel energy. Oil and coal became cheaper and cheaper during the twentieth century thanks to increasing mechanization and ever increasing use of those (ever cheaper) fossil fuels: a helpful upward spiral. In this abundant world, energy costs, even of energy producing devices, didn't really matter. When oil was abundant and accessible, oil production consumed only 2% of the energy the oil contained. Most other energy resources (coal, tar sands, photovoltaics) require several times that energy investment and, partly because of this, will require a far larger energy infrastructure. (A 1970's criticism of nuclear power plants was that more energy went into building them than they would ever produce; this turned out to be not true.)

Some sectors of the economy, such as the service sector, are less dependent on natural resources. The medical industry (15% of the United States' economy) or the educational industry (another good-sized chunk) are examples of wealth creating industries that don't require such high levels of resource exploitation per unit of output; at any rate less than mining, logging, metal smelting or agriculture. Of course, modern physicians and teachers depend on the already high level of resource exploitation that the society itself requires (some of that mining, logging and agriculture). Computer software development is an almost purely intellectual effort, but behind it lie customers and computers, power plants, chip manufacturing facilities, heated or air-conditioned buildings, trucks, tremendous quantities of polluted groundwater. In 20 years computers have risen to consume 10% of the American electricity supply. In no time the person at the desk becomes part of the whole petrochemical stream.

Our world is constantly recreated by living things. Climate is partly regulated by the amount of carbon dioxide in the atmosphere and thus by the storage of carbon in vegetation, soils, peat bogs and carbonate rocks. Plants absorb carbon dioxide from the air, turn some of it into plant tissue and transfer some of it to soils, where it breaks down rocks into useful plant nutrients (so called, in-place weathering). Soil invertebrates, bacteria and fungi immobilize the carbon left behind in dying plant roots. (A considerable percent of the fine roots of trees and perennial grasses die and regrow every year.) Physical processes in the oceans would precipitate out dissolved carbon dioxide as calcium carbonate (limestone), but living things do it first, so limestone is mostly formed of shells and corals. Uplifted limestones from the sea form 10% of continental crusts. One day they will be subducted into the earth's fiery mantle and their vaporized carbon returned to the atmosphere through volcanoes (a long term carbon cycle). Fossil fuels are probably the product of one major past storage event of tens of millions of years' duration during a warm wet period especially favorable to vegetation. (Coal certainly is; some writers claim oil and gas are not, but are continually regenerated from inorganic matter by physical processes in the earth's mantle.) At present, peat bogs and grasslands are both better than forests at long term storage of carbon and tropical forests are better at storing it than temperate ones.

The main absorption band of carbon dioxide corresponds with the earth's peak thermal emissions, which explains its role as a greenhouse gas. Carbon dioxide levels in the atmosphere have been declining for tens of millions of years, and as a result, the earth has been slowly cooling, despite the slowly increasing radiance of the sun. (The increasing radiance of the sun is a very long-term process.) A cooler climate tends to be a drier one, because less water evaporates from the oceans, and so declining carbon dioxide levels favor grasslands over forests, which need more moisture. The rise of carbon-storing grasslands and grazers over the last several million years has corresponded with our cooler glacial ages and, by absorbing more carbon dioxide into the soil, creates a positive feedback for a cooling earth. Earth's temperature depends on the placement of the continents, which influences the ability of the oceans to circulate heat, and of glaciers to form on land near the poles (no land, no continental glaciers); on the amount and seasonality of sunlight (long mild snowy winters and short cool summers in the land-rich Northern Hemisphere favor the formation of glaciers); on the level of volcanism and formation of new rock (volcanoes release carbon dioxide to the atmosphere and also form new silicate rocks that take it up); on levels of the different atmospheric gases; and on the functioning of the biosphere (whether carbon absorbing or not). Under present planetary conditions, carbon dioxide (and the related gas methane, which oxidizes to it, and, to a lesser degree, the other heat absorbing gases) acts as a natural thermostat. Carbon dioxide is stored as a gas in the atmosphere, as a weak solution of carbonic acid in the oceans, as clathrate hydrates near the poles, as carbonate minerals in the earth's crust and in oil and natural gas. Cooling of the earth causes the reactions that store carbon dioxide in the oceans and soil to slow down, thus moderating the cooling. Volcanic activity continues to release carbon dioxide, which slowly rewarms the earth. (A volcano is erupting somewhere on earth every day.) But (apart from the influence of people) under current planetary conditions carbon storage slowly increases overall, so the activity of the carbon thermostat is superimposed on a slow cooling. The Milankovitch cycles of solar radiance, which are caused by the periodic wobbles of the earth on its axis and the slow changes in its orbit around the sun, influence the seasonality of incoming solar radiation (warmer or colder winters, longer or shorter summers, in the north or the south) and are thought to determine (more or less) the onset and waning of glacial cycles. Without the carbon we have been putting into the atmosphere, we would be entering another glacial age; and perhaps be on the way to a much cooler earth overall.

All the warming gases (carbon dioxide, methane, nitrous oxide) are recycled by the biosphere, which helps maintain their level in the atmosphere. The level of oxygen is maintained by photosynthesis, largely of cyanobacteria and blue-green algae, but also by higher plants. Carbon dioxide, methane, nitrous oxide and dimethyl sulfide (important in cloud formation) are all produced by bacteria or algae; methane and nitrous oxide absorb heat at different wavebands than carbon dioxide (which increases their warming effect); dimethyl sulfide, produced by marine organisms, provides the nuclei around which the water vapor in clouds condenses (as do many bacteria, which migrate between sea and clouds and use the atmosphere for dispersal, and whose DNA is found in rain and snow). Clouds are also important in regulating climate. Marine organisms in coral reefs may increase their output of dimethyl sulfide on warm days, creating clouds that shade the sea. Because of such facts some writers argue that earth's climate is, within limits, self-regulating. It is inarguable that without life, the composition of the atmosphere would be very different: the amount of carbon dioxide would be much greater, the turnover of nitrogen gas 10-20 times slower, the amount of oxygen very small. Atmospheric methane would not vary with the expansion and contraction of termite habitat or of methane-producing tropical wetlands (their size controlled by the strengths of summer monsoons, those a function of plate tectonics, ice cover in the Arctic, ocean currents, earth's orbital cycles, and the intensity of the summer sun.)

———

The particular state of our world is maintained by its ecosystem services, of which the living world's influence on climate (a sort of regulation) is the most dramatic. Others include maintaining the natural water cycle (regulating the heights of ground water tables and floods; regulating the amount of rainfall); cleaning the air and water (forests filter the air passing through them; ecosystems scavenge nutrients and chemicals from water passing through them and influence its temperature, its siltiness, its chemistry, its rate of flow); maintaining the relative composition of the gases in the atmosphere (such as oxygen and carbon dioxide); creating and maintaining soils (the structure, water-holding capacity and nutrient levels of soils are all functions of the living things in them as much as of their mineral composition and their location in the landscape); the cycling of nutrients essential for life, including human life (among the atmosphere, oceans, plants, animals and soils); detoxifying pollutants (the workers here include the many organisms of decay, including those microorganisms that can reduce human hormones and chlorinated hydrocarbons to more benign chemicals); the pollination of plants and the regulation of plant and animal populations. The yearly value of such services has been put between $3 trillion and $33 trillion dollars (the latter about equal to the world's gross domestic product, the former—given a yield of 5%—implying an investment of around $60 trillion). While such calculations are economically useful, ecosystem services are essentially irreplaceable. Both estimates are likely way too low. We can't create systems that perform the services natural ones perform. Even on small scales, as in space ships or sewage treatment plants, this is difficult. Partly this is because of their size (and therefore their cost, both in wealth and in energy consumption) and partly because we don't fully understand how the systems work.

We all share in our effects on the natural world. Any permanent human settlement changes plant cover and the local water cycle, and modern settlements require energy use and chemical output far above those of a few centuries or a few decades back. The footprint of a modern person—the acreage of farm, forestland, mines, manufacturing plants, roads, oil wells, waste dumps needed to sustain his or her life—is much larger, about 24 acres for the average American. Our combined demand requires more land than the United States contains. Is this necessary? Certainly not for human life. Energy use in a hunter-gatherer society, which is mostly food, plus some fuelwood, is little more than a person's caloric requirements, about 2000 kilocalories per person per day, all of it recyclable. Agriculturalists, who often grow a surplus of food, and also may use draft animals, irrigation water, and derive power from wind and water, use 5-10 times as much, 10,000-20,000 kilocalories per person per day. While such societies are powered by the sun and potentially sustainable, they may not be so in practice. (They may overcut their forests or overexploit their soils.) Industrial peoples of the nineteenth century with their railroads, steam engines, piped water and gas light, used about 70,000 kilocalories per person per day, much of it from coal, a fossil fuel, that is, a fuel whose energy comes from stored solar energy; burning a fossil fuel returns its carbon to the atmosphere from which it came, in general at a rate too fast for the land and oceans to absorb and neutralize. Late twentieth-century people in the developed world, with their automobiles, large houses, electric lights, airplanes, televisions and computers use about 120,000 kilocalories per person per day, or 60 times that of hunter-gatherers, and twice that of industrial populations in the nineteenth century. North Americans use twice that; but live no better than Europeans. The difference is thought to be partly extravagance (bigger cars, larger, warmer houses, a larger military establishment), partly more energy-intensive industries, and partly North America itself, with its more extreme climates and longer driving distances. Astronauts, because of the energy demands of escaping earth's gravitational pull, use 2.7 million kilocalories per person per day. The footprint of the developed world, if applied to a population of 6-9 billion, is not sustainable: that many people living a western life would require several earths. Most of the energy used by the developed world comes from the combustion of fossil fuels.

Fuel use is not the only way to measure the environmental impact of a society; synthetic chemical production and patterns of human settlement are others. The size of the American corn crop is a measure of the environmental impact of American agriculture. (The picture is one of relentless growth: corn yields are up 31% since 1995, 72% since 1975, that is from about 4 billion bushels in 1970 to 10 billion in 2000, largely because of genetic improvements, even as the land planted to corn shrank; each bushel requires 1.25 quarts of oil and one or more bushels of eroded topsoil to grow.) However extraction and combustion of fossil fuels constitute one of the most important causes of land degradation and water pollution, as well as a leading source of man-made greenhouse gases. A writer has remarked that the more carbon that passes through the human economy, the more the non-human world declines. Compared to natural energy fluxes on earth, human fossil fuel use is small, amounting to .11 calories per square meter per day. Input from the sun is 4900 calories per square meter per day, primary production by plants 7.8 calories, weather 100 calories. The outsize effect of fossil fuel use on natural systems derives from several factors. Water use by electric power plants constitutes the largest use after irrigation. While it is not a consumptive use like irrigation (water is pumped through the plants to cool them), it warms rivers and changes their flow regimes, simplifies their flora and fauna, and ruins their fisheries. (New plants can cut water use by 90%, at a small increase in the cost of power.) Obtaining fossil fuels takes up land for wells, roads, mines and pipelines, and through spills or disposal of wastes, contaminates land and water. Burning them puts the products of combustion into the atmosphere: carbon dioxide (the gas of global warming), heavy metals like arsenic and mercury, dioxins and other hydrocarbons, sulfur and radioactive materials. The heavy metals, sulfur and radioactive materials are natural contaminants of the fuels. While arsenic, lead, mercury and radioactive elements are present in small concentrations in the natural environment, current man-made fluxes of them are comparable to, and often greater than, the natural ones. For instance, the flux of mercury produced by people is 10 times the natural flux. Thus the chemical soup we inhabit.

It is the production of carbon dioxide by the burning of fossil fuels whose effects may turn out to be the most dramatic, and perhaps the most catastrophic, for us. Carbon dioxide from the clearing of woodland, the cultivation of soil, and the burning of fossil fuels has apparently reversed the slow fall in the atmospheric concentration of this gas during the last several million years. Sunlight falls through the earth's atmosphere with little absorption (a little is absorbed as heat) to strike the earth, where it is absorbed or reflected. When reflected (re-radiated) from the earth as infrared radiation (radiant energy of a different wavelength), it is captured by several gases in the atmosphere, including carbon dioxide, methane, nitrous oxide and water vapor. Water vapor is the most abundant and most important greenhouse gas. By capturing a small amount of additional heat, man-made carbon dioxide raises the temperature of the atmosphere and increases the evaporation of water vapor from the oceans; this creates a positive feedback that raises the temperature of the atmosphere further; the climate warms. (And warms; and warms.) Since the ability of the atmosphere to hold water vapor increases rapidly with temperature, the small rise in temperature caused by carbon dioxide ends up having a large effect on atmospheric temperature.

The gases that absorb visible or infrared radiation are not the only factors that influence the earth's balance between incoming and outgoing solar radiation. The amount of dust in the atmosphere, clouds, the reflectivity of the earth's surface (that is, whether it tends to reflect or absorb sunlight: ice reflects sunlight, water absorbs it) also alter that balance, warming or cooling the planet. Clearing a landscape alters its reflectivity, much of the planet lies under a man made haze of dust, sulfur dioxide and smoke, jet contrails coalesce into clouds (cooling the earth by day, warming it by night), soot falls out on the Arctic ice and melts it. The global temperature is currently up 0.74° C. (1.33°F.) from the pre-industrial level, though it is now thought that the pre-industrial level had been raised slightly above the "natural" level by human influences beginning several thousand years ago. These include forest clearance for agriculture and the cultivation of rice in paddies (rice paddies, man-made swamps, produce methane). Clearing forests and cultivating soils release carbon dioxide that would otherwise remain stored. To keep global warming below 2° C. (considered a "safe" level) carbon dioxide in the atmosphere must stay below 450 parts per million; this corresponds to an average carbon emission of 1-1.5 tons of carbon per person per year. (More recent studies put carbon emissions at zero, in order to stop ongoing warming; 2° C. is still a level at which the permafrost boundary will move 400-600 kilometers north and sea level rise several feet.) A person in India currently produces about 1 ton of carbon dioxide, an American 20 tons, a European 12 tons. Thus stabilizing the climate implies large cuts in carbon dioxide emissions in developed countries, as countries like China and India develop; 70% in Europe, 95% in North

America. This is not necessarily a disaster. Carbon dioxide from stationary sources (such as power plants) can be immobilized, at costs estimated at 2-4 cents per kilowatt hour, or 10-20% of current electricity costs (some claim double that). This would come to about $15 a month on the average American bill. (What we would do with several billion tons a year of magnesium carbonate blocks is another matter.) Norwegian offshore oil and gas production is currently taxed at $38 per metric ton for the carbon dioxide stripped from the methane gas it produces; to avoid this tax the state oil company reinjects the carbon dioxide stripped from its gas wells into salt formations under the North Sea. Renewable sources of energy work; and efficiency gains in the use of energy of far greater than 50% are possible, probably 1000% in the case of cars. The French and Japanese in the early 1960s supported themselves on energy usage one-seventh that of the United States today. They certainly lived modern lives, and because of tremendous increases in energy efficiency to date, would lead even better ones on the same amount of energy today. The world today produces more than enough energy; the problem, as with food, is one of distribution (and use). Even lower energy consumptions than those of France or Japan in the 1960s will support societies with low infant mortality, high life expectancies, varied diets, good medical care and good educational opportunity. People in the Indian state of Kerala are said to live such lives on per capita incomes of less than $400 a year. (Since other writers claim that an adequate diet, that is, one that allows full expression of one's genetic potential for growth, arrives only during the early stages of modernization, at per capita incomes of about $4000 a year, ten times that in Kerala, one must wonder at such claims; or perhaps assume that being slim and short is no fundamental disadvantage. In photos from the 1940s American men seem half the size of moderns.)

The direction of societies ruled by markets is upward; their purpose is growth; their citizens constantly strive to increase their incomes. The notion of constant material improvement in human life is recent and unsustainable. It makes people miserable. What is clear is that our current way of increasing our wealth is too much for the biological world to bear. All high civilizations fail; and their people die or disperse to lead simpler lives. In the past such failures have occurred when climates changed, rains failed, trade routes disappeared, soils grew less fertile. Since people took up agriculture and began to settle in villages 10,000 years ago, societies have created wealth from the extraction of virgin resources and the clearing of new land. For the most part, such societies occupied relatively small areas of the earth and used relatively small amounts of easily available natural resources (timber, soil, coal, copper, iron ore). While all societies fail in the end, none will take as much of the world with them as we will.

Chapter 2
The Land Trip

In general, the more that is invested in a landscape, the more it's worth. Put another way, the more human manipulation a landscape undergoes, the greater its market value, and the less its value as an ecosystem. Capitalist manipulation of a new landscape begins with harvesting the natural wealth that has accumulated over time: furbearing animals, fish, mussels, timber, or (over a very long term) topsoil, iron ore, coal. The 80,000 beaver pelts exported yearly from Dutch settlements in New York State in the 1650s financed buildings in 1600's Amsterdam. Virgin forests in the eastern United States hold 3-6 times the standing biomass of second-growth, and yield 10-20 times the board footage of lumber; so the first cut was extremely valuable. Eastern aboriginal societies used the forest differently than early modern Europeans—for harvesting acorns and hickory nuts and the animals that feed on them, for firewood, building material, medicines, basketry fibers, material for making canoes. For the most part large old trees remained. These peoples manipulated the landscape by hunting large herbivores, which affects plant succession; by horticulture; and by fire. Some of their uses were not sustainable. But no current capitalist use is sustainable. One can point to a few groups of farmers, and a few wild fisheries, such as the salmon fishery in Alaska and perhaps the lobster fishery in the Gulf of Maine—though its success depended on the fishing out of the cod; but in general in a capitalist world, resources are harvested as fast as the market will absorb them, with no thought to the landscape from which they are taken.

So capitalist economics influences the biological as well as the man-made world. Defining organisms like bison and trees become commodities. Formerly economic landscapes that conserved their soils are abandoned because they become too labor intensive to exploit, such as hillside terraces in Greece or Peru, or tiny vineyards on slopes along the Rhine. Industrial landscapes are abandoned when exploitation is no longer profitable, such as Roman silver mines that still pollute rivers in Spain. The products of human effort (fields, mines, factories, downtowns), having repaid the capital invested in them, erode, breakdown, decay; all were once (before development) functioning natural landscapes. During the settlement of the United States, when a landscape's natural wealth was gone (the furs, timber, game; the beaver that also supported the colony at Plymouth; the live golden eagles one could buy in the game markets of American cities in the 1820s), the land might be worth very little. Or, depending on its soils and location, it might be worth a great deal as building lots or farmland. During the boom in settlement of western New York State after the Revolution, it was said that unimproved forest land increased in value 2-3 times over 10 years, while land cleared and converted to farms increased 5-20 times. The firewood cut off an acre would pay for the clearing of another four acres; and the cleared land could be sold at a profit. One could buy a farm for a year's labor. Many people "proved up" land with the first crop, sold it, and headed west, where the flat lands along the Ohio beckoned. In Europe in 1790 a year's agricultural labor would buy one's family food, the right to collect firewood, and some sort of shelter. Most of the land east of the Mississippi was wooded, but until the railroad made shipment of bulk commodities profitable, much of the timber removed in clearing land was not marketable as timber. It cost too much to ship: 50-75% of timber's retail value consisted of transportation. Timber was more marketable as converted ashes or potash, whose value per unit volume was much higher than sawn lumber or firewood. Potash was an important nineteenth century industrial chemical used in the manufacture of glass, soap, gunpowder, and in bleaching linen and scouring wool. About 5% of the several tons of hardwood ashes produced by burning an acre of old growth forest would convert to potash. The sale of potash would pay for the land and its clearing. Converted wood ashes were among the first goods shipped east on the New York State's Erie Canal (the waterway through the Appalachians that, from 1825, let the agricultural Middle West ship goods to the U.S. east coast. The canal cut transportation costs by about 95%).

A landscape's annual return determines its value. In general, a landscape's return is also a measure of how much has been invested in it. The manipulation such investment implies is usually inversely related to the land's natural value; that is, to its function in the biological world. For instance, land used for forestry is worth less than land used for agriculture;

and forestland tends to have a higher natural value (in preventing runoff of surface water into streams; in maintaining streamflow; as habitat for plants and animals) than farmland. But agriculture yields a much higher annual gross return. Agriculture's higher return depends on more yearly investment and manipulation, as the farmer puts more labor and money into his land. Grazing land, which requires less human input than land used for row crops, is worth less than cropland, and yields less gross return. The biological effect on the landscape from grazing, where domestic animals do the work, is considerable, but in general pasture land, if not overgrazed, retains more of its natural functions than land in row crops; and logged forestland more than pasture. (Most human uses fall along a continuum and the ecological effects of grazing and logging can be extremely large.) Gross income per acre is both a measure of the landscape's degree of manipulation and of its importance for the larger economy. Development usually creates pressure for more development. Farmers for instance, need machinery, chemicals, fuel, transportation, housing, credit, insurance and services; and much more of all this per acre than loggers. More farmers require more of all this, and more people benefit. Local governments, banks, oil companies, farm equipment dealers, insurance companies, real estate agents all see a rise in their incomes and all push for more development: more farms, more subdivisions, more downtowns, more infrastructure to support all that.

If it works, land development is always profitable to the landowner, though it may not be profitable to the society as a whole. In the middle of the nineteenth century, unclaimed wetlands in the Middle West (river overflow land, swamp forest, swampland, lands still owned by the federal government) was granted to the states, who were supposed to sell it to raise money for drainage, in order to turn it into farmland. Drainage is expensive. Land speculators bought much of the land, began drainage schemes, then resold it. Many private drainage schemes failed. Some were taken over by the government. Some were completed and the investors never paid (agriculture couldn't pay the costs). Drainage eliminated the useful public functions such lands performed, such as storing flood waters and reducing the height of floods along the river, and the removal of silt and nutrients from river water. Overflow lands provide spawning areas for fish and greatly increase the productivity of the riverine fishery. The silt and nutrients the floods spread on the swamplands grew trees, game and fish. All these services were free as long as the land was left undeveloped. After development, such functions were replaced with more saleable goods in the form of tax paying farmland, towns and industrial installations. With drainage, forest clearance, and the conversion of so much land in the watershed to farmland, floods along the Mississippi became more and more pronounced in the twentieth century. Siltation from farmland widened and shallowed the river beds, which made floods more frequent and severe. By the 1920s the flood peak as far north as Minneapolis had risen by 40%, and the new landscape resulting from riverside drainage had to be supported by greater and greater expenditures of government monies for levees, dredging, and other channel work. While dwellers in the floodplain profited from the development of their landscape, they could not afford to replace the natural services their development destroyed (in this case, flood control), and had to ask the federal government (that is, every inhabitant of the United States) for help. In 1927 floods along the Mississippi destroyed 160,000 homes and every bridge for 1000 miles upstream of Cairo, Illinois. This flood led to the federal government taking over the flood control system and developing modern schemes for controlling the river. Currently the costs of anyone living in a floodplain in the United States are paid for through federally supported flood insurance and river control projects; that is, by everyone.

In some cases, now, the human conversion of natural landscapes such as prairies and forestland to something of more immediate human value (logs, farmland) no longer seems economically reasonable. In the 1920s sportsmen argued against draining and diking swamplands along the Mississippi using the same arguments used today: that the value of the natural river was greater than that of the drained land created by levees. African pastoralist systems of keeping cattle are several times more productive in cash and protein per acre than western style ranches consisting of fenced private property in similar dry climates; the unpredictable weather makes nomadism a more reasonable economic strategy. But wild animals, such as gazelles, will produce three times the edible flesh of cattle on the same grasslands, and buffalo, together with their commensals mule deer, elk, pronghorn antelope and prairie dogs, will produce more flesh per acre than grass-fed cattle on the American plains.

Nowadays the net income from forestry on good forestland can equal that of a farmer on the same land. If prices are good, the income from the furs and mushrooms taken from an acre of Middle Western oakwoods will equal what the land nets if cleared and planted to corn. Gross income from the forest will be less, but the forester is less heavily capitalized than the farmer, and so has fewer costs. Sustainably harvesting mature timber (one white oak per acre every few decades) doubles or triples the yield per acre, letting the forester earn considerably more than the farmer. The scheme works now because farm commodities are abundant, while wild mushrooms and good quality timber are scarce, so the price of forest mushrooms and timber is high and of farm crops low; and because modern transportation systems allow perishable goods like mushrooms to be marketed. (As for furs, the price varies; some people consider wearing fur unethical and their influence, waxing and waning with the fashion world, can lower the price considerably. Fur boycotts helped put trappers of wild animals like the Inuit and Cree out of business. Most fur now comes from farmed animals.) Row crops net $50-$100 an acre, one-tenth the value of a large white oak log. Since the log takes 150 years to grow, and will increase in value for another 150 years, and one might have 50-100 trees, or 100-300 logs, per acre, the potential incomes from sustainable forestry and row crop agriculture are not that different. However, since the gross income per acre of the forest owner is so much less, his contribution to the total economy is also less. His needs are fewer, he supports fewer people in other industries. He also receives no return for the clarity of the water that flows off his land, nor for his land's value in storing carbon, nor for its value as habitat for plants, animals and pollinating insects. A problem with any scheme of sustainable forestry in a capitalist society is that a stand of timber takes a long time to mature and that at any time the stand can be cut for cash.

Medieval Europeans had more or less sustainable lives. For most of European history, peasants were poor; economic life in Europe went up and down (one can read such long-term cycles in the price of wheat) and the material life of an agricultural laborer in Europe in 1700 was little different from that of one in ancient Rome. One gets a hint of such lives from current peasant agricultures. In the western foothills of the Sierra da Estreda in Portugal, people still farm the sides of steep valleys, whose soil consists of an infertile clay eroded from Precambrian shale. The hillside soil is held in place by dry stone walls. Planting terraces are irrigated with water held in pits dug in the shale slopes above. The terraces are fertilized with shrubs cut from the upper slopes of the hills. The material is used as bedding for goats for 2-4 weeks, then piled up and let compost, and mixed with the soil of the terraces at planting time, at a rate of 3-4 tons (dry weight) per acre. Thanks to its time in the goat pens, the material's decay is rapid. The compost provides nutrients and improves the water-holding capacity of the soil. The high shrublands (the mato), which provide the fertility for this system, are cut on rotation every four years. One heath species is dug out by the root, and the root used in distilling alcohol from wine. Unless it were removed, this species would tend to dominate the heath, so removing it is important, since the different species of shrubs have different uses (besides compost; and probably in the ecology of the shrubland itself, which I assume is also grazed by the goats). Most of this work is done by hand.

Such a perspective makes it clearer how French and Scottish fur traders in the 1600s and 1700s could settle in to live with the natives with whom they traded; their material lives were not that different. Settlers in Massachusetts found the Wampanoag wickiup warmer and more weathertight than their wood-framed houses. Venting smoke through a hole in the roof was no surprise to people coming from a land where the chimney was just coming into use, and who built their chimneys from sticks and mud. Indian moccasins were more comfortable and watertight than English shoes, their bows more accurate then English blunderbusses. The Indians' diet was better: they were taller than the Europeans. The Spanish in 1500's Florida could not pull the bows of the Timucuans. It was said the bows were the thickness of a man's arm and shot arrows that traveled 200 yards and split oak logs six inches thick. (I assume some of this is exaggerated.) The bowmen could reload at a rate six times faster than a harquebus or a crossbow. In the 1500s the natives in New England would not allow the Europeans to settle permanently, but only to trade, and after a few incidents of kidnapping by Europeans, the Wampanoag killed most of the survivors of a French shipwreck on Cape Cod and enslaved the rest; and killed everyone on board a ship in Boston Harbor and burned the ship. European diseases defeated them. An epidemic in 1619, probably of a viral hepatitis spread by food, perhaps from a survivor of the French shipwreck, killed 90% of the people in coastal New England, in a band 200 miles long by 40 miles wide, from southern Maine to Narragansett Bay. The Wampanoag thought their deities had failed them. The Pilgrims, arriving a year later, saw in the empty land the good hand of God. In 1633 a smallpox epidemic killed 35-50% of the remaining Indians in New England.

While their material lives may have been similar, the paradigms of European and Native American lives were utterly different. Most of the Native Americans in New England were horticulturists, but made much of their living (so to speak) from the forest's mushrooms and furs. They weren't that interested in accumulating wealth, especially at first, when their societies were still intact. (Status mattered but was based less on wealth than in capitalist societies.) Wealth for Native Americans was found in the existing landscape. The Europeans' advantages included their diseases (which in succeeding waves eliminated most of the native peoples); their metal goods, which the natives found useful enough to trade for; their domestic animals, which helped transform the landscape; and an uncompromising ideology of spiritual superiority and financial gain. The English in New England found the native men shiftless, since they spent so much time hunting, and not in the fields, which were worked by women. (A cultural difference: for the English hunting was merely sport and agriculture men's business; but in many cultures women give birth to children and plants, so till the fields.) The English philosopher John Locke argued that the farmer, by making the land more productive, usurped the rights of the aboriginal inhabitants; his labor gave him a superior moral right to the land. A difficult argument: should farmers then surrender their land to industrialists, who would further increase its value? (When paid for it, they did.) Most of the eastern Indians were horticulturists and crops formed an important part of their diet but their fields were a small part of the landscape. Lacking domestic animals, they hunted the animals of the forests, rivers and seashore and gathered wild plants. The English soon adopted native crops and cultivation methods. Certainly clearing more of the land to grow corn made it possible to feed many more people, and once they had a foothold, the Europeans rapidly increased.

For a hundred years, until late in the twentieth century, the principal value of the redwoods and Douglas firs of California and the Pacific Northwest was found in cutting them down. These forests contained a greater mass of living tissue per acre (up to 500 tons) than tropical forests. Little thought was given to how the trees formed their particular riparian ecosystems, influencing the salmon streams that flowed through them, and the beaches and marine fisheries at their mouths. Logging the forest created value in the form of jobs, tax monies, profits for the timber companies. On private land, property taxes were partly determined by the value of the timber, and fell if the timber were removed. On government lands, some of the stumpage paid by logging companies funded local school systems. The cutover forests were replaced with planted forests of a single species, usually the fast growing Douglas fir. After cutting the trees and burning the slash, the former woodland was sprayed with herbicides to eliminate the nitrogen fixing broadleaved trees and shrubs that form the natural succession in these forests. Eliminating them and planting conifers speeded up the rotation. (Under natural conditions, the nitrogen fixing plants built up the stores of nitrogen in the soil, which would be exploited by the firs and spruces that succeeded them. The trees would live on the banked nitrogen until nitrogen-fixing lichens appeared late in the succession.) If possible, especially as the population grew, cutover private land was sold for building lots, because even with short rotation times (80 years for planted Douglas fir, rather than the 400-1000 of a natural forest), and generous government subsidies, timberland, with a return of 2-3% a year, cannot compete with other investments, such as real estate or stocks. Money made from timber or land sales and invested in the stock market would yield 10-11% a year over the long term; and the long term in the stock market is 10 years, not 80-150 (or 300-800, for sustainably managed old growth trees).

Various modern scenarios point out that the salmon, sea trout, mushrooms and other renewable products of the natural forest, together with some timber, harvested so as to maintain the watersheds and the structure of the forest are worth more over several rotations than the timber alone, clearcut when it reaches economic maturity. Sustainable rotations are very long in economic terms, with considerable valley bottom land and most steep slopes left in lightly thinned old growth: ideally, one would remove only trees that are not likely to survive until the next cutting cycle, while leaving some large dying trees to become snags. (John Muir proposed harvesting only dead trees.) The sustainable forest would produce a more or less stable yearly income, rather than the boom and bust of the capitalist forest, with its clearcuts and periods of regrowth. It would probably store carbon. In general, sustainable models of renewable resources produce no net overall growth but a low steady flow of income; while capitalist models produce large bursts of growth followed by decline, and often end with the extinction of the resource. Growth in the capitalist economy as a whole comes from investing the profits of resource extraction elsewhere. With forests, prices for timber can rise and new uses found for forest products, so the value of what is produced can rise, but the amount taken from the forest remains the same.

A major difference between the capitalist and the sustainable models is time. The long term in the capitalist model is the time between quarters; or a decade of market returns; or the time taken to liquidate a resource, whether oil or timber (a month, 5 years, 20 years). Factories such as paper mills are built to cover the lifetime of the resource. The longest term for an individual investor is a human lifetime (80 years, perhaps 40 or 50 of that spent working); after that, as Keynes pointed out, we are dead. The long term in a natural environment is the life of the ecosystem. What matters here are things like the rate of soil formation and erosion, the likelihood of volcanic eruptions, the speed of climatic change. The lifetime of a north temperate landscape like the Pacific Northwest in our glacial age is probably 10,000-20,000 years. (Perhaps less: tropical ecosystems last longer; while the geologic structures of mountain ranges average 70 million years.) The current ecosystem in the Pacific Northwest is 7,000-10,000 years old. This is the age of its organic soils, created out of glacial dirt by generations of trees. (Say, 70-100 tree generations at 100 years each, or perhaps fewer: the age of senescence in old growth Pacific Northwest trees ranges from 300-800 years for Douglas firs, pines and Western red cedars, and up to 2000 years for redwoods and sequoias; but the trees start reproducing themselves at younger ages, and so generation time is less than this.) This is longer than any human "high civilization" has lasted.

Apart from timber, the renewable products of an old growth forest include salmon, trout, furs, mushrooms and recreational space; its services include water storage, water filtration and carbon sequestration. All this might be worth $100-$200 an acre a year. The lower number is too low, but unless the landowner runs a hotel or collects mushrooms himself, most of the potential income from the land goes to outfitters, bed and breakfasts, gas stations, motels and restaurants. In a world that valued sustainable uses, the landowner would receive income from licenses and leases. Farmers in Costa Rica receive annual payments for restoring and maintaining their forests. The payments are related to the forests' value in reservoir maintenance (less silt to fill the reservoirs, more even water flow for power generation), carbon sequestration and tourism. Over a rotation of 300 years, $100 an acre comes to $30,000. During this time, the timber, harvested regeneratively—I am assuming this is possible—has a stumpage value of $15,000- $30,000. The total ($45,000-$60,000) is in the range of the stumpage value of an acre of 300 year old trees, clearcut. The numbers are comparable, but the money comes from different sources, and at different rates. Because of the hand labor involved in sustainable management, many more people would be employed. (Sustainably managing a fully stocked, 100-200 acre woodlot in the Canadian Maritimes, including sawing the logs onsite, would employ 3-4 people fulltime.)

Numbers change as markets vary. Both old growth timber and wild fish have increased in value as they have become scarce. In 2000, some acres of old growth timber were worth more than $100,000 in stumpage value (stumpage is what the logger pays the landowner). On the same acre the annual mushroom harvest might be worth $60,000. Truffles worth $60,000 per acre wholesale (600 pounds at $100 a pound) are harvested from some conifer forests in the Pacific Northwest. (Every acre doesn't have harvestable mushrooms but in the 1990s the value of the commercial mushroom harvest in Washington and Oregon was about $40 million annually, approaching that of the tree harvest.) If the truffle harvest is renewable at half this level, licenses to harvest truffles, or the truffles themselves, would be worth a fortune. (It's the same story, though on a lesser level, with the mushrooms, ginseng and other collected wild plants in eastern forests.) Truffles are characteristic of old growth forests. They are a major food of forest voles, which maintain the planting by spreading fungal spores in their droppings; the voles are eaten by the endangered spotted owl. If it were clear that managing forests for a steady source of income would produce a wealthier, more stable local economy, managing public lands sustainably would be justified. Private lands, insofar as their management interfered with a public good, that is, the health of the ecosystem, could be forced into sustainable management; or sustainable management could be made more worthwhile by compensation for the value of the water that runs off the land or for the land's storage of carbon. The time involved in a sustainable rotation is a problem; perhaps landowners could get loans against the streams of revenue from their mushrooms, water and trees, at least 50 year ones: tree futures. Then one would have to thin the forest every so often. However the fact that costs and prices now work out cannot be the chief argument for using landscapes sustainably, because both may different tomorrow.

———

Modern human occupation of a landscape almost always results in its degradation. Ecosystems are landscapes characterized by conservation of nutrient flows; nutrients are valuable and ecosystems evolve as organisms figure out how to use each other, or each other's wastes. So the flow of nutrients within an ecosystem (that is, the number of times each element of nitrogen or water is reused) is several times greater than the flow of these nutrients across its boundaries. The notion is a bit hazy in nature; where to draw the boundaries may be unclear. Movement of nutrients occurs mostly through the enveloping fluids of water and air, but animals also move between ecosystems, carrying nutrients, seeds and other propagules with them. Bacteria and other single-celled organisms, fungi, and small invertebrates in the soil and water for the most part manage the webs of nutrient conservation and re-use. Such organisms live in the shelter of the large organisms that define landscapes: trees, grasses, kelp forests, coral reefs. The structural organisms are affected not only by the microbial and invertebrate life of the soil or sea (whose habitats they create), but by other organisms that live on or about them, such as the burrowing rodents that reshuffle soils, grazers like elk or buffalo, tree-breakers like elephants, nitrogen-fixing termites, browsing moose, seed-eating mice, leaf-munching and pollinating insects, insect-eating birds. Large herbivores create their own landscapes. In some African ecosystems, elephants alternate with grazers like wildebeest, or more recently, Masai cattle, the grazers leaving behind developing stands of brush and trees, the brush eating elephants leaving behind grasslands. All these organisms also have their own predators and parasites. The water that flows off an undeveloped landscape has a chemistry determined largely by the soils and biology of that landscape. Biological life changes soils; the mineral earth, deposited in much of the temperate region by melting glaciers 15,000-20,000 years ago, was only the beginning of the story. The biological landscape also helps determine the total runoff of water (forests intercept 50-66% of total rainfall and return it by transpiration and evaporation to the atmosphere); the water's temperature (ground water is cooler, shading influences the temperature of streams); and its timing (whether streams rise quickly, or more slowly, how long perennial flows last).

Human settlement interferes with all this. The effect of something like farming on a landscape is clear; crops replace forests and prairie; insects of cropland (orders of magnitude fewer in variety) replace those of the forest; mammals, amphibians, reptiles, birds are different, mostly gone, replaced by those that can survive in fragmented habitats, and by farm animals and pets. The nutrient-conserving web of organisms is broken apart and the land sheds nutrients and soil into ditches and streams. Even under ideal conditions, farmlands shed nutrients; during heavy rains or with cool spring temperatures the simple biological systems of a farm let nutrients slip through. In the Caribbean Sea over the last 40 years, coral reefs have gone from 20% flat corals and 80% tree-like corals to 75% flat corals. The branched corals create a more varied habitat for fish and better protect the coast from waves. This change is largely due to the increasing human population in the basin, with the accompanying fishing and silt and nutrient runoff into the sea.

Settlement also changes micro-climates. Without the shelter of the trees, the land's surface warms more quickly, freezes more deeply, dries out more thoroughly. Wind speeds near the ground increase. Water tables fall and springs dry up, the latter a common complaint of later settlers in the Northeast and Middle West. Rainwater on the prairies filtered down hillsides into sloughs, then into perennial streams; after the slopes were plowed, the water ran off quickly, the sloughs dried in the sun, and the streams ran only after rains. Cleared ground holds less rain or snowmelt, thus less water goes into groundwater recharge, which feeds springs. The exposed snow melts more rapidly, and so water runs off faster and in larger amounts, carrying more soil particles and nutrients into streams. Summer flow is lower. Deforestation in the eastern United States typically increases total water yield 20-40%. In the Rockies, forested watersheds are cut in specific patterns so as to yield more water for storage dams. So the spring pulse of meltwater is larger and earlier, summer flow lower and warmer. The bacteria, algae, zooplankton, invertebrates and fish that inhabited the stream were adapted to the temperatures and flows of the old environment; they may survive the new regime or they may not.

Virtually every physical, chemical and biological aspect of surface water affects fish and their insect and planktonic prey. It affects their habitat among the rocks, logs, mud, and gravel of the stream bottom. Addition of more material with more settlement (nutrients, hydrocarbons, metals, sewage, topsoil, fertilizers, pesticides) increases the stress on the organisms in the streams. Dams change rivers' flows as well as their water temperature and chemistry. Riverside wetlands store much of the spring floodwaters and release them non-simultaneously, thus maintaining stream flows. Such wetlands also trap about 80% of the sediment entering a river and turn it into land. Drainage ends this. All changes in the riverine landscape propagate downstream, where in the flatland reaches of large rivers the size and timing of the spring pulse and the level of summer and fall flow are crucial for the breeding success of many species of fish. Further downstream, in ocean estuaries, the size and timing of the water pulses and the strength of seasonal flows, along with nutrient levels, help determine the breeding success of many species of marine fish, about two thirds of which spawn in estuaries. For each dam built, and each riverside farm that helped destroy a natural river (and the estuary it helped feed), the human construction seemed worth so much more than the fish; but as all the rivers were dammed, and as more and more farms and cities were built along them, the fish disappeared.

The changes in rivers, lakes and the ocean associated with the development of modern landscapes are not small. In one of those brutal, useful scientific demonstrations of the middle years of the twentieth century, two scientists clearcut a mountainous New Hampshire watershed, treated the vegetation with herbicides to prevent regrowth, and studied the effects on the movement of water and nutrients through the system. Flows following summer rains increased 3-4 times, and the warm, respiring soils of the watershed released enough nitrogen into the streams to kill trout. Siltation increased enormously. Erosion rates in logged forests are about 500 times those in undisturbed ones. Rates drop slowly as the forest recovers. Landslides occur on the steep slopes of western forests 3-5 years after clearcutting as the roots that held the slope together rot away. Insignificant rains set them off.

All this of course has long been known, if not in a proper scientific sense. In the eighteenth century, wharfs at river mouths along the Connecticut shore of Long Island Sound had to be extended 1000 feet or more into the sound because of siltation caused by land clearance upstream. The wharf extension reached 3900 feet in New Haven harbor in 1821. Some extensions were the result of larger ships needing greater depths of water, but nearby bays and marshes were also filling in. Tidal areas of San Francisco Bay were raised several feet by silt washed downstream from gold mining in the Sierras in the 1850s. Hydraulic mining, which used water pressure from channeled streamflow to remove whole hillsides, generated enormous quantities of material. Some Sierra streams accumulated 100 feet of sediment and are still cleaning themselves out today. The sediment filled the bay's coastal marshes and ended the bay's oyster fishery. The bed of the Sacramento River was raised 10 feet and the city of Sacramento flooded. Riverside farmland also flooded. About 6000 square kilometers of farmland along the Yuba River were buried in mine spoil. The destruction of private property led to the banning of hydraulic mining, though only after the gold was gone. Much of the salmon rearing habitat in the Sacramento basin was also destroyed. In the 1880s canneries put up 200,000 cases of salmon a year, about 10 million pounds of fish, from the Sacramento watershed north of those destroyed by mining. That much salmon would be worth $100-$200 million as fresh fish at retail today. (The value of the sport fishery would be more.) Miners removed 700 tons of gold from the Sierras, worth $1.5-$2.0 billion at today's prices, that is, 7-20 years of salmon. They used 7000 tons of mercury to do this. Much of the mercury ended up in San Francisco Bay, where 150 years later it still makes the fish unsafe to eat.

In northern oceans, single-celled plants called diatoms make up 80-90% of the spring algal bloom and 50% of the annual phytoplankton production. They leave a record of their lives in their siliceous shells, which accumulate in sediments. Cores taken from Chesapeake Bay in modern times indicate that the bay's diatom population was changing by the 1760s. Diversity was dropping and the organisms left were characteristic of turbid water. The rivers that fed the presettlement bay were said to run clear, even in high water. (This may be an exaggeration, but it is something one would think Europeans, used to the turbid farmland streams of the 1600s, would notice.) All the same, native Americans had thousands of acres of cornfields in the river valleys and tens of thousands of acres of fire-cleared grassland. Early Euro-American agriculture in the Chesapeake region (the later 1600s) mimicked the native hoe culture, but was a commercial agriculture using slaves and indentured servants for labor. Corn and tobacco were planted in small fields among the trunks of girdled trees. Fields were not squared off, but followed drifts of suitable soil. Crops were fenced to keep out the cattle and hogs that roamed the woods. Fields were used for three years, then let return to forest. Only the best land was used; perhaps 2% of the land was in cultivation at any one time. Cultivators were aware of the dangers of erosion. Since crops were shipped out by water, and dredging was expensive or impossible, siltation and shoaling of rivers was a serious matter.

Tobacco was the first cash crop. It was extremely profitable. In Spanish colonies in the West Indies, tobacco, grown with slaves, returned 6 times the value of any other crop. In Virginia the tobacco tax returned to England 4 times the revenue the Crown received from its colonies in the West Indies. But tobacco needed a constant supply of new land. After the 1720s, as settlement increased and more land was cleared, wheat began to be grown for the New England market. Cultivation of wheat meant draft animals and plows, permanent fields, permanent pastureland, shorter fallows, use of manure from penned animals on the crops, more water and soil running into the bay. Pasturing animals in the forests, in the old style, also resulted in erosion and soil compaction. Using the forest as pasture will eventually turn it into a woods of scattered trees, with very little reproduction. (Some protected ancient woodlands in England are of this sort.) By the 1750s, rates of water runoff into the bay following rainstorms were several times natural levels and summer anoxia in the bay was growing. Some summer anoxia in the bay is natural, caused by warm fresh water from the rivers flowing over cold salty water brought in on the tides from the Atlantic. The warm fresh water sitting over cold salty ocean water reinforces the stratification of the water column caused by the heating of the summer sun and prevents mixing and reoxygenation of the bottom water, where most organisms live.

Agriculture had caused widespread siltation of streams by the 1770s, with shoaling in the rivers feeding the Upper Bay. Riverbeds rose by tens of feet. By the early 1800s perhaps 40% of southern Maryland was in cultivation. Sedimentation in the bay probably reached its height in the late 1700s and early 1800s. The silt brought by the rivers reduced the bay's water clarity, a critical characteristic of the presettlement bay. The additional nutrients from runoff fed planktonic algae, which also reduced its clarity. Before European settlement, much of the bay's productivity was channeled through its seagrass meadows, which removed silt and nutrients (silt settled out among the grasses, nutrients were absorbed), provided shelter and breeding habitat for fish and invertebrates, and served as food for waterfowl and other wildlife. The meadows depended on high levels of light for photosynthesis. By the 1780s water runoff rates, accelerated by forest clearance and agriculture, from parts of the bay's drainage are estimated to have been 20 times natural levels, while sedimentation (the build-up of mud on the bottom) had probably risen 2-4 times. Over the next two centuries the watershed of the Potomac River went from 80% forested to 20% forested and the soil carried by the Potomac increased 8 times. (While the total watershed of the bay has recovered from 20% forest to 40% forest at present, much of the forest is very fragmented. Sediment carried by the Patuxent River to the bay went from 160,000 tons in 1950 to 710,000 tons in 1980, corresponding with a 21% further removal of forest in that river basin.). Nitrogen entering the bay increased 4-8 times, phosphorus 10-30 times; the nutrients fueled tremendous algal blooms, which blocked light from the seagrass meadows. When the organisms died, they sank to the bottom of the bay, where their decay reduced the supply of oxygen. Silt also smothered the oyster reefs and the seagrass beds, which both eventually declined by 90%. Oysters are filter feeders that remove plankton from the water. In 1500, oysters were sufficiently abundant to filter the bay's water every 3-7 days. They were a major factor in keeping the water clear. Oysters were also overfished. Dragging for oysters slowly removed their reefs, which were useful as settling points for young oysters. Reefs were also dredged to clear shipping channels.

Fish runs were declining by the middle 1700s, probably largely from habitat change caused by settlement. Silt from eroding fields infiltrated the spawning beds of river gravels and smothered developing eggs. Higher stream flows after storms flushed eggs and larvae downstream. Higher water temperatures and lower summer flows made the rivers less habitable for juvenile fish. Chesapeake Bay is the drowned estuary of the Susquehanna River. When the Susquehanna Basin was fully forested, peak flows in the river would have been 25-30% lower, summer flows higher. Much of the summer river would have been cooler. Many of the bay's fish are anadromous, that is, sea fish that spawn in fresh water, go through a juvenile stage there which lasts from one to several years, and return to the sea to mature. These include river herring, sturgeon, striped bass, perch, alewives and shad. Before the rivers were dammed, river herring ran up the Virginia rivers to the Shenandoah Valley, and up the Rappahannock and James to the Blue Ridge. Small mountain streams had fish returning from the sea to spawn. Shad ran up the Susquehanna to Binghamton, New York. There was a shad fishery on the north branch of the Susquehanna in Wyoming, Pennsylvania. (The Susquehanna had perhaps 1500 miles of shad spawning habitat.) Commercial fishing began in the Chesapeake in the early 1800s and peaked in 1833 with an estimated 750 million river herring and 25 million shad caught and salted. Both catches would be reduced by 99% by 1878 (and then recover slightly), partly from overfishing, partly from dams, partly from the deteriorating condition of the rivers, which would no longer support spawning fish or their juvenile offspring. Such losses would have repercussions throughout the habitat. (In Long Island Sound, the population of alewives, a major forage fish, is now at 3% of former levels, largely because dams on the rivers and streams feeding the sound have eliminated its spawning habitat. Predatory fish like striped bass and swordfish and birds like ospreys all depend on fish like alewives, another species of herring.)

The abundance of plants and animals in the new world bewitched the Europeans. Thomas Morton, an affable English opportunist, described the gastronomic pleasures of the turkeys, geese and swans of early New England, which he shot more or less from his porch. (The image of a man with a gun, setting off into the wilderness, so quintessentially American, seems to have been a European one. German immigrants in Pennsylvania made the rifles; the Scots and English headed off into the woods.) Morton scandalized the Puritans by setting up a maypole and dancing around it with the few remaining natives, and trading rum and firearms to them for furs. In the end the Puritans fined and jailed Morton and burned his house to the ground. Back in England, he wrote his memoirs.

The abundance of wildlife lasted for some time. Markets in American cities were full of game until the 1890s, when game was gone from the West. Most game had vanished from the East forty years earlier. Center-fire cartridges, repeating firearms and fast refrigerated transport by railroad disguised the falling abundance of western game animals and made the

1880s the golden age of the market hunter. As market hunting declined, hunting continued at a local level. White-tailed deer were virtually extinct in much of the Northeast by 1900 (in Vermont by 1840). The abundance the Europeans saw may have partly been the result of the depopulation of the country by European diseases several decades before Euro-American settlement took off (thus the rebounding of populations of game animals no longer hunted), partly the creation of a more varied landscape in the early days of settlement (more environment of the edge), partly the enduring emptiness of a large part of the continent. European fisheries had been as abundant several hundred years earlier, when fish were caught in traps or with hook and line; the invention of the trawl, along with the boom in population and settlement in Europe after the year 1000, began their decline.

In North America, the Europeans were interested in what was marketable; their new landscapes produced not game, fish and clean water, but crops like tobacco, corn and cattle; goods like firewood, potash, sawn lumber; power from flowing water for mill dams. Dried fish and salt beef were sold by New Englanders in the 1700s to markets in southern Europe and the West Indies. The excess fish catch was fed to pigs and used to fertilize fields. Oak staves for making sugar casks were also in demand in the Caribbean. The most profitable use of New England timber was for building ships, used in the triangle trade for slaves and rum. (African slavery was not new in 1500 but centuries or millennia old. However plantation agriculture in the New World increased the worldwide demand for slaves. Some writers suggest that the introduction of peanuts, manioc and maize to Africa from South America in the 1500s let African populations increase sufficiently to meet the new demand for slaves.) In the capitalist world of the 1600s and 1700s resources were turned into marketable products and capital, which turned into other goods. As profits from an enterprise fell, investment moved elsewhere. Fields were abandoned as yields fell; or went into long-term fallow. Tobacco and corn farming were always considered migratory occupations in the South, where animals were rarely penned, making their manure unavailable. In Virginia, low ground or bottomland was planted to corn, for food for animals and slaves. The land under upland hardwood forest, regarded as more fertile, was used to grow tobacco. Trees were girdled and tobacco grown among the dead trees for three years, then wheat for two or three, then the land was abandoned; such land usually succeeded into pine forest. (Thus the "piney woods" of the South, a modern creation.) By 1800 much of coastal Virginia and Maryland was gullied, abandoned farmland. By 1817 North Carolina had abandoned land equal to that under cultivation. Breeding slaves for the new lands to the west became a more and more profitable agricultural enterprise and led inexorably to the Civil War. Tobacco is a nutrient-demanding crop, needing 10 times the nitrogen and 30 times the phosphorus of other crops, and cannot be manured (manure ruins the taste). It was thought for a long time tobacco needed the nutrients in newly cleared ground. While loss of nutrients is one problem, after a few years the build-up of plant sucking nematodes and a fungus in the soil also stunt the tobacco plant's growth. Controlling the fungus requires a very long fallow.

Many attempts at using the new lands went awry. Some landscapes, such as upcountry New England, or the American plains beyond the line of reliable rainfall, should never have been settled with traditional European crops and animals. The hilly uplands of the Piedmont South had a boom in cotton after the Civil War. Newly discovered phosphate deposits along the South Carolina coast, whose cost was half that of Peruvian guano, together with wide-row cotton culture (the crop was cultivated with mules), led to excellent crops, but also to enormous erosion from the Piedmont's coarse soils. Cotton responded well to phosphorus but its use led to nitrogen depletion, which reduced yields, so more and more land was brought under cultivation. While the use of fertilizer led to good crops (German potash and nitrate was used to supplement the phosphorus), the cost of the fertilizer made the crop barely profitable to most farmers. Some farmers rotated cotton with corn interplanted with cowpeas. The cowpeas helped maintain soil fertility, improved soil structure and reduced erosion in the corn; people and hogs ate the corn and peas. This would have made the whole system more sustainable. But few farmers performed the rotation, since it took land away from cotton, the cash crop.

Erosion from hilly cotton lands was horrific. The eroding soil raised streambeds, creating swamps behind the rivers' natural levees on former river bottomland. Fertile riverside fields thus became swamps, especially in areas of low gradient. The sediment depth in some Georgia bottomlands reached 3 feet, while riverbeds rose from 3-17 feet. In general American agricultural practices have caused the loss of about 6 inches of topsoil per century (say, 12 inches to date), with the soil ending up in lowlands, rivers and the sea. The loss in the Piedmont was similar but more rapid. Many Piedmont soils gullied spectacularly. (Well known gullies inspired tourist visits.) Piedmont soils in Alabama and Georgia had probably been forested for hundreds of thousands of years and at the time of American settlement had evergreens 2.5-4 feet in diameter, 120-320 years old, which grew without fertilizer or care. The trees could have been sustainably harvested indefinitely. But one can tell a similar story about much American farmland. By 1979 about 100 million acres, 17% of U.S. cropland, had been ruined for commercial agriculture by soil erosion.

Economic development of the landscape during the settlement of the United States focused on creating additional real estate value. Much development was anachronistic or pointless. Rivers were destroyed by canals going the same way as the railroads that would replace them. Rivers today continue to be turned into waterways, since with the state providing a straightened, dredged stream, dams and locks, river transport is cheaper then rail. Large barge tows can carry more than fully loaded coal trains (the most massive land transportation device), and cheap transportation makes low-value goods such as timber and grain marketable. Once river ports develop, the value of their real estate must be protected, which leads to more development of the river. In many cases, settlement continued in river floodplains long after there was any economic purpose in being near the water; as new development supported previous development. Politically it was easier to control the river than to move the people. The capacity of developed and abandoned landscapes to produce timber, game, fish, row crops or grass, or to conserve and recycle nutrients, was permanently (in human time), reduced. Given enough time, and especially in more humid regions, such lands would recover, though now set on a different trajectory. (Degraded cropland and abandoned strip-mines can also be regarded as an economic opportunity: places to set up arrays of solar panels, or electricity-generating windmills, to build factories, to rejuvenate soils with carbon-storing crops.)

Abandonment and rebuilding is part of the capitalist cycle; it is the response to changes in demand. Some of this change is driven by technology. Thus the automobile made all previous commercial construction in the United States obsolete. Nineteenth-century downtowns were built to handle people who walked or took public transportation. The trams or cabs of public transportation were always moving; city construction didn't allow for room to store the private vehicles of the same number of shoppers. When public transportation systems were abandoned in the automobile boom of the 1950s (many of them bought up and let deteriorate by automobile companies), urban downtowns couldn't deal with the number of cars. Commercial activity moved out to the suburban edge, and urban downtowns, more and more inhabited by the less well off, were left to decline. As urban housing typically takes many decades to decay, and falling rents make such places desirable for a long time, urban centers turned into slums, a trend rising childless populations and the lure of the city sometimes reversed.

New development always moves toward cheaper land, which is less developed land. And cheaper land tends toward development, in an attempt to make itself more valuable. Development is partly pushed by property taxes, a major cost of holding land, and partly by the return from land trying to keep up with returns from money invested in manufacturing or the stock market. Unless they are protected, it is the ineluctable fate of all private lands in a capitalist economy to be developed. So one know the fate of an abandoned apple orchard in Fairfield County, Connecticut, which birders identify as a stopover site for hawks on their long slide south, where the birds rest and pursue grasshoppers, dragonflies, meadow voles and starlings; or of the scattered clumps of pine trees among farmland in the narrow neck of southern Mexico that local farmers point out as the places migrating hawks roost. With development, lands lose much of their biological value; if polluted, they acquire a negative biological value. Abatement of a public nuisance is the owner's responsibility, so polluted lands are expensive to redevelop, since they first must be stopped from leaking toxic materials to the atmosphere, into peoples' lungs, and to waterways. (Abandonment is the simplest solution, though more difficult nowadays.) Developed, polluted landscapes are the natural result of capitalism left on its own. Capitalism is not concerned with virtue and always seeks the lowest-cost solution. There is no economic benefit in taking account of the value of a natural ecosystem; none in not using dangerous materials if they are the least expensive (especially if one has constructed a factory to make use of them); none in not disposing of waste products as cheaply and as shortsightedly as possible.

Values change with development and also over time. Everyone knew that building dams on the rivers of the Pacific coast would eliminate the salmon. The effects of logging, farming and mining were perhaps less obvious. But in the 1860s the Chimariko Indians, who depended on the salmon runs, fought battles with local gold miners over the color of California's Trinity River, its clarity gone thanks to mining operations. Currently the income that would be generated by the run of salmon and sea trout in a river like the Elwha on Washington's Olympic Peninsula is probably worth more than the electricity generated by its dams. Dams commodify rivers, turning the energy of the moving water into private or state property, and making the stored water saleable, while eliminating the fish that were free to everyone. The first dam on the Elwha was built on speculation. It was built without fish ladders, which was illegal: to compensate for the loss of spawning area the owners built a hatchery. The hatchery never worked and was soon abandoned. The Elwha flowed through old growth forest and the electricity from the dam made it possible to set up a paper mill. Another dam followed, also for the mill. At that time, the company payroll was probably worth more than the fishery. (The Elwha's run consisted of perhaps 250,000 fish, of all 5 species of Pacific salmon plus sea-run trout, say 2.5 million pounds of fish, a third to a half harvestable on an annual basis. The dams eliminated much of the fishery and violated the spirit of a treaty with the local S'Kallam tribe, which guaranteed them rights to the Elwha fisheries.) But numbers change. The electricity generated by dams on the main stem of the Columbia is now worth about a billion dollars a year. In the eighteenth century 50,000 Native Americans along the Columbia caught 20-40 million pounds of fish, which amounted to 5-20% of the run of 11-16 million fish (100-800 million pounds of fish). Taking up to half a salmon run is considered a sustainable harvest. If one could harvest, on average, 100 million pounds of fish, the salmon run would be worth $1-$2 billion at fresh retail, half that wholesale, a third of a billion dollars at retail if canned. The fishery might be worth more than the power. Dams of course have to be rebuilt, while a natural river maintains itself. (Dams are designed to last 250 years but need major renovations every 25-50 years. Such maintenance costs must be subtracted from the value of the power.)

Electricity generation however subsidized the river's development for irrigation, ship traffic and water supply. Without cheap power, and an aluminum industry to use it, and aircraft manufacturers to use the aluminum, how would the Pacific Northwest have developed? (I'm sure there were other ways.) What would have happened in World War II? One could also argue that it was economic development, made possible by cheap electric power, that made fresh salmon worth so much. We could also ask why electric power and aluminum should be so cheap; and whether rivers and their landscapes must be developed so as to destroy their natural functions. Could we not use rivers in a different way?

Sailing west to Greenland in the year 1000, the Norse recognized they were passing south of Iceland by the pods of blowing whales. Similarly, the Pilgrims sailed into Cape Cod Bay through schools of whales, arriving in a land that depended on corn, beans, beaver, white-tailed deer and fish. Runs of Atlantic salmon were smaller than those of the Pacific Coast. Perhaps 2.5-5 million Atlantic salmon returned to rivers in the northeastern United States and Canada. But the Atlantic Coast of North America had several other abundant species of anadromous fish. Fish provided a third of the calories for many tribes and much of their protein. Early observers may have mistaken the silvery, spring running shad for salmon. By the late 1600s the blocking of fish runs by dams was an issue. Riverside farmers put up fish for the winter; food was a major expense. The modern world lay in the future. Colonial legislatures passed laws requiring dams to be removed during fish runs or modified to let spawning fish pass. (Existing dams were exempted.) Such laws were known from England, where dams on salmon rivers were required to leave gaps wide enough to accommodate a well-fed 3-year-old pig. These laws were widely ignored, though less in New England than in the Chesapeake region, or later, on the Pacific coast. In the early nineteenth century Boston capitalists dammed the main New England rivers to provide year round power for their textile mills. No Victorian businessman was going to open a dam for a month to let spawning fish pass. So-called salmon stairs, which were invented at about that time in Scotland, and which would have worked with such low dams, and were not considered expensive, were not installed. So the fish runs ended. By 1850 half the potential salmon habitat in eastern North America had been closed off by dams. Today 900 dams block New England salmon rivers. In the latter part of the nineteenth century logging was also fought over as its effects on rivers became clearer. New York State's Adirondack Park, the largest park in the United States east of the Mississippi, was established partly to preserve a potable water supply, the Hudson River, for New York City. The upper Hudson was already shoaling from two centuries of agriculture in its watershed. (Of course the park was only established after the most valuable timber had been cut.) Preserving the Hudson's flow did not stop the State from allowing its pollution. New York has never tapped the Hudson, but uses the nearer Catskill drainages, also turned into a state park, for much of its water.

A general definition of a capitalist society is one that is able to mobilize capital, or its equivalent, to accomplish some remaking of the world (an irrigated field, a dam, a mine), that produces a profit, that is, a surplus, for the society. Socialist and communist societies and chiefdoms are capitalist in this sense, differing in how they collect and allocate the profits. (Economists will not like this loose definition.) In general, capitalist societies have not managed renewable resources well. They are not to be blamed for this; their focus is on the human. Nature is a commodity to be used. Most capitalist economic doctrine essentially eschews renewable management, and favors managing resources to economic depletion as fast as the market allows. (Modern so-called stakeholder capitalism tries to take the needs of the owners, the workers, the local community, the nation and sometimes the larger ecosystem into account.) Virtually no federal forestland in the United States has been managed so as to recreate, over time, its original biology and structure, or to protect the streams and fisheries which its landscape shelters. The timber has been cut as fast as the market would allow, or as the timber companies demanded, in the interest of maximizing short-term profit and community development. When the trees were gone, the timber towns (like river or mining towns) were finished, and their real estate values declined. Timber companies followed the trees elsewhere and the loggers (stuck with real estate of little value) blamed environmentalists for their problems.

Lots of schemes have been put forward for the renewable use of tropical rainforests: collecting gums, nuts and fruits; collecting medicinal herbs; "sustainable harvesting" of the timber (which may or may not be possible in that environment); raising iguanas for meat on the forest edge. But the value of the trees is too great. Estimates of the value of the timber in the Amazon basin range from $3-$5 trillion. This may be too high, given the amount of clearing to date, and that Brazil's legal cut in 2004 was only worth $1 billion (though illegal cutting might double or triple that), but a value of $1-$3 billion a year, some of which goes to feed very poor people, means the timber will be cut, even if, as seems likely, cutting the forest will change not only the soils but the rainfall regime (one half or more of the water that falls on the Amazon basin is recycled from the forest itself), and thus make regeneration of the forest, especially in a warmer, drier world, impossible. (Elimination of the forest will also reduce rainfall in the savannahs of the Cerrado, the immense and profitable agricultural region south of the forest. Such considerations make it seem Brazil would do better to try to eliminate poverty by other means, perhaps by redistributing the wealth of its citizens more vigorously.)

The world is now capital rich. Capital could be found to preserve the Brazilian, Indonesian, and African rainforests (all with huge stores of carbon in their plants and soils), as well as those remaining on the west coasts of Canada and the United States. The aboriginal population of the Amazon has been estimated at 4 million people (larger than the current population), but such estimates continue to grow. Amazonian communities were dense, and vulnerable to European diseases, and most of the population disappeared shortly after European contact. It is thought their survivors formed the scattered hunting and gathering tribes now found in the basin. While settlement was concentrated along the rivers, more signs of dense settlement are being found in places between the rivers, together with more areas of *terra prieta*, the fertile black earth created by human manipulation (mostly by the addition of slowly charred woody material). The Amazon rises 40-60 feet during the wet season and its falling waters leave tens of thousands of hectares of land available for cultivation. Some of this has been developed for rice cultivation. Much is used on a small scale for grazing water buffalo. Whether riverside and *terra prieta* agriculture, sustainable fishing, and the gathering of tree gums and fruits in a manipulated forest, in an economy connected by river transportation, would support a modern rural population as large as the aboriginal one, and one with a reasonable per capita income, of say, $2,000-$4,000 a year, is another matter. (Currently 80% of the population of the Amazon Basin is urban. Two million people live in Manaus, a duty-free port with a thriving industrial sector.)

In the case of the forests of the Pacific Northwest, over the remaining life of the ecosystem (perhaps 12 Douglas fir lifetimes) the return from the landscape would be greater (2-3 times greater?) if it were managed as a whole functioning ecosystem, for timber, mushrooms, hunters, naturalists, tourists, electricity, irrigation water and fish, rather than for timber, river navigation, irrigation water and electricity alone. With most of the old growth timber now gone, such schemes become possible. The short term return from newly purchased forestland (no clear cutting) would still be less. 'Sustainable management' is only suitable for an economy that does not grow. Rather than maximizing the return per acre over the short term, so as to efficiently deplete the resource base and accumulate a large amount of capital, sustainable management manages a variety of interdependent sources of income over the long term so as to provide a steady annual income, with reliable employment.

Sustainable development anywhere would be preceded by a period of growth, as the infrastructure for it (solar panels, selectively harvested forest farms) is put in place. Then such growth disappears. A difficult idea, especially if the human population continues to grow, and if the point of life, as an old French text claims, is to become rich. (Without that goal, the authors state, man loses hope.) Indeed, without growth, how will capital grow? The yield from a sustainably managed landscape might be 2-3% per year, about equal to inflation. The stock market deals in minutes and quarters, and returns on investment, say, in land, are supposed to be paid back in a length of time considerably less than a biologically appropriate timber rotation: 150-500 years for many eastern forests, 300-1000 years for those of the Pacific Northwest. Even short-lived trees live longer than people. There have been 100 tree generations in the black spruce forests of eastern Canada since the glaciers retreated, but several hundred human generations. Business loans are short-term (3-10 years, sometimes 20). Government supported loans to farmers with low interest rates and long repayment times are a concession to the limited returns that are possible from cropland, even under modern chemical management, where the soil lost from the fields outweighs by several times the grain, compared with the returns from investments where the rate of output is under better human control, such as mines, factories and office buildings.

Part of the problem of capitalist land management is its success; and therefore its scale. Developed landscapes now dominate many environments (the urban east and west coasts, the agricultural regions of the Mississippi drainage). So-called "high civilizations," faced with growing populations, continued on the same paths and tended to create environmental problems that were outside their control. The Sumerians knew that irrigation was causing saltation of their fields; under the pressure of a growing population and the military rivalry among the cities of Mesopotamia, they substituted more salt-tolerant crops and brought new land under cultivation. The Hohokum who lived along the Salt River in Arizona had similar problems. In a more limited landscape, their solution was to keep rotating their fields, using only 10% of their cultivable land at any one time; fields were used for a maximum of 10 years. When the water table became too high and salts too pervasive, the land was retired for a period. Perhaps they had fewer enemies. The Hohokum supported 100-200,000 people in the Salt River valley for 1500 years on irrigated agriculture in the flats and plantations of agave in the uplands, and were probably eliminated by European diseases. (Floods that dug riverbeds below the intakes of their canals are another possibility.) Ancient peoples were no less intelligent than we; moderns are abandoning irrigated land as fast as it is being created for the same reasons (saltation and waterlogging) as the Sumerians. As long as new land is available, reclamation is not thought worth the investment. The Sumerians lasted 1000 years, until the new land ran out, and a long drought lowered the level of the Euphrates, and probably increased its salt content, and their problems overwhelmed them. We have used the fossil fuels that make our lives possible for 150 years. Economically recoverable oil was probably half gone in 2000. Coal should last 250-400 years more, or perhaps 1000. (Or perhaps 60: the time coal will last depends on the rate of global economic growth, or rather, on the rate of coal use.) We do very little about the problems fossil fuels are causing. The reasons to shift to other energy sources are not only ecological.

Chapter 3
Climate Problems

From 100 to 40 million years ago the planet had no permanent ice; sea level was 70 meters (210 feet plus) higher than now; trees grew in Greenland and Antarctica in the midwinter darkness. About 35 million years ago, as the earth cooled, the polar ice caps formed. The cooling was probably the result of a slowing in the movement of continental plates and thus less volcanism and less emission of carbon dioxide from subducted seafloor limestones. Ecosystems also continued to remove carbon dioxide from the atmosphere. Perhaps 20 million years ago large herbivores shifted from eating leaves to grass, grasses developed the ability to reproduce vegetatively by tillers and rhizomes, and carbon-storing grasslands and grazers became major components of the earth's landscape. For the last 2.5 million years ice ages and interglacials have alternated on the planet. During the last million years the periodicity of the glaciations has shifted from approximately 40,000 to 100-125,000 years. Eighty percent of the last 2.5 million years have been spent in glacial periods. The primary reasons for the alternation between glacial and interglacial periods are thought to be astronomical: variations in the earth's orbit about and angle to the sun and thus in levels of solar radiation reaching the planet, especially the Northern Hemisphere, where most of the land and land-based ice, is, in summer (the time of snowmelt; periodic variations in the strength of the sun correlated with sunspots have also been suggested). That such small seasonal changes in sunlight (the variation is on the order of 8-9%) causes such dramatic climatic changes has to do with the amplification of the effects of solar radiation by the carbon dioxide and other greenhouse gases in the atmosphere. The effect of the radiational changes is also amplified by the present position of the continents, which affects the amount of heat absorbed by the northern and southern hemispheres. The position of the continents affects the movement of ocean currents, which are a major factor in moving heat from the tropics to the poles (and so also affects the strength of the monsoon rains in parts of India, Africa, the southwestern United States, and Australia). But the current glacial climate depends on slowly falling atmospheric levels of carbon dioxide.

Carbon dioxide is absorbed by the oceans and stored as limestone in the shells of sea creatures. Much carbon dioxide was locked up by plants 200-300 million years ago in coal and oil. Some carbon dioxide has been absorbed during the last 100 million years by the oxidation of the material exposed by the uplifting of the Himalayas and the Tibetan Plateau. Plants continually transfer carbon to soils, which also store it. Bacteria change carbon dioxide into methane, which is stored in permafrost under frozen northern bogs; and in clathrates (icy lattices of methane and water) in cold ocean muds. The recent increase in grasslands may have been related to their more efficient use of carbon dioxide (there being less of it in the atmosphere) versus most shrubs and trees; grasslands also transfer much carbon to their soils (one reason grasslands make such good farmland). Carbon dioxide levels in the atmosphere during past glacial periods have been about 190 parts per million, during interglacials about 280 parts per million. In our industrial, fossil fuel burning world the level is currently over 380 parts per million (about 390 ppm in 2009), and increasing by 1.5-3 parts per million every year. The combined effect of all the warming gases, expressed as the warming potential of carbon dioxide, amounts to a carbon dioxide level of about 430 ppm. A rise in carbon dioxide levels of 100 parts per million at the end of the last glaciation caused a rise in global temperature of 9° F. Why our temperature rise so far has been so much less than expected is uncertain. Partly, it is because the climate system hasn't had time to respond to the increase in forcing gases (for instance, it takes ocean waters, whose temperatures are a major determinant of climate, 30 years to equalize in temperature with the atmosphere), partly because sunlight reaching the earth is blocked over much of the planet by a man-made haze of dust, smoke, sulfur dioxide and clouds (from jet contrails, from condensation about fine man-made particulates). This reduces sunlight reaching the earth's surface by 10-20% and so reduces heating.

The difference in atmospheric concentrations of carbon dioxide in different times reflects reallocation of carbon among the oceans, land and atmosphere under different environmental conditions. While the reasons for the reallocation are poorly understood, lower carbon dioxide levels are associated with cooler conditions. A simple explanation for the fall in atmospheric carbon dioxide as the temperature cools is that a cool ocean absorbs more carbon dioxide from the atmosphere than a warm one. (The oceans are a very large sink for carbon dioxide, containing about 50 times that in the atmosphere.) Absorption of more carbon dioxide by the oceans thus reinforces the atmospheric cooling. (This feedback relationship will raise the warming predicted for rising carbon dioxide levels by 50% as the warming ocean absorbs less carbon dioxide.) Iron rich dust rich stirred up from deserts falling on iron poor seas during dry, windy glacial periods (the windiness a result of the greater temperature contrast between poles and tropics) increases the growth of phytoplankton, which die and sink (or end up as fish poop and sink: the poop pump), increasing carbon storage in the deep ocean and keeping temperatures low. Atmospheric iron increases growth in tropical forests, which also store carbon.

As the ice age continues and continental glaciers become larger however, they also become unstable. Their weight compresses the earth below them (as much as half a mile) so their tops sink to lower, warmer altitudes. Heat rising from below melts them; and their bottoms sink below sea level, making them liable to catastrophic collapse. As continental ice caps begin to melt with an increase in the northern summer sun at the beginning of an interglacial period, the flood of freshwater and icebergs from them puts a lid over the salty northern North Atlantic, suppressing the deep ocean circulation (which depends on the sinking of cold salty water near Iceland). Eurasia cools as the surface of the North Atlantic freezes, reducing the asian monsoons (which depend on the heating of inland plateaus to draw moist sea air inland, an effect one can read in stalagmites from Chinese caves). No longer able to export heat north, the tropical ocean warms, expelling some of its carbon dioxide to the atmosphere. As the seas rise slightly, coral growth expands, both upward in existing reefs (to keep up with the deepening sea) and along newly flooded shorelines. The formation of coral skeletons also releases carbon dioxide to the atmosphere. If the overall global warming caused by the release of carbon dioxide to atmosphere is sufficient, the ice caps will continue to melt. As the ice retreats further, carbon dioxide and methane are released from newly uncovered northern peat bogs, reinforcing the warming; as the seas rise further, the ice caps float and rapidly collapse.

Then, as the climate warms further (and summer insolation in the north changes to one less favorable to summer melting), this process reverses. Carbon's rate of storage in peat bogs, and also in forests and grasslands increases, pulling carbon dioxide out of the atmosphere and setting the stage for a new glaciation. The 15 millennia of the last interglacial exceed any interglacial of the last 420,000 years in duration, stability, amount of warming and concentration of greenhouse gases. If orbital geometry were the sole factor in control, our climate should have begun a slow cooling several thousand years ago. Thanks to earth's feedback mechanisms, carbon dioxide levels were slowly falling until 8000 years ago. The current rapid warming (about 0.75° C. in a century) is thought to be caused largely by the burning of fossil fuels, but clearing forestland for agriculture, growing nitrogen-fixing crops, fertilizing crops with artificial nitrogen, raising cattle and growing irrigated rice also add to rising temperatures.

Agricultural activities contribute through the addition to the atmosphere of carbon dioxide, nitrous oxide and methane; like carbon dioxide, nitrous oxide and methane are greenhouse gases. Nitrous oxide is released by bacteria from the breakdown of nitrogen compounds in water or soil (whether from nitrogen fixing legumes or fertilizer); methane through the bacterial breakdown of carbonaceous material in the guts of cows, and also in rice paddies and water reservoirs. (Swamps, lakes, volcanoes, the guts of termites, and perhaps forests, are natural sources of methane.) Carbon dioxide is released by clearing forests for agriculture (perhaps 20% of current man-made warming gases) and by cultivating soils. (Cultivation speeds up oxidation of the soil's stores of carbon.) If this interglacial had followed the course of the previous ones, reglaciation should have begun 4000-5000 years ago. It is now thought that land clearing for agriculture in Eurasia, which started 8000 years ago, produced enough carbon dioxide to compensate for the declining radiance of the sun, so evened out a natural decline in carbon dioxide, and produced our moderate and benign climate. If so, our civilization, which depended on a climate favorable to agriculture, is the result of a lucky accident. (On the other hand, there was another long interglacial about 400,000 years ago, when the earth's orbit around the sun was more nearly round, as it is now.)

A global warming of more than a few degrees Celsius will make most existing human infrastructure obsolete. Current thinking is that if we can hold carbon dioxide levels, or their equivalent, under 400-450 parts per million, we can keep the warming under 2° C. and escape its worst effects. Since this level is essentially here and we are doing essentially nothing to control the gases that cause warming, or to control human population, and changes are taking place in the Arctic and Antarctic with a large potential for positive feedback, escaping climate change seems extremely unlikely.

Some infrastructure will be made obsolete by changes in the water cycle. Condensation and rain power the earth's weather, releasing heat to, and removing it from, the atmosphere. A warmer world is a wetter world, since more water will evaporate from the oceans; and in fact moisture in the upper trophosphere (the lowest level of the atmosphere) is up 6% since 1982. Warming will affect humidity levels, evapotranspiration from vegetation (a source of moisture for convective rainfall), sea levels, soil moisture levels, rates of storm runoff, of groundwater recharge, snowfall and snowmelt, river flows, lake levels, rates of leaching and erosion in soils. Half the United States' population lives within 10 miles of the beach and is vulnerable to rising sea levels. Rising temperatures themselves become a problem, as people cannot survive more than a few hours of skin temperatures of 35° C. (94° F.). In dry climates sweating now lets us survive ambient temperatures of 45° C. but in high humidity sweating doesn't help. A 35° "wet bulb" temperature is the maximum for human survival. One writer argues that a temperature rise of 2° C. is too much and that we should try to return to a carbon dioxide level of 350 parts per million (from the current 390 ppm, or 430 ppm if one includes all warming gases). He reasons that a rise of 2° C. gives us the temperature of 3 million years ago, when sea level was 20-25 meters (70-80 feet) higher and the coastline of the east coast of the United States south of New York followed the Orangeburg Scarf, a formation about 60 miles inland of the present coast. Such a rise would imply more than the meltdown of the Greenland glacier, which is estimated to occur at a rise of about 2.7° C. and would raise sea levels 6-7 meters. The meltdown of the Greenland glacier is supposed to take 1000-3000 years— or 300-1000 years, sources differ—but with glacial meltdowns going on worldwide, and sea levels also rising from thermal expansion, sea level rises of 1-4 meters per century are very possible. The earth is currently absorbing enough extra heat to cause sea level rises of a meter per decade if all the heat went to melt ice.

Changing the amount of water that falls as rain or snow creates other problems. California depends for its summer water supply on thawing snowmelt from the Sierra Nevada. If more of the snow falls as winter rain and the snow melts a month or two earlier, the water available in summer falls. (It flows through the reservoirs in winter and spring. Estimates range up to an 80% decline in available water over the next 50 years.) Mountain glaciers stabilize the summer flows of the great South Asian rivers like the Ganges, and also the flows of many South American rivers, by capturing the variable monsoon rains as snow (that compacts to ice) and providing a strong regular pulse during the summer melting season. Meltwater from Himalayan glaciers provides 60% of the flow of South Asian rivers, whose waters help feed over a billion people. At present rates of melting, such glaciers will be gone in 40 years.

Severe storms, those with more than 2 inches of precipitation in 24 hours, have increased 20% over the last 30 years in the eastern and midwestern United States. One hundred year storms now occur every 9-10 years in the Schoharie Valley of New York State. This has required New York City to strengthen Gilboa Dam, one of its water supply dams. Holes were bored down through the dam and rods anchored in the bedrock to hold the dam in place. Water levels reached 6.5 feet over the dam in 1996; overtopping by 8 feet could cause failure of the dam. In general, small changes in reservoir inflow can cause large changes in reservoir water yields: that is, too much or too little water. Large inflows increase siltation, which shortens the reservoir's life. (They also increase its production of methane.) Warmer seas also mean stronger hurricanes (and stronger windstorms in general); whether hurricanes will be more numerous is still debatable. Rising sea levels, and more frequent floods and extremely high tides, driven by more powerful storms, will force the abandonment of much housing along seacoasts and in river floodplains. Saltwater will infiltrate lenses of fresh water near the coast, already depleted by pumping, turning them undrinkable. Losses in the United States will amount to many trillions of dollars. (Florida alone has $3.5 trillion in development at risk from storm surges.)

Warming shifts ecosystem boundaries, usually north (south in the Southern Hemisphere), or upslope (about 20 feet upslope per decade during the twentieth century). Warming in parts of the American tropics has made the days cloudier and cooler, and the nights warmer, creating favorable conditions for the spread of the chytrid fungus (imported to most of the world on African frogs) among amphibians. In Central America, population after population of frogs has been wiped out as the fungus spread up the valleys. (The fungus kills frogs by interfering with the absorption of sodium and potassium through their skin.) In a similar way, tropical diseases have begun to move into temperate regions. The climate will probably become less favorable for agriculture in hot continental interiors, such as the American Corn Belt, or the Ukrainian Black Earths, where rising temperatures and more sporadic rainfall will reduce yields. Each rise of 1° C. will move the habitat for cultivated grains 100 to 200 kilometers further north, but into subarctic regions where soils are poorer (the southerly regions will become too hot to grow grain; and as temperature and humidity rise beyond a wet bulb temperature of 95° F., some will become too hot for humans to survive). Rising global temperature will cause the release of more carbon dioxide from subarctic forests and methane from Siberian bogs. In the last decade, the Alaskan tundra has changed from a carbon sink to a carbon source; and much of the boreal spruce forest of northwestern Canada and Alaska has been killed by bark beetles no longer controlled by the length of winter. Now bark beetles are killing forests throughout the western United States. The carbon dioxide released by the decomposition of the dead trees will add to that in the atmosphere. In parts of Siberia, methane now bubbling up from thawing tundra lakes keeps them from refreezing during the winter. Other lakes are draining away as the permafrost beneath them thaws. One quarter of the land surface of the Northern Hemisphere currently remains frozen year round. Northern permafrosts contain something like 400 billion tons of methane, a greenhouse gas 20 times more powerful, if shorter lived, than carbon dioxide, equivalent in warming potential to several centuries of human production of carbon dioxide. Thawing will release it.

———

Warm tropical waters flow north with the Gulf Stream and moderate the climate of Europe and eastern North America. Winds push and pull surface ocean currents. The Gulf Stream (equivalent to the flow of ten Amazons) is drawn north by the sinking of cold salty water in the North Atlantic. Surface water from the Atlantic contains 2-4 pounds more salt per ton than water from the Pacific. Weather and geography let the Atlantic export moisture from its basin. Temperate westerlies send Pacific moisture against the high mountain ranges that run along the west coast of the Americas, where most of it condenses out as rain and fog and flows back to the Pacific in rivers and springs. Westerlies carry Atlantic moisture across the Eurasian plains, where some of it falls in the basins of rivers that empty into the Atlantic, the North Sea or the Mediterranean, and some, after several cycles of evaporation and precipitation, reaches the Pacific. Eastward blowing tropical trade winds carry Atlantic moisture directly across the Isthmus of Panama to the Pacific, while the easterly Pacific trades drop their moisture on the mountains of East Africa (from where it returns to the Indian Ocean). So the Atlantic loses 3-4 inches of water a year (the flow of one Amazon) to the Pacific and its surface waters are saltier.

Salty dense Atlantic waters become denser as they move north and cool. Water reaches its maximum density at 4° C. (39° F.). North of Iceland cold salty Atlantic surface water begins to sink through the warmer water below it. Sinking is helped by the annual freezing of the ocean surface, which takes up fresh water, leaving a pool of cold, salty water behind. The water sinks to the bottom of the ocean and flows south in a return flow towards Antarctica, perhaps pulled by the circumpolar winds about the Southern Ocean. Another sinking pool of water develops in the Weddell Sea near Antarctica. Atlantic water thus fills the abyss under all the oceans. The return flow of warm tropical water from the Pacific and Indian oceans around Africa to the North Atlantic transports heat from the tropics to the poles and returns the lost moisture.

Heat (or the lack of it) and salinity drive this current (the "ocean conveyor") by affecting the density of water. Any warming of the ocean's surface near Iceland (which reduces the water's density) or any inflow of fresh water that lessens the water's saltiness (also reducing its density) slows the sinking of the water. Discharges of the six largest Eurasian rivers into the Arctic Ocean rose about 7% from 1936 to 1999, probably because of increased rainfall. Arctic sea ice has thinned by 40% since the 1980s and its extent has shrunk dramatically. Ice reflects sunlight, while seawater absorbs it, so melting sea ice produces more warming, and more melting. It now seems the Greenland icecap is also undergoing net melting: another huge potential inflow of fresh water to the North Atlantic. Measurements indicate the salinity of the Arctic Ocean has fallen slightly 1965 to 1995. Measurements also indicate the sinking of surface water over the underwater ledges east of Iceland has slowed by 20% since 1970, but the measurements are few. About Antarctica, the extent of sea ice shrank by 25% from the 1950s to the 1970s, a fact that escaped notice for some time in an era without satellites. (Then curious scientists started examining whalers' logs.) So sinking may also slow in the Southern Ocean.

A slowdown in the ocean circulation has various feedback effects. Seas around Europe and north of Iceland would cool. Sea ice would probably increase. Temperatures in Europe would fall by 5-10° F., half of their current advantage over similar latitudes in North America. Wine growing in much of northern Europe would retreat toward North Africa. Nutrient upwelling, which is driven by currents, would slow throughout the Atlantic Basin. This would lead to a decrease in ocean biomass. Less capture of carbon by plankton and slowing of the ocean's circulation would reduce the carbon storage capacity of the ocean, increasing the atmospheric warming overall. Sea levels would rise about a meter around the North Atlantic Basin from a lessening of the plughole effect, caused by the sinking of several billion gallons of water per second. The Indian monsoon would weaken and Central and South America would lose a considerable amount, perhaps half, their rainfall.

The global circulation of ocean water is also the main method by which the deep sea is kept oxygenated. A substantial reduction in water exchange between the surface and the depths would result in partial stagnation of the deep sea. This would cause a severe decline in deep ocean biomass, extinction of many marine organisms, and perhaps releases of carbon dioxide and hydrogen sulfide, chemicals toxic to most land organisms. (Hydrogen sulfide, a natural product of decomposition, is released from undersea sediments when oxygen levels fall near zero. Its release turns the water column toxic.) A stalled ocean circulation at the end of the Permian period 251 million years ago wiped out 90-95% of marine life; 70% of land plants and animals also died. The temperature rise at that time was caused by rising levels of carbon dioxide from volcanism. Sea temperatures rose 8° C. at high latitudes, but less in the tropics, reducing the temperature differential between them and thus the deep ocean circulation. No one has much idea of the time scales involved in such events, but one cycle of deep ocean circulation takes several thousand years (figures range from 1000-10,000). The last such ocean-mediated cooling, the Younger Dryas Event, began 12,700 years ago and was probably caused by the draining of remnant glacial lakes (primarily Lake Agassiz) in mid-continent North America down the Saint Lawrence River. The cooling began with a drop in average temperature of 27° F. near Greenland. Winters became several months longer and the atmosphere stormier. The Younger Dryas lasted about 1300 years and ended abruptly, probably within three years. It was associated with drought in Southwest Asia. The drought may have helped push people like the Natufians, who were living in villages near the Jordan Valley and supporting themselves by gathering acorns, pistachios and some wild cereals and by hunting gazelles in a landscape managed by fire, into domesticating cereals and becoming herders of goats and sheep: that is, into becoming agriculturalists. The drought would have reduced the production of acorns and pistachios and made the growing of grain more attractive. During the Younger Dryas Event, tundra returned to northern England. (The petals and pollen of an Arctic flower, *Dryas octopetala*, are its markers.)

It is now thought that the effects of global warming will not be enough to stop the North Atlantic Circulation but will just slow it (estimates are about 25%), and that the warming of the atmosphere will more than compensate for any cooling the slowdown in the circulation causes. Another feedback effect of man made warming is the release of methane from permafrost in the Arctic and from clathrates (ice lattices) in shallow Arctic seas. Permafrost is already releasing its methane. The last time oceanic clathrates released their methane temperatures jumped 6-8° C. in a few thousand years. A 1-2° C. rise in temperature over shallow ocean shelves is sufficient for such releases to start.

Another troubling scenario is the collapse of the Amazon rainforest, with the release of the carbon in its vegetation and soils. Trade winds blow Atlantic moisture west over the Amazon Basin. As one heads west, more and more of the water that falls on the forest is transpired from the trees. (Three quarters of rainfall in the western Amazon is recycled from plants).) Cutting enough of the trees in the eastern parts of the forest (some calculations say 20%) break the cycle, then droughts begin to dry the western forest, fires open it up, and the forest collapses from west to east through drought and fire into a savanna, emitting the carbon in its trees and soil (500-900 metric tons of carbon per hectare) to the atmosphere as carbon dioxide. The releases of carbon dioxide lead to a further warming of the global climate. Brazil is already the fourth largest emitter of anthropogenic carbon, much of it from forest clearance. Indonesia is third. (Both countries are relatively poor.) Indonesia's emissions come largely from the clearing and burning of its tropical peat swamp forests for paper pulp and palm oil plantations. Because of this, its per capita emissions of carbon are close to those in the developed world. The release of all the carbon stored in the soils and vegetation of Amazonia would raise global carbon dioxide levels by about 300 parts per million. The clearing of tropical forests is completely under human control and could be stopped tomorrow if the parties involved could agree on a scheme for compensation.

In the modern world, surprises abound. Fifty years ago no one suspected that refrigerants released into the atmosphere would cause an increase of ultraviolet radiation at ground level. But depletion of ozone levels in the stratosphere by the reaction of ozone with the refrigerant molecules has caused an increase of ultraviolet-B radiation of 10-20% in northern latitudes over the last 30 years and greater increases in the far south (of southern Australia, southern Africa, southern South America). Ultraviolet-B causes skin cancer and cataracts; ultraviolet-B unfiltered by the atmosphere would sterilize the surface of the earth and the upper layers of the ocean. (The creation of the ozone shield as a byproduct of the creation of an oxygenated atmosphere by early photosynthesizers allowed the colonization of land by life.) It is now suspected that the warming of the lower atmosphere, the trophosphere, that is, global warming, reinforces the rate of ozone depletion. The lower atmosphere warms by absorbing more of the heat radiating out from the earth, thanks to the presence of elevated amounts of greenhouse gases and water vapor. Since a limited amount of heat is available, increased warming of the lower atmosphere cools the upper atmosphere. Cooling favors the reactions that destroy ozone, which take place at very low temperatures. The upper atmosphere is also contracting (a thermal contraction, one associated with its cooling) at a rate of about a kilometer every six years. This is not a small amount, and no one knows its implications. But such facts abound. The building of very large dams in the Northern Hemisphere over the last 70 years, by concentrating so much weight toward the North Pole, has slowed the rotation of the earth by a fraction of a second a day; so has the warmer atmosphere, which has an increased angular momentum, or drag. Modern settlement is associated with drainage (of farm fields, of building sites) and the subsequent lowering of continental water tables has raised ocean levels about three centimeters over the last century (something more than an inch, but another rather large number).

Anthropogenic carbon dioxide has made the oceans more acid. (About half of anthropogenic carbon dioxide emitted stays in the atmosphere, a third goes into the oceans and a sixth is taken up by plants. The oceans are the largest reservoir of planetary carbon dioxide, while soils are second.) The pH of the ocean was 8.3 after the last ice age; 8.2 before the Industrial Revolution; 8.1 now: a 30% increase in acidity. As carbon dioxide in the atmosphere becomes more abundant, more of it dissolves in ocean water. Some is removed from the water by living things. A large proportion of the spring and summer blooms of coccolithophoric algae (those with calcium carbonate shells) sink to the bottom of the ocean, thus locking up carbon dioxide from the atmosphere in calcium rich ocean sediments. But some of the additional carbon dioxide remains in solution and raises the water's acidity. As the acidity of the ocean rises, carbonate becomes less available, calcium carbonate skeletons harder to form, and less carbon is stored by living things. Corals and oysters grow more slowly. As the water becomes more acid, their shells and skeletons begin to dissolve. Oxygen also becomes harder to extract from seawater as the water becomes more acid; this affects the growth and reproduction of organisms with high oxygen demands, such as squid. Toxic metals become more available in a more acid ocean. The ocean is buffered by calcium carbonate deposits on its bottom and the circulation of acidic surface water through the deep ocean will eventually reduce its acidity but this takes tens of thousands of years. At the present time humans are releasing carbon dioxide too quickly for the buffering mechanism to work.

Modern people have other effects on climate. The shading effect of aerosols from the burning of fossil fuels, especially coal; from burning grasslands; and from burning dung or wood, reduces heating from the sun at the earth's surface. Off the east coast of the United States the reduction is about 10%. The aerosols include sulfates, nitrates, sooty organic chemicals and fly ash. Similar reductions are found over the Indian Ocean, from the so-called "brown haze" that hangs over the Indian subcontinent; and over the Yellow Sea, from the cloud that hangs over China. Such aerosols probably reduce global temperature but, settling out over the Arctic, increase the melting of snow (which the Inuit claim now looks yellow rather than blue). Too small to make good condensation nuclei, the aerosols may reduce rainfall. (This seems to be the case in eastern China, where rainfall has fallen from 1965 to 2000, as the countryside industrialized.) They probably change the speed and place of jet stream winds. Despite the phasing out of most chlorofluorocarbons (CFCs) and related chemicals by the Montreal Protocols, ozone depletion has lessened very little, probably partly because of anthropogenic cooling of the stratosphere, partly because of cheating, partly because some CFCs are still allowed, partly because of resistance to phasing out some chemicals like the bromine used to sterilize agricultural soils, partly because more chemicals than first suspected are involved. Much is still unforeseen. Fluoroform (HFC-23), a waste product of manufacturing the new refrigerant (HCFC-22), turns out to be an extremely efficient greenhouse gas that is rising in concentration in the atmosphere by 5% a year. Its current volume has the warming potential of 1.6 billion tons of carbon dioxide. Nitrogen trifluoride, used in the production of semiconductors and liquid crystal displays, is 17,000 times more potent than carbon dioxide and in 2008 was causing more warming than the carbon dioxide released by coal-burning power plants. Like the Sumerians, we face a bind: fossil fuel use is causing some of our worst environmental problems, but use of ever cheaper fossil fuels will let us solve the problems global warming is going to cause, such as rebuilding infrastructure, opening up farmland to the north, producing food by bacterial fermentation, and so on. This will lead to a warmer, more unstable climate, and the need for more use of fossil fuels. A major investment in non-carbon energy sources such as solar cells will also cause a spike in the use of fossil fuels (but a temporary one).

The exceptions to the land trip tend to confirm it. Humans now appropriate 30-40% of net global terrestrial primary productivity, that is, net annual growth on land, for their own use; this growth is taken through harvested crops, felled timber, grazing by animals, land clearing for agriculture or housing, through forest and grassland fires. Most of the biomass is not used directly. Only a few percent of most crops is used for food; and a small percent of logged trees end up as fiber. We harvest about 35% of the productivity of the oceans. Such use does not slow. In the 1990s the United

States converted about 2.2 million acres of open space to developed land, 50% more than in the 1980s. The United States uses about 24% of the biomass its land produces, Western Europe uses 72%, South Asia 80%, South America 6%. (So South America remains wild). As human occupation of the globe increases, empty, more or less wild landscapes rise in value. Such unused, or used and abandoned lands (the drier, rougher American plains, or cutover but partly regrown forestlands) are seen as a retreat from the real, industrial landscape, and are affordable by those who were more successful in creating that industrial landscape. Empty, wild lands rise in value because they are now scarce.

Chapter 4
The Natural World

All forms of life influence their surroundings. Photosynthesizing bacteria and blue-green algae are thought to have created our oxygen rich atmosphere about 2.5 billion years ago. Before that the atmosphere was a reducing one, carbon rich and oxygen removing. The creation of an atmosphere dominated by oxygen was a disaster from the point of view of most organisms; some retreated far underground to rock formations from whose chemical elements they could derive energy, some to the anoxic muds of wetlands, ponds and the deep ocean. (The oxygenation of the atmosphere also created a layer of high altitude ozone, which, by absorbing ultraviolet light, allowed the colonization of the land by plants and animals. Energetic ultraviolet light tends to disrupt cell chemistries.) The current global concentration of carbon dioxide is low thanks partly to the burial of so much carbon in fossil fuels 200-300 million years ago, and thus its effective removal from planetary chemistry; and partly from reduced volcanism, which fell as the movement of continental plates slowed. Large amounts of carbon are also stored in frozen northern bogs, in ocean sediments and ocean water, in soils, and in standing vegetation. Much of this carbon was sequestered by living things, some by simple chemical processes, and much of it can be released to the atmosphere under the right conditions. We, for instance, free the carbon in fossil fuels by burning them. Eventually that carbon would have been put back in the atmosphere as the continental plates in which it is buried are drawn down into the earth's hot mantle and the carbon vaporized and carried back into the atmosphere through volcanoes; but that is a very long term process.

The oxygen content of the atmosphere also is falling proportionately to the burning of fossil fuels, but because the oxygen content of the atmosphere is so large, the fall is inconsequential. Like that of carbon dioxide, the oxygen content of the atmosphere is seasonal, growing during the northern hemisphere's growing season, when its plants photosynthesize, and falling during the northern hemisphere winter; while carbon dioxide falls during the summer, when it is converted into plant tissue, and increases during the winter dormant season. The percent of oxygen in the atmosphere is maintained, somewhat mysteriously in the face of constantly exposed, oxidizing minerals (exposed by geological activity and digging humans); fires; and of oxygen-respiring living things, by bacteria and blue-green algae similar to those of 2.5 billion years ago, which live in soils or as photosynthesizing plankton on the ocean's surface; and by rooted green plants.

It is now known that bacteria live in rock formations up to 3.5 kilometers below the surface of the ground and several miles deep under the ocean. They can tolerate temperatures up to 113° C., 13° C. above boiling. The bacteria in deep ocean muds derive their energy by reacting hydrogen with nitrate to produce ammonia. Some of the organisms that live in the deep, hot underground biosphere also depend on hydrogen fuel: these form the so-called subsurface lithoautotrophic microbial ecosystems; or, SLIMES, the rockeaters. SLIMES were first discovered in drill holes 1000 meters down in Columbia River basalts. It is thought these SLIMES use hydrogen generated by the reaction of water with the ferrous silicates in the basalt for an energy supply. As with the case of the sulfur reducing bacteria about hot water deepsea vents, other organisms show up to eat the first ones; or to use their wastes; and others to feed on those; so an ecosystem develops. The mass of the deepwater organisms, because of the immense size of their habitat, may be greater than that of life on the surface of the earth, much of which is in the first, well oxygenated foot of soil, the first hundred feet above the soil, and the first half inch of ocean surface. Whether the underworld life has much influence on the surface one isn't known. Bacterial life in deep ocean muds has some effect on life on the surface, because what goes on in the muds influences the chemistry of the oceans, and ocean water circulates at greater speeds than crustal rocks (thousands of years rather than tens or hundreds of millions: the carbon from fossil fuels wouldn't have turned up in the atmosphere any time soon). At any rate, the living influences on the planet's surface and atmosphere are set among more general planetary and gravitational ones (such as the output of the sun, the speed of the earth's spin, earth's aspect to and distance from the sun); and those of plate tectonics. Rocks circulate too, but slowly: the burial and rebirth of plates not only help recycle planetary carbon but also reposition continents and oceans; the position of the continents on the globe influences heat distribution between tropics and poles, and thus rainfall and other aspects of global climate. It has become more and more clear that nothing—no climate regime, no ocean shoreline, no level of solar output, no stand of trees—is permanent. Things like the structure of a particular forest are only partly replicable. Ecosystems have histories: different histories explain differences among otherwise similar landscapes—landscapes similar in microclimate, soil, aspect, hydrology, but different in animals and vegetation. Their different histories from, say, a fire, a period of heavy browsing, or windthrow, nudge them in different directions.

From a more local perspective, trees create conditions that favor forests: they transpire water from their leaves, cooling the atmosphere and creating mist and rain; their roots penetrate the soil, letting water and oxygen in, and secreting carbon dioxide which dissolves rocks; the yearly turnover of root hairs and rootlets (approximately one third of the total root mass) adds carbon and other nutrients to the soil. These nutrients are recycled by soil organisms and the trees. The damp, shaded, more temperate forest soils (not so hot in summer, not so cold in winter) become a home for insects, rodents, soil arthropods, bacteria, fungi. Fungi are important in all forests but especially so in evergreen forests, which are noted for their survival in poor soils, and whose needle-fall tends to resist recycling by arthropods and bacteria, thus locking up nutrients for long periods of time. In some Pacific Northwest forests, soil fungi constitute half the soil biomass. Some evergreens won't survive without specific fungi in the soil. The fungal hyphae provide water and other nutrients to the tree and increase its effective root area 1000-10,000 times. They secrete enzymes that extract nitrogen and phosphorus directly from humus, eliminating the bacterial food chain. Hyphae secrete chemicals that break down the surfaces of stones and transport trace metals to the tree. They help protect tree roots from acid soils. So firs will grow on piles of mine spoil; and clear-cutting, which causes a dramatic decline in soil arthropods and fungi, makes dry sunny sites in the arid west difficult to reforest.

Fungi secrete growth hormones to increase tree root growth and antibiotics to protect tree roots from bacterial infection. Hyphae connect trees of the same and different species, and by providing nutrients from overstory to understory trees, help make forest succession possible. They turn the forest into a single organism. Such services are not free; approximately 40% of the photosynthate produced by the leaves seeps out of the roots to the mycorrhizal and bacterial ecosystem that lies within a few millimeters of the root hairs (the so-called rhizosphere); the photosynthate includes nutrients (sugars) and vitamins. Processes like this are the sort that commercial foresters (or farmers, in another situation) try to short-circuit, by using herbicides and commercial fertilizers to alter ecosystem function, and thus allocate more of the products of photosynthesis to plant growth. This approach ignores the place of the tree in the whole soil and forest community. By providing trees with water, fungi help them better withstand droughts. Polysaccharides secreted by the bacteria and fungi of the rhizosphere also change the soil's structure, making it better aerated and better adapted to the movement and absorption of water (thus better adapted to tree growth). In test plots in a tropical forest in Guinea, trees responded better to a mulch of wood chips than to chemical fertilizers. In general, trees with mycorrhizal connections grow more vigorously than those without them.

The rest of the soil community is also important. The arthropod communities of Pacific Northwest forests include the grinders of successively smaller size that break down litterfall (5 tons per acre per year in old growth forests), until the material, successive levels of bug poop, finally reaches the scale of the plant cell. Bacteria scavenge the available nutrients, die, or are eaten by other organisms, which excrete them, their nutrients becoming available to the trees; the nutrients are also transferred by fungal hyphae to the trees directly. Logs are broken down in a similar fashion. Total incorporation of a Douglas fir log into the soil can take 400 years. While 5% of a tree consists of living cells, 20% of a rotting log is living tissue thanks to the log's inhabitants. Such soil arthropods can be used to define a site. From the structure of the arthropod community one can tell the time of year, the altitude of the collecting site, its aspect (north or south slope), the understory plants, the forest's successional stage, whether it is old or second growth, its mix of trees. The arthropod community has a turnover time measured in weeks (compared to days for the soil bacteria and centuries for the trees) and so is a much better indicator of subtle changes than the trees, whose long lifetimes put them at a distinct disadvantage in terms of short-term changes. Along the same lines, Douglas fir needles have fungi that live in them and manufacture chemicals that suppress the growth of needle grazing insects. The fungi, unlike the tree, can evolve at a rate similar to the insects. As with the soil fungi, these fungi live on nutrients from the needles as well as on nutrients scavenged from killed insects. That is, the service isn't completely free.

Forests intercept and soften rainfall. Rainwater splatters into a mist as it hits the canopy, drips off leaves and branches and runs down trunks, enters the soil along channels left by rotted roots, dead fungal hyphae, rodent burrows, the smaller burrows of soil invertebrates, cracks left by drying soil or ice. Rainwater is absorbed into the spaces that occupy 50-60% of a healthy forest soil. A considerable percent of rainwater evaporates from the trees. In grasslands, where much the same thing happens but on a smaller vertical scale, more water evaporates and runs off into streams, and less enters the ground, than in forests. In older forests, those over say, 150 years in the Pacific Northwest, which is about average for the beginning of maturity in temperate forests, the changing microclimate in the forest begins to favor the growth of lichens on the branches and trunks of the trees. In time, the lichens and other epiphytes on an old-growth tree may weigh four times its foliage and a tree may have 1000 species of affiliated fungi. The lichens gather their nutrients from the air. Rainwater leaches nutrients from them down to the forest floor, where they become available to the tree (pieces also break off and fall to the floor). *Lobaria* lichens are the largest single source of nitrogen in old-growth forests of the Pacific Northwest and fix several times the nitrogen they need for their growth and maintenance. Roughly speaking, the excess is what is available for the rest of the forest. *Lobaria* appear in Douglas fir forests at 100 years (20 years after the last of the planted trees in a plantation is cut) and become abundant at 200 years. A full range of all species develops at 400 years. Thus old-growth forests begin a new nutrient cycle, harvesting nutrients from the air to supplement what is available from the soil, much of which by then has been locked up in the trees. The lichens are a food source for arboreal rodents, and the rodents food for predatory mammals and birds, creating a new loop on the new nutrient cycle. More growth means more carbon stored in forest soils. Flying squirrels and red-backed voles, among whose main foods are the fruiting heads of the fungi of the forest floor, those valuable edible mushrooms, are the prey of spotted owls, a bird whose precipitous decline has led to the controversy over cutting the remaining old-growth forests of the Pacific Northwest.

The spongy forest soil, clumpy with polysaccharides secreted by tree roots and soil organisms, draws in the water it receives. Rainwater is naturally slightly acidic because of carbon dioxide in the air. In the soil it is enriched with more carbon dioxide exuded by respiring tree roots and soil microorganisms. This increased acidity accelerates the process of subsoil erosion (or weathering), releasing nutrient minerals from rocks and rock particles under the tree. Water also carries rock-dissolving fungal enzymes. Most of the nutrients are taken up immediately by the forest community. Some of the soil water, stripped of its nutrients, sinks through the vadose zone to ground water, some is pulled up by the trees and transpired. The 50-65% that evaporates or is transpired from the canopy to the atmosphere will fall on the landscape nearby in the convective rainfalls of spring and summer. Trees grow tall not so as to produce straight sawlogs, but to compete for sunlight, their source of energy. Another function is to intercept water. Precipitation under coastal redwoods in California is 2-3 times the yearly rainfall measured above the trees, thanks to the redwoods' fog-catching abilities. Trees high on mountainsides may get half their moisture from cloud water condensing on their leaves and branches. Some of the moisture runs into the ground and feeds perennial streams. (The process supports forests on oceanic islands whose main source of water is fogs.) The great mass of the trees of the American West Coast may be related to water storage (in tree tissue and soil), in a region of summer drought. For the most part, the dead heartwood tissue of trees is thought to be useless, and hazardous as the tree gets larger and its great weight, relative to its root area, increases its liability to fall. Some biologists have theorized that heart-rot fungi, which help the heartwood decay without harming the living outer shell of the tree, are adaptive mechanisms on the part of the tree to reduce its weight and convert the nutrients stored in the heartwood to something the living parts of the tree can use; thus the tree continues to produce seeds, which is the tree's goal—surely a horrible notion to foresters. Thus some old broad-leaved trees send adventitious roots from their branches down into their hollow cavities or into the soil that accumulates along their branches.

Forests are part of the larger hydrologic and biogeochemical cycles of earth, and of the earth's energy relationships with the sun. Through photosynthesis, evapotranspiration, and the absorption, reflection, and re-radiation of sunlight, forests affect the flow of solar energy through them and determine the forms in which this energy is dispersed. They change the form, amount and chemistry of the water that flows through them, the chemistry and speed of the wind-driven air, and help determine the amount and kinds of dust and sediment released to these fluids. They govern in some degree the behavior of the interconnected streams, rivers, lakes and ocean estuaries. They affect local, regional and global climate. In temperate regions, forests, through the evapotranspiration of water from their canopies, have a cooling effect on the summer climate. In snowy regions, certainly in winter, and probably overall, forests have a warming effect on the climate, because they absorb more heat than bare ground or ground covered with snow. Locally, this warming is thought to exceed the global cooling effect caused by their absorption of carbon dioxide. So the movement of the taiga forests north as the climate warms will likely reinforce the warming. Cutting the European forest seems to have raised midwinter temperatures in Europe by 2-3° C., while cutting the forest on Vancouver Island in British Columbia raised summer temperatures there by 2° C.

The storage of carbon in the tissues and soils of forests connects them to the global carbon cycle. One scheme to lower global carbon dioxide levels is to plant millions of square kilometers of new forests, which while young and growing absorb much carbon dioxide. Existing older forests would be cut down to do this. Cutting down an existing forest releases large amounts of carbon dioxide from the logs, the soil and from rotting vegetation. In temperate forests, it takes 10-40 years before the carbon dioxide taken up by the growing trees equals that released by decomposition. (Tropical forests take much longer.) Once the forest has grown to the point of maturity (that is, to the point where its increase in mass has slowed, and thus the storage of carbon in trunks and branches slowed), the problem is how to log the forest without releasing a high percentage of the carbon. Most of the carbon stored in the wood of a logged forest will return to the atmosphere within a decade. Much of it goes into paper or other short-lived products, and maybe a third is waste, which is burned or decomposes. Since it now appears that forest soils continue to absorb considerable carbon dioxide even as the trees themselves increase in mass more slowly, the likely answer is never to log the forest, or to log it very slowly, so that the soils themselves continue to absorb, or at least limit, their release of carbon. (Cut only trees unlikely to survive until the next cutting cycle.) In this case, the choice of lands to reforest becomes important. We should choose more environmentally vulnerable or valuable ones, such as river floodplains, very steep slopes, watersheds used for water supply, and aquifer recharge areas. We should reforest connected areas of sufficient size for the birds and animals characteristic of a region. We should try to recreate natural forests. Tropical forests cool the local climate and continue to store carbon over their lifetimes. They have a large influence on local rainfall and should not be cut, or be cut very carefully, without much disturbing the canopy, and thus kept from releasing their carbon.

Of course, new forests on deforested or destroyed lands (such as worn out agricultural lands or strip mined lands) constitute an advantage from the start; and are extensive; establishing them may be difficult but involves no initial loss of carbon. In this context, restored degraded grasslands, such as the American plains or African savannahs, are other good candidates for carbon storage.

Forests are connected to the larger world by the water and air that move through them, by the animals that graze and browse on them, and by the plants and animals that otherwise take advantage of the habitat. Insect depredation is usually kept below 20% of leaf mass, partly by insect diseases, partly by chemicals secreted by the trees (24 hours after a caterpillar bites into an aspen leaf, the leaf has increased its poisonous phenols by 5 times; Ponderosa pines flood the tunnels of attacking bark beetles with sticky resins; some leaf responses to damage are measured in minutes), partly by predatory insects, and partly by insect eating birds. (While losses to insects are a serious matter, much of the energy in the grazed leaves returns to the trees through the insects' dung and bodies.) Summer bird densities in the forests of the eastern United States can be startlingly large. Eastern deciduous forests average 863 pairs of songbirds per square kilometer (in effect, about one pair per 1200 square meters, a square 30 by 40 meters on a side). Of these, 587 pairs migrate here from the tropics. Eastern coniferous forest averages 644 pairs, 412 of which are neotropical migrants. Riparian woodland in California averages 1596 pairs per square kilometer,

399 of them neotropical migrants. (California has a more moderate climate, with more insect and plant food available year round, and thus more year round resident birds.) The neotropical migrants come to take advantage of the summer burst of insect life in the temperate and arctic regions. Birds of one species divide up the suitable areas of the forest into separate territories. A given plot of forest is however shared by many different species; the different species divide it into distinct feeding zones, which are gone over daily and intensively. Intensively, since the birds are feeding not only themselves but several growing young. Birds specialize by focusing on different parts of the forest (the forest floor, the mid-level, the canopy), in how they search those areas for food (scratching in the leaf litter, searching under fallen leaves, probing the crevices of trunks, hawking for flying insects), in what they choose to eat. Among wood warblers that forage in the conifers of the northeastern forests, for instance, one species forages in the outer surface of the canopy, one near the trunk, one along a vertical axis, one horizontally, one chooses the upper third of the tree, one the lower third; some are specialists in specific insects, such as spruce budworm. The winter territories of the neotropical migrants are thousands of miles away. Between are the lands that are traversed during migration, slowly in the spring, where, in Central America, the birds' northward movement corresponds with the ripening of some species of tree fruits, whose seeds the migrating birds disperse. Along the east coast of North America, spring migration follows the hatch of geometrid moth larvae, at bud break, up the Appalachian Mountains. The leisurely spring migration, during which the birds feed on abundant, protein-rich, nontoxic leafeating caterpillars, is quite different from the birds' abrupt autumn departure from the Atlantic capes for flights over the ocean to the Caribbean islands or the northern coast of South America. The demanding autumn flights require abundant food supplies concentrated in small areas, that is, easy-to-find fruits, insects, or arthropods, as well as shelter.

To be effective in their job of insect control, the birds must be numerous; ideally they must flood their breeding territories, so that their summer densities are limited primarily by competition for food. This means their "winter" territories must also be optimal, as well as those lands in between used during migration. All these areas (summer territories, winter territories, migratory corridors) have been considerably degraded in the last

50 years. The Atlantic capes have been developed, reducing their supplies of food and shelter. Breeding areas suffer from fragmentation. Most neo-tropical migrants do best in large areas of unbroken forest. Smaller areas reduce the amount of their insect prey, as the drying effect of forest edges reaches into the forest and reduces its invertebrate life. Smaller woodlands also make neotropical migrants, most of which are small, brightly colored, and build rather visible cup nests, more vulnerable to nest predators and parasites. Most neotropical migrants are too small to deter nest predators like crows, raccoons, blue jays, opossums, house cats and skunks, all of which are abundant in forest edges, and may destroy 75% of nests. Cow-birds, the most important nest parasite in North America, lay their eggs in other birds' nests and let those birds raise their young. As they grow, the young cowbirds, which are larger, push the natural offspring out of the nest. Cowbirds, once a bird of the prairie, have increased tremendously in the last several decades. They are a bird of open country or forest edges. The clearing of the forest three centuries ago let them spread east and the more recent fragmentation of the forest, along with the decline in small bird eating raptors, has let them increase. (Other prairie birds, such as the clay-colored sparrow, seem to be following them east.)

At the same time more and more winter range is being lost in south-ern Mexico, Central America and the West Indies to agriculture, log-ging and development for tourism. Species that winter in tropical second growth, that is, cut-over woodland (these include northern parula warblers, indigo buntings, rose-breasted grosbeaks), are increasing in number, while those that winter in tropical primary forest (such as wood thrushes, black-burnian warblers, chestnut-sided warblers, hooded warblers) are declining. The quality of winter habitat makes a difference. American redstarts that winter in tropical forest in Jamaica and Honduras gain weight in winter and return early in spring, while those that winter in cutover dry scrub lose weight and return later.

Hazards of industrial settlement are many. Lighthouses and lighted towers (TV towers, cellphone towers) kill 100-300 million migrating birds in North America each year, lighted windows 100- 1000 million (dim-ming the lights in downtown skyscrapers during migration reduces deaths by 80% and saves millions of kilowatt-hours of electricity); cats 500-5000 million (a modest 5-10 per cat); ingested lead shot 300,000; hunters 120 million; cars perhaps a million a week. (There are perhaps 20 billion birds in North America in late summer.) The night sky through which song-birds fly south is no longer an empty space, lit by the moon and stars, but probed by beacons, saturated with radio waves, swept by thundering jet engines, lit by the urban glow from below. Development removes meadows and shrublands rich in food, groves of roosting trees and coastal forests from migratory corridors. Habitat degradation, along with pollution of the Northern Hemisphere by heavy metals and hydrocarbons (such as mercury, oil, benzene, DDT), and the depletion of calcium from forest soils by acid rain (which reduces the abundance and nutritional quality of their insect life), helps explain the falling numbers of neotropical migrants in North America. One count puts numbers of migrating songbirds at 5% of pre-Columbian populations, with a fall of perhaps 50% since the 1950s.

Migrating birds extend a forest's reach into a larger realm, as does the water transpired from the trees (which enters the hydrological cycle, both local and global); and the chemicals exuded by the trees into the atmo-sphere. Some of these chemicals are part of the trees' cooling mechanism, some function as warning signals of the presence of grazing insects to other trees (which also begin to secrete them, spreading the warning). Predatory insects are attracted by these deterrent and warning chemicals, and eat the grazers or lay their eggs on them. (Some species of bean plants secrete chemicals attractant to predatory insects when the eggs of insects that eat foliage are laid on them.) Such chemicals are secreted in sufficient concen-tration to effect local atmospheric chemistry; in the presence of automobile exhaust and ultra-violet light, they form low level ozone and add to photo-chemical smogs. At least 120 chemicals are found in the air of California's high Sierras. Among the most abundant are monoterpenes, which have a role in preventing and curing cancer. Such chemicals enter the bloodstream through the lungs or the limbic system through the olfactory nerves. So a breath of mountain air is a good thing.

A forest's watery connections lead to the sea, whose nutrients and evaporated water may return to feed it. After Pacific salmon spawn, they die, and the nutrients in their bodies leach out into headwater streams and lakes. Up to 30% of the nitrogen and carbon in algae and aquatic in-vertebrates in the steep, nutrient poor streams where coho salmon spawn come from dying coho. The coho also contribute up to 18% of the nitrogen in streamside vegetation (up to 50% of the nitrogen in some old growth trees). Trees grow 3 times faster along salmon streams. Spawned out fish contribute 25-40% of the nitrogen and carbon in juvenile coho, a direct parental contribution to growth. Most of these nutrients come from the sea, so the coho are part of an accumulated pool of nutrients shared among the sea, streamside forests and the fish, and to a lesser extent with the bears, martins, otters, beetles, fly larvae, white-tailed deer, red squirrels and eagles that eat the salmon carcasses, and which spread the bounty from the sea further into the landscape through their own bodies and dung. In the salmon fed rain forests of the Pacific coast, ancient murrelets nest among the roots of the trees and marbled murrelets in the tops: both are seabirds which contribute the nutrients in their droppings to the forest. East Coast streams also had abundant runs of anadromous fish (herring, sea-run trout, striped bass, shad, Atlantic salmon, the catadromous eels) and so took part in a similar exchange of nutrients; 90% of Atlantic salmon also die after spawning. Fish that didn't die contributed their voided wastes (and their milt and eggs), which correlated in amount with their body masses, much of which derived from the nutrients of the sea.

Forests are also influenced by the presence of large predatory animals. Plant eating ungulates such as deer and elk are capable of rapid increases in population. If unchecked by weather or predation, the animals overshoot their food supplies, their populations crash, and then cycle up and down at a much lower density, in a degraded habitat. Deer at densities greater than 10 per square mile inhibit regeneration in favored species such as hemlock, yew and white cedar. Browsing that eliminates the understory reduces the value of the forest as habitat for many animals, including nesting birds (which reduces the birds' effect on insects). White-tailed deer densities in the eastern United States range from 5 per square mile in mature deciduous forest to 20-30 per square mile after logging; with protection, populations can reach 70 per square mile. High densities prevent forest reproduction and result in a permanent reduction in food supply for deer. In Yellowstone Park, before the wolf was re-introduced, browsing by elk had prevented regrowth of cottonwoods along streams. Most of the cottonwoods were 60 years old or older and dated from the time wolves were eliminated from Yellowstone. Reintroducing wolves cut the elk population from 5-30%. (Elk numbers are more influenced by food supply and weather than predation. Elk will continue to increase during a period of mild winters even as the wolves increase.) But the presence of wolves also changed the elks' behavior. Elk now spent much less time in stream bottoms, where they were more vulnerable to ambush by wolves. This reduced their browsing there and let young willows and cottonwoods regrow. The regrowing forest improved habitat for nesting songbirds and also for beaver (another food animal of wolves: so, here, more wolves meant more beaver). The trees stabilized stream banks and shaded the water. Trout became bigger and more abundant. The ponds the beaver built on small streams provided more habitat for birds and amphibians, slowed water flow, raised ground water tables, and let stream beds build up (aggrade) rather than downcut. Wolf predation also halved the coyote population, so numbers of smaller predators like red foxes and raptors that prey on small rodents rose.

Wolves control deer and elk by taking animals too sick, too hungry, too old or too young (or whose mothers are less powerful or aware) to defend themselves. (This is the usual story. In prairie dogs predation also falls on males obsessed with sex and on pregnant females, who are probably slightly slower. Predation on young deer or elk can become a problem if the prey population is reduced or otherwise stressed.) Thus animals like wolves exert different evolutionary pressures from those of human hunters on their prey. Human hunters from the Stone Age to now take healthy animals of breeding age. Large males are favored as trophies, young females for their meat and hides. To stabilize deer populations human hunters must take 35-45% of the animals annually, not a small number. The difference in predatory behavior is thought to be one reason why prey populations tend to fall in body size under human predation (a smaller size lets them breed earlier) but remain the same or rise under predation by wolves (larger animals are more difficult to capture). Predation influences evolution. Healthy wolf populations depend on migration of wolves from other regions, and thus on extensive connected areas of plains or forest.

The plants, animals, fungi and the invisible life of the soil are influences on a landscape as important as its topography, climate and its soil's mineral composition (all of which, on a local and a global scale, are modified by life). Natural landscapes are connected largely by the movement of water and its associated nutrients, but also by the flow of the atmosphere and the movements of nutrients and energy through seeds, migrating fish, birds and other migratory and wandering animals. Through such connections, changes in one habitat (such as logging a part of a forest) can affect the whole landscape's cycling of water, soil and nutrients, that is, affect the flow of these materials through the larger landscape. When the landscape was edible and its major purpose was to provide food, people were in most cases a less obtrusive part of this landscape. They altered it less, or in less disturbing ways than now, partly by influencing the abundance of large grazing animals, partly by small-scale horticulture, mostly through the use of fire. People were another part of the biota, comparable to an efficient predator like a wolf or a cougar, or to a transformer of soils like the prairie dog, an important but not an overwhelming influence on the landscape, not like now, comparable to a volcanic eruption or a cometary impact.

Chapter 5

A Little History of the Edible Landscape in North America

Where did the modern system of resource exploitation come from? Before the world was marketable, it was edible, and humans, like all plants and animals, were part of that world.

About 13,000 years ago, amidst the retreat of the glaciers that had covered much of northern North America, signs of human occupation appear. In some places, signs appear much earlier: in southern Chile perhaps 30,000 years ago, somewhat less in a cave in western Pennsylvania (the age of the Pennsylvania remains is still disputed). Some writers claim the diversity of North American Indian languages would have taken 50,000 years to evolve. In the present-day United States undisputed signs of people appear as fluted stone points embedded in the bones of the so-called Pleistocene megafauna, the larger (greater than 100 pounds) grazers and browsers and their predators (also large) that inhabited that slightly earlier earth, 400-600 human generations removed from us. So-called Folsom points have been found with the bones of ice-age bison in Folsom, New Mexico, and with those of woolly mammoths and mastodons in the Ohio Valley and New York State. The eastern landscapes would have been just south of the glacier's edge. During the next few thousand years most of these large animals would become extinct. Some writers claim the extinctions occurred much faster, over a period of 300 years, from 13,200 to 12,900 years ago. In North America the Pleistocene megafauna included species of hairy elephant like the herb-eating mammoth (an animal of the tundra) and the forest-browsing mastadon, a fierce galloping carnivorous bear (which some claim kept people from crossing the Bering Plains), 4 genera of giant ground sloth, a large wolf, a tiger, several kinds of horse and donkey, a camel, a moose-elk, a tapir, a cheetah, 2 types of llama, a yak, a giant beaver, 3 kinds of ox. All these animals disappeared. Most of them were larger than those that succeeded them.

In the eastern half of North America, a belt 50-100 miles wide south of the glacial front was open tundra of grasses and sedges, with scattered clumps of balsam poplar and black spruce. Modern deciduous trees like oak, hickory, beech and birch grew near the glacier in favorable locations. White-tailed deer browsed there. The tundra was more productive than the modern tundra because of its more southerly location and warmer summers. The grassland extended south along the Appalachian highlands to Georgia. Lower elevations along the highlands supported an open and parklike boreal forest, with deciduous trees in south-facing coves. In Georgia this forest merged into a closed oak and hickory forest like that of the Middle West today. West of the Mississippi, the plains and Southwest were wetter. Much of the plains was a humid forest; the water in the Ogallala aquifer was still accumulating. Sea levels were 300-400 feet lower and grazing and browsing land extended some ways out onto the continental shelves, though less far off the coasts of North America than off those of Europe, whose shallow continental shelves were extensive. What would become cod banks in the Gulf of Maine were forested islands.

The climate that favored the tundra-steppe was created by the mile-high ice, which held a constant high pressure system over it, altered the flow of the jet stream and kept the climate cool and dry. As the glaciers had advanced, trees had retreated south, finally reaching their so-called southern refugia. As the ice melted, the climate moderated and the trees began to move north. Trees move a few hundred to a thousand feet a year, depending on how they are dispersed (gravity, wind, the guts of seed-eating birds, the beaks of blue jays, the mouths of squirrels). Oaks in Europe returned north at 500 meters a year, in North America perhaps two-thirds of that, or 200 miles a millennium. (While jays will carry an acorn several kilometers in a trip, it takes several decades for that tree to bear acorns.) The seeds of North American wild ginger are dispersed by ants up to 35 meters a year, so ginger could theoretically move north only 10-11 kilometers (6 miles) in 16,000 years; but ginger probably moved north 450 kilometers (250 miles) in that time. Many herbs and trees seem to have moved faster than observation would support, an indication that uncommon occurrences (such as seed transport by floods, windstorms or tornadoes) become relatively common ones over long periods of time. Fish also retreated south along favorable drainages as the glaciers grew and disappeared from rivers and lakes under the glacial ice. The north-south alignment of the Mississippi is one reason that river is so rich in fish species. But oak, hickory and white tailed deer were found within a hundred miles of the glacial wall in New England; and oak, elm and ash (the last two nutrient-demanding species that require good soils) amidst clumps of black spruce within 15 miles of the glacier in southcentral Illinois. Such trees would re-occupy the landscape when the climate ameliorated.

In North America, as the ice began to retreat, and incontrovertible signs of people appear, the Pleistocene megafauna began to disappear. Its disappearance was worldwide, except in Africa, where modern people and large animals evolved together. The extinctions took much longer in northern Eurasia, where people had also lived with the animals for a long time. Were the extinctions caused by people or the changing climate? People may have been new to the Americas but people had been hunting mammoths in central Europe and Russia for tens of thousands of years. In Australia and South America large mammal species also disappeared when humans arrived. In Australia this happened 50,000 years ago and corresponded with a long-term drying of the climate and also with the modification of the habitat by fire. Fire set by humans changed large portions of Australia's semi-arid zone from a mosaic of trees, shrubs and grassland to fire-adapted scrub, and the animals adapted to the former environment disappeared, among them a large flightless bird and the marsupial lions and carnivorous kangaroos that preyed on it, and a 7 meter long lizard.

One of the more ingenious theories to explain the extinction of the Pleistocene megafauna in North America puts animal behavior, climate change and human predation together. As the glaciers retreat and the climate grows wetter (and more favorable to trees), the large grazers help maintain the grass, through the fertilizing effects of their dung and the constant re-creation of new grass shoots, whose high transpiration rates dry and oxygenate the soil. Mastodons, like modern elephants, fed on trees, breaking off branches and stripping bark. They kept forests open. Mammoths and mastodons recycle nutrients quickly, eating 250-300 pounds of vegetation a day, much of it of poor quality (that is, woody: only large animals with large fermenting stomachs can eat such poor quality vegetation). Thus they maintain the habitat. Continual hunting pressure reduces the number of large grazers somewhat. A relatively small rate of predation may accomplish this since the large animals mature slowly and have widely spaced young. While in Asia, these animals had evolved with increasingly skilled human hunters, in North America game animals had lived without people for tens or hundreds of thousands of years and had little fear of them, and so, despite their size, may have been easy to hunt. As grazing pressure falls, parts of the grassland begin to be replaced by forbs and shrubs, then with willow, aspen, birch and black spruce. With a climate less and less favorable to grassland, and continuing predation on the large animals, such changes accelerate. At the same time a rapidly rising ocean eliminates more of the grazers' habitat. So this scenario turns into a downward spiral for the grassland and its inhabitants. (The replacement of open steppe in the north by trees would have warmed the climate worldwide by 0.1° C. So if the mammoths were in fact eliminated by humans, this would have been the first time human activity affected global temperature.)

For the most part, the large animals were replaced by smaller modern grazers and browsers, all adapted to predation by man: the modern bison, elk, big horn sheep. Many of these animals came with people across the Bering Strait. Because of the extinctions, some niches are still open. The Southwest could support more browsers on its thorny scrub. The fruits of some species of large fruited trees in the American tropics are eaten nowadays only by a few species of large beaked birds. Once they were eaten by elephants and their relatives (and several other genera of megafaunal fruiteaters), the seeds that survived passage through the animal's gut getting a good start in a pile of dung. (This trick was used by a biologist to re-establish dry tropical forest in Central America; he fed the tree fruits to horses, who shed the seeds in their dung; the seeds sprouted in the manure.)

While the demise of large, tame animals under human hunting has happened in many places (California Indians eliminated large sea mammals from their coasts, New Zealand Polynesians the moas, the Australians their large mammals), matters may have been complicated in North America by a comet that struck 12,900 years ago. The explosion caused immense wildfires, and would have reduced or eliminated both the peoples of the Clovis culture and the large mammals on which those peoples depended. This would explain why Clovis points disappear abruptly from the fossil record about the time the comet struck, while stone points of a different make (reflecting another wave of immigration?) appear some time later. The comet may also have triggered the massive release of glacial meltwater from the center of the continent to the North Atlantic that slowed the Gulf Stream, causing the Younger Dryas cooling in Europe 12,800 to 11,500 years ago.

Deglaciation took 8000 years. It was a time of flood stories, biblical or Abenaki. The melting ice raised sea levels about 360 feet. During some periods the sea rose as much as 10 feet in 70 years; depending on the slope of the coastline, this could mean the sea moved inland hundreds of yards during a man's lifetime. In northern Europe, which is bounded by shallow seas, rising waters between 11,500 and 6000 years ago covered 40% of the inhabited area, driving people and animals inland. But there were more catastrophic events. One reading of the geological data indicates that approximately 7600 years ago, the rising waters of the Mediterranean broke through a natural dam in the Bosporus and flowed into the basin of the Black Sea. At that time the lands around the Black Sea were occupied by Neolithic farmers. The refilling of the sea happened in a year or two, the rising flood waters moving upslope over the grainfields and villages of the lakeside plains at a quarter mile a day. (An alternative theory claims that the rise in the sea was much slower and was caused by glacial meltwater draining into it from the north; and that the Black Sea then broke through the dam and flowed into the Mediterranean. At any rate a flood, and almost within sight of Ararat.)

Rising sea levels invaded Hudson's Bay and floated the remains of the Laurentide ice sheet about 12,800 years ago. Icebergs sailed out through Hudson and Davis Straits and drifted 1900 miles across the Atlantic toward the coast of France, dropping behind them over the ocean floor the rocks and mud of the bay that had been frozen to their bottoms, thus letting us read of their passage (not the first such armada of ice). The melting of the ice sheet changed the air circulation pattern over North America, and let a modern climate develop. The glacial grassland of eastern North America became an open woodland of boreal conifers, with paper birch, balsam poplar, alder and willow; and then a forest of northern hardwoods, such as sugar and red maple, quaking aspen, and yellow birch, together with hemlock, red and white pine, and red spruce. The more nutrient-demanding species such as white ash, elm, and basswood arrived last. Some species, like red oak, continue to move north and upslope. The boreal conifers continued their move north, overshooting the modern treeline by

175 miles during the Hypsithermal, 9000 years ago, when the climate was warmer and drier than today. A fully modern climate was dependent on the return of the Gulf Stream, which was shut off by massive meltwater drainage into the North Atlantic during the final collapse of the North American ice sheet.

With the return of the Gulf Stream came the establishment of a modern air circulation pattern over North America. East of the Rocky Mountains the climate is controlled by three air streams: warm, moist air from the Gulf of Mexico; cold, dry air from the Canadian Arctic; and moderate maritime air from the Pacific. Precipitation east of the Mississippi for most of the year depends largely on the converging streams from Canada and the Gulf, one cold and dry, one warm and moist, meeting along the axis of the jet stream. Summer precipitation is often convective: moisture rising from landscape warmed by the sun and transpired by trees and grasses condenses into clouds as it rises into the cooler upper atmosphere. Condensation occurs about marine salt particles and methyl sulfate nuclei produced by marine algae, as well as about other natural and industrial aerosols, and falls as rain. (Condensation droplets that form around sulfur dioxide particles from the burning of fossil fuels are usually too small to fall as raindrops and remain as bright clouds.) The Pacific air loses its moisture over the mountain chains of the western United States (the Sierra Nevada, the Rocky Mountains): thus the broad arrow of modern grassland that pushes east from the rain shadow of the Rockies over the plains of the Dakotas and the prairies of Illinois and Ohio into western Pennsylvania, and along the shores of the Great Lakes into New York State. The modern climate let a modern forest develop in the eastern United States by 6000-8000 years ago, though more warmth loving trees (oaks, hemlock, hickory, chestnut) kept moving north and upslope; and red spruce, a slightly more boreal species than its companions in the northern forest, and a signature tree of the higher elevations of the Appalachians during Euro-American settlement, only became abundant there 2000 years ago.

People had inhabited this landscape for at least 11,000 years. They had hunted the large animals and watched the glaciers retreat. They had begun to influence forests and grasslands long before the plants came into equilibrium with their soils. Were the prairies, newly created, with their buffalo and elk, entirely natural? Some argue that by helping to eliminate the Pleistocene megafauna, human hunters also helped create the modern grazers (and thus the modern grassland), partly through hunting pressure, partly by opening the ecosystem to animals that migrated with the people from Eurasia. Hunting pressure by humans tends to favor animals that reach sexual maturity at a younger age and smaller size: the modern bison, as compared to the larger earlier species, whose bones are found at the bottoms of the boneyards beneath buffalo jumps.

The grasslands of North America were once thought to be a creation of climate and topography alone; of atmospheric circulation patterns, a mountain rain shadow. Now it seems more likely that in parts of the landscape climate and topography only set the scene. Much of the plains are too dry and windy for trees, and the wet prairies of Illinois were too wet for too long in the spring and then too dry in the late summer and fall; these wet prairies were inhabitated paradoxically by drought-resistant plants. But grazing (by buffalo, elk, bighorn sheep, prairie dogs) and browsing and grazing (by antelope and by jackrabbits in the drier short-grass and sagebrush steppes) as well as burning (by lightning strikes and Native Americans) maintained and extended the reach of the prairie. In part, both the landscape and the large grazers that inhabited it were man-made. Former prairies in Illinois or Minnesota that are not burned today are invaded by trees. Burning was a major tool of landscape modification by gathering and hunting peoples. Modern studies indicate that burning every 2-5 years in spring maximizes prairie productivity. The Kentucky barrens (called barrens by the Americans because they were treeless) consisted of 6000 square miles of buffalo and elk pasture kept open by grazing and browsing by the animals and by burning by their Native American hunters. Some eastern prairie landscapes still existed at the time of European contact, when most of the natives who maintained them were already dead of European diseases. LaSalle saw buffalo grazing along the south shores of the Great Lakes. These would have been the Pennsylvania buffalo, larger and blacker than the Plains buffalo. There were still an estimated 12,000 of these animals in Pennsylvania in the 1700s. (Since the modern buffalo crossed the Mississippi in their movement east only 1000 years ago, these animals may have been there only a few hundred years.) Buffalo, New York, was named for a buffalo trail that went along Lake Erie. Such trails, following the easiest route between two points, may have been quite old; perhaps migrating mammoths laid out the original paths. At any rate, during a period of damper climate, burning would have held back the forests in places where they would have otherwise invaded. Thus the "oak openings" of Ohio and western Pennsylvania; the buffalo along the southern shores of the Great Lakes; the elk in Michigan and New York State; the populations of buffalo along parts of the Atlantic coastal plain. The oak scrub and blueberry barrens used by prairie chickens in New England and New York State were also maintained by fire, and may have been (like the vegetation on mountaintops) relics from glacial times, kept open by humans for their plants and animals. All these animal pastures, which had existed for thousands of years, many since the melting of the glaciers, would be cleared of wild animals in no time by the Euro-Americans, who had no intention of pasturing communally owned wild animals. They were capitalists and agriculturists, with visions of a privately owned landscape of barns, fields, woodlots, iron plantations, gristmills, cattle.

Edible landscapes provide food, shelter, clothes, tools and fuel in a renewable way, and in amounts that support a more or less constant human population. While the human population seems to have increased steadily up to the time of the adoption of agriculture 10,000-11,000 years ago, most of this increase is thought to have been because of the occupation of new habitats. For most of human time (perhaps 160,000 years), the landscape was renewable and edible; edibility was its function. A buffalo hunter sitting on the edge of the Plains in Wyoming a thousand years ago would have been aware of animal traps as far as he could see. A trap might consist of a log enclosure at the foot of a low bank. The animals would be driven over the bank into the pen; the injured, frantic animals caught inside were then speared, shot with arrows, battered with stone mauls, slashed with hatchets. Converging lines of piles of stones (the stones for the piles and the logs for the trap had to be gathered and carried by hand, perhaps for miles, from wherever they lay) extended back from the brink of the bank for a thousand yards or more, forming a chute up which the animals would be driven. People would jump out from behind the stones waving hides to drive the beasts on. The problem for hunters on foot was getting the animals into the chute. Buffalo have poor vision but are curious and approachable upwind; one might decoy them by pretending to be buffalo, or wolves. In the early spring the grass near a chute might be burned: this would provide new grass to lure the animals. A tall enough bank or cliff wouldn't need a pen since the animals would be sufficiently hurt in the fall so as not be able to escape. Some sites are quite old; some cliff drives have bones from the larger, Pleistocene bison that preceded the modern "dwarf" bison of the plains.

Hunting in this way was a group endeavor. Once dead, the animals had to be tended to immediately or their flesh would spoil. The meat from four adult bison (about 1600 pounds) would yield about 160 pounds of dried meat (depending on how dry one got it, the amount could be double that), which would keep for several months, and feed a family of six for two months on a ration of half a pound a person a day. (Fresh meat was consumed at 6-8 pounds per person per day.) Most of the animal was used. The Plains Indians were buffalo gourmands: delicacies included the stomach contents (a sort of salad), unborn fetuses, the milk from nursing cows, bone marrow, nose gristle. Hunters in a hurry took only the hump and tongue. A green soup was made from drowned and rotten buffalo. What was left at a kill site, which would vary with the number of animals killed and the number of people and the state of their food supply, turned into bait for scavenging bears, wolves, golden eagles, coyotes, skunks, badgers and bobcats, all of which fed and were ambushed at kill sites. The buffalo returned to the prairie not only through their dung and remains but through the dung and bodies of their predators. Cliff drives undoubtedly sometimes resulted in more dead animals than the people could use.

Animal traps varied with location and prey. The Pit River, the longest tributary of the Sacramento River of northern California, was named for the abundance of pit traps for deer (steep-sided holes in the ground, lightly covered with brush) in its valley. In historic times mountain sheep were trapped in pens in Rocky Mountain pastures by the Shoshone. The drives exploited the animals' tendency to run down from a look-out area and then up again; and from that height off a bank and into the pen. Nursing herds, which formed more tractable groups, were usually driven. Animals in the pen were killed with clubs, which have been found in caves. Eight thousand years before this, when the climate on the plains was wetter, and some anthropologists think the hunters came by sailing canoe from Pacific islands rather than by foot across the Bering Strait (or by boat along the kelp bed that extends from northern Japan along the Aleutians and down the Pacific coast of North America), mountain sheep were netted with nets of juniper bark cordage. Nets 50-60 meters long and 1.5-2 meters high have been found folded in mountain caves. In the Great Basin, pronghorn antelope, which can be restrained by a visual fence, were lured into brushwood enclosures and driven round and round until exhausted, then clubbed to death. Their numbers in the Great Basin may have been kept artificially low by this method. On the North American taiga, frozen lakes were used to set up traps for caribou. The chute for the drive was made from small evergreens set out on the ice, the trap constructed in the forest on shore. Samuel Hearne reports trap enclosures reached a mile around and the chute would reach out for a mile or more. Openings in the trap's brush fence were fitted with snares. Animals that weren't snared were shot with bows and arrows or speared. If possible, in all these cases, no trapped animals were spared, as people thought that escaped animals would warn the others. Modern peoples of the eastern woodland, such as the Huron, drove white-tailed deer into similar pens, on land.

People who depended mostly on hunting and gathering moved often. Some of the plains hunters of the time before the horse moved every second or third day. Because of the lack of water, only the edges of the plains, along the rivers and creeks, were habitable. (But seasonal pools, seeps and wallows were more abundant in the presettlement plains, letting individual hunters travel further.) The technology of drying meat was one of the things that made the lives of hunters on foot possible, as were dogs, used for dragging tent poles and carrying household goods. Whether such cultures had any effect on the numbers of game animals is uncertain. Before the plains hunters began trading with the Americans, their effect on the buffalo was probably negligible. Antelope may have been suppressed in the Great Basin, but probably not in California (where the Central Valley had a herd of a million or more animals). In the East some writers think the population of white-tailed deer was kept artificially low. The Iroquois take of deer has been estimated at 11,000 animals, while the annual deer kill in New York State by automobiles is now about 60,000 animals, and the herd numbers over a million; the kill by hunters is around 200,000. The farmed, logged-over and developed habitat of the State is now much more suitable for deer, which like most game animals prefer early successional (some would say partly destroyed) landscapes. (White-tailed deer are one of the few large mammals in North America more numerous now than at European contact.) But villages could keep the populations of game animals near them low.

Most hunters and gatherers get the bulk of their calories from gathering and the buffalo hunters also gathered wild plants: young stems and leaves of *Balsamorrhiza* and *Chenopodium*; bitter-root roots; sego lily bulbs; wild onions; fruits of chokecherry, buffalo berry, gooseberry. Such wild fruits, bitter-sweet and rich in vitamin C, would be beaten up with fat and dried meat into pemmican, a calorie-rich, easily transportable winter food. In the Great Basin, seeds of yucca, Indian ricegrass, wild rye, limber pine, saltbush and *Chenopodium* were collected, the small-seeded grasses beaten off the stem into baskets, the pine cones with their large seeds collected before the seeds had fallen and let open in baskets or on mats in the sun or around the fire; some cones, like those of the sugar pine, were set alight so they would open and shed their seeds. Sometimes rodents collected the food: the Pawnee and Winnebago stole caches of ground beans from meadow voles, the Northwest Indians took caches of aquatic arrowheads from muskrat nests, the Seri and Yaqui robbed caches of mesquite pods from pack rats. In the Great Basin, Mormon crickets, an abundant, large insect, were driven into shallow trenches, grass was set over the top of the trench and lit, the roasted insects then collected and stored. A woman could collect several bushels in a day, the caloric equivalent of a year's supply of pizza.

Harvesting can increase the abundance of some plants. Grizzly bears create lily gardens above timberline by digging and eating the lily bulbs in the fall. Digging loosens the sod in which the lilies grow. Some of the sod is left exposed and rots, releasing nutrients to the ground. The lilies reproduce by bulb scales on the sides of the bulb, which fall off when they are handled, thus starting new lily plants; the bears also leave many bulbs (especially smaller ones) in the ground. The result is a thicker growth of lilies in the dug-over ground (a lily garden), more forage for animals like mountain goats (which eat the blooms), more spectacular scenery for the hiker, a greater resource for the bears. Bitter-root and sego lily (as well as roots like camas and groundnut in other regions) probably responded this way to harvesting by the Native Americans, as ramps (wild onions) are said to increase yearly when gathered by modern commercial collectors. The point of course is not to take all the bulbs. Australian aborigines replanted the tops of the wild yams they dug, so the plants would regrow. In general, harvesting seeds and wild fruits tends to spread the plants around. Some plants however require other techniques: California Indians burned stands of wild rice grass in the fall to increase the yield. Burning volatilizes some nutrients but cycles others through the system more rapidly, increasing yields. With some species of perennial plants, lightly harvesting young stems for food, as the Californians did with their native clovers, increases the size of the underground parts, which then produce more stems. Using the landscape by gathering thus subtly changed it, often for the better, from the point of view of the gatherers.

Other landscapes demanded other forms of accommodation. In the boreal woodland south and east of Hudson's Bay, a more difficult landscape fed the Cree. The Cree ate mammals and fish, greens, berries. One of their prey animals, the varying hare, has a population cycle with a 7 year period. Such cycles are typical of many northern animals. In this case, the rise and fall of the hare was followed, at a slight remove, by that of the lynx that ate them. The lynx is a furbearer, so records of lynx skins sold to traders provide a record of varying hare populations. The Cree would eat more hare when they were available, taking hunting pressure off the moose, their other major meat resource and the big game animal of the snowy regions south of the caribou range, and north of the deer. Beaver were a reliable winter food, if not over-exploited. A hunter knew where to find them, how to trap them. It was said a winter camp smelled of balsam boughs and boiling beaver. The animal's fur was an added benefit. Anadromous fish like Atlantic salmon, blue-back herring and sea-run trout inhabited the rivers of the St. Lawrence drainage up to Niagara Falls and north around the coast through much of Labrador (when, far enough north, trout and salmon are replaced by Arctic char). Other fishes lived in the drainages of the large inland lakes. Fish were relatively easy to capture in spring and fall, when they made their spawning runs. They could also be caught through the ice in winter. Waterfowl and shorebirds were seasonally abundant. Some bands depended largely on muskrat, which were extremely abundant, and also somewhat cyclical, in large river deltas. The small lakes scattered inland from Hudson's Bay held large lake trout. They could not be fished every year. Growth of the fish was too slow, so the lakes had to be fished in rotation. Game management was handled by the tribes through the assignment of individual hunting territories. (Individual territories probably replaced those of small hunting bands, a change that came some centuries ago with commercial trapping for fur.)

Further north, caribou were taken where their migration paths took them across rivers or through lakes, where the animals could be ambushed and shot with bows and arrows as they approached a river crossing, or speared from canoes while swimming a lake. Berries were often abundant. Blueberry barrens were burned, which increased their yield tenfold, making more food available for people, for foxes and bears, for the chicken-like birds of the barrens (grouse, ptarmigan, prairie chickens). Fires often got away and regrowth of the forest could take a long time. The Cree of northern Alberta set landscape manipulating fires to produce a mosaic of mature and early successional landscapes attractive to game: berries to lure birds and bears; herbs and grasses for rabbits and mice, and thus martens and foxes. The Cree might renew cattail marshes for muskrats or aspen stands for beaver with fire. Life under the gray winter sky of Quebec or Maine was perhaps not so hard as one supposes: one old Cree lady told an anthropologist that her rabbit skin winter clothes were so warm that as a girl she sometimes slept outside in the snow.

Perhaps this was an exaggeration. In the late 1700s Samuel Hearne records finding a young Western Dog-rib woman living alone on the taiga. (Hearn, an employee of the Hudson's Bay Company, made several exploratory walks of a thousand miles or more north and west from the Company's trading settlement on Hudson's Bay.) She had been captured by a war party of Athapascans and adopted into that tribe. The Athapascans had killed her infant child when she was captured. She had not forgiven them this and had escaped. Her tribe lived far to the west and she didn't know where she was, so once free of her captors she set up camp on the taiga, certain that sooner or later some of her people would come her way. She lived by snaring ptarmigan, hare and ground squirrels, along with a few beaver and porcupines. She had made fire from two stones and had sewn a new set of clothes from rabbit skin. She also had fashioned a hundred fathoms of line from the inner bark of willow for a fishing net, which she was planning to use when the fish began to run that fall. She had lived alone for seven months. What is impressive about the woman is that she could survive so easily and so well; and she apparently didn't mind being alone. The landscape was her home in a way it could not be for the Europeans; their own natural landscape was not home for them in this way. For Europeans of that time, human life took place on a divine earth, remade by men in god's image, in an act of redemption: the cultivated, domesticated landscape of Europe, a landscape whose history went back several thousand pre-Christian years. The natural post-glacial landscape of Europe, the European forest (as impressive as that of nineteenth century Ohio or Kentucky), once as familiar to its hunter-gatherers as the taiga to this woman, had become an alien one to its present inhabitants, the forest a home of spirits and devils (a hairy wild man with a club was a feature of village festivals); home of the chaos of nature. Europeans took their domestic landscape with them, and so could live far from the apple orchards and barnyards of Europe as long as wheat could be grown, fruit trees planted, and communication with Europe (the spiritual center of the world) existed. Being marooned on an island like Robinson Crusoe was one of the mythic fears of the time. But the whole colonization of the earth by people had consisted of journeys something like that woman's; with small bands of people who could make do with what was around them, and whose spiritual lives were connected to the landscape about them.

Chapter 6
Improving the Edible Landscape: Horticulture

At the time of European contact, the most complex civilizations in America north of Mexico were the chiefdoms of the lower Mississippi valley. The Mississippians depended on the cultivation of corn, beans and squash. Corn, or maize, originated as a tropical grass. Corn became a dietary staple in Mexico about 3000 years ago, but *Teosinte*, the grass from which it came, which has small cobs and small, hard kernels of little nutritional value, was beginning to be domesticated 9000 years ago in tropical Mexico. *Teosinte* stalks are sweet (modern maize stalks are up to 16% sugar) and *Teosinte* may have been used to make beer. It probably took thousands of years for *Teosinte* to become productive enough to cultivate as a grain. Anthropologists reason that corn in Mexico was competing with mesquite pods, a gathered crop, as a food, and that grain yields had to reach 200 kilograms per hectare for corn to surpass the calories yielded by mesquite pods and so be worth cultivating. Whether the native tribes made the same utilitarian judgment is another matter. (Grinding implements with maize residue on them date to 8700 years ago.) Two hundred kilograms of grain per hectare is what modern peasant farmers in Oaxaca expect and the yield reached 4000-3500 years ago, when the first evidence of large scale clearing in Oaxaca appears. Daylength is important to plants and it took more time for corn to make its way north from the tropical Mexican highlands, with days of more or less equal length year-round, to the regions of long summer days in temperate North America. Along with beans and squash, corn became a food crop in eastern North America about 1100 years ago. One can read its spread in the increase in river sedimentation in the southeastern United States 1000 years ago, as river lowlands were cleared for crops. The combination of corn, beans and squash is good nutritionally; the proteins in corn and beans compliment each other, and squash seeds are high in protein and fats. It also works horticulturally, with the three crops grown together in the same field. One hectare of corn, beans and squash grown together yields as much food (calories) as 1.75 hectares of corn grown alone: the combined crops use water, sunlight and soil nutrients more efficiently.

The corn growing chiefdoms visited by De Soto in the Southeastern United States in the early 1500s consisted of collections of large villages occupying favorable sites along the Mississippi and its tributaries. The major villages had plazas, ceremonial platforms, and mausoleums. They were connected to other major villages by roads. Satellite villages surrounded the central ones. Large animals like elk and buffalo were scarce. (De Soto never saw a buffalo.) There had been earlier advanced civilizations in the Mississippi Valley. The early mound-building cultures of the Ohio and Illinois valleys had been based partly on North American domesticates, partly on hunting and gathering. Their mounds are now understood as astronomical constructions that mark moonrise at the northernmost point in the moon's orbit about the earth (a cycle with a period of 18.6 years.) Their crops included a squash grown for its shell and seeds; the oilseeds sunflower and sumpweed; and the small seeded grains maygrass, little barley, knotweed, and goosefoot. These crops were more difficult to grow and less productive than maize, and the villages of the Mississippians were larger: Cahokia, a city of farmers at the junction of the Mississippi and Missouri rivers had 15,000-30,000 people. As in Amazonia, both cultures used the river overflow lands (the bare ground between the rise and fall of the river) for cultivation. They also cultivated uplands. The Mississippians ditched wet upland prairies and grew rows of corn in the mounded earth.

The earlier cultures relied heavily on wild mammals such as deer and squirrels, and also on fish, shellfish, migratory waterfowl, and wild nuts. Riverine resources were important. Forests were probably managed with fire to encourage the growth of nut trees, which are somewhat fire resistant. Wild nuts in the Middle West include hickories, chestnuts, oaks and pecans. Two bushels of hickory nuts yield about 28 pounds of edible nutmeat, which would feed a person for a month. Two-thirds of a bushel of white oak acorns yields the same amount of nutmeat (in fewer nuts), but hickory nuts are a more balanced source of amino acids. Nut production is related to crown development, that is, to tree spacing, so forests can be "managed" by fire, which over time produces a park like forest of large trees.

Such management incidentally manages them for the squirrels, turkeys, deer, bear, elk and buffalo that compete with people for the nuts. Besides increasing the yield of nuts, controlled burning, by increasing the yield of browse and herbaceous forage in deciduous forests, can increase the animal biomass by 4 times. The yield of white oak acorns in southeastern Ohio is 0-156 pounds per acre, of the less edible red oak 5-179 pounds per acre, of hickories 5-46 pounds per acre. (Three hundred pounds per acre seems to be the average yield of modern oak forests in the North American East and Midwest, though other writers claim a single mature white oak will produce 1000 pounds of nuts and an acre of mature oaks in good habitat up to 6000 pounds of acorns per acre per year: this compares well with grains.) These numbers compare with the yield of mesquite pods in the Southwest, or of Oaxacan corn, though probably less than a half to a third of the nuts would have been gathered. The years of low yield are a defense against seed eaters, are more or less predictable and may coincide over large areas. Luckily nuts are storable. (Female red squirrels in the Yukon can predict high cone years in firs, probably through chemical signals in the buds—leaf buds and cone buds differ chemically—and in years of good cone yields produce two broods.) The corn growing chiefdoms collapsed soon after De Soto's passage, most of them without ever making direct contact with Europeans. Their relatively dense populations made them susceptible to European crowd diseases, which cause high mortalities in populations without immunity. When La Salle descended the Mississippi River a century after De Soto, he saw few people and many buffalo and elk.

The native population of the Americas before European contact will never be known, but estimates keep going up. A recent one claims 43-65 million people for both continents and Central America (or, perhaps, double that), with 3 million people in North America. Other estimates for North America alone go up to 18 million. These people were killed less by warfare than by European diseases. The smallpox epidemic of 1520-1524, originating in Mexico, is thought to have exposed most of Central and South America, and perhaps parts of North America, to a disease with a mortality rate of 30-70%. The people De Soto came near may have been killed by diseases carried by his pigs. Recurring epidemics of smallpox reduced the population of central Mexico 70% in the 30 years following the Spanish conquest, from 25 million to 6 million people. The population finally bottomed out at 700,000 people, a 97% reduction, after 100 years. By 1650 perhaps 5.5 million people were left in the Americas, with 1 million in North America. A hundred years later probably 90% of the native population of North America had been eliminated by successive waves of European diseases. These included smallpox, typhoid, bubonic plague, influenza, mumps, measles and whooping cough in the century after Columbus; typhus, yellow fever, scarlet fever, diphtheria, malaria, amebic dysentery, later. Many of the diseases were childhood diseases in Europe, some of which had been affecting Europeans for thousands of years. Despite some immunity (some genetic, some passed on with mother's milk), children in Europe suffered heavy mortality from them. Until the 1780s the mortality of European children from childhood diseases was 10-30%. But when such diseases affect adults who have no immunity to them, the effect on the population is much more severe, since the reproductive population collapses, and children and old people die without those in their vigorous years to support them. The result may have been that by the 1700s and 1800s the Americas were wilder (less managed by people) than for a long time; the forests in 1750 thicker and more extensive than they had been for 1000 years.

At the time of European contact, agriculture was a sideline for many tribes, but the new crops of corn, beans and squash were grown to the limits of their range, and traded further north. The three were almost always grown together, in the same field, a tradition that may have developed from the observation that wild beans and squash often grow with *Teosinte*, the ancestor of corn. Their cultivation was regarded as useful, even if they provided less than a third of a group's calories or, because of poor growing conditions in difficult areas, were not always reliable from year to year. Because most North American tribes depended heavily on gathered and hunted foods; did not have state-like ambitions; did not keep large herbivores (goats, pigs, sheep, cattle), whose effects on forest reproduction, as well as on wild animal populations, can be large; did not burn lime or brick to build monumental cities; and did not produce large agricultural surpluses, I would call them horticultural rather than agricultural. Their agricultural lands occupied only a tiny percentage of their territories. Eighteen million people in America north of Mexico is about 6% of the current population, 2 million (a current low estimate) less than 1%. These people required more land per person for food than we but modified the landscape much less.

Horticulture changes the effect of human occupation on the land-scape. Cultivating plants begins a process that will eventually result in the elimination of the natural edible landscape, and its replacement with a man-made one. At the northern limits of corn culture the Huron, an Iroquoian tribe, got an estimated 80% of their calories from cultivated crops, 65% of this from corn. (Corn constituted 50-75% of the diet of the southern New England tribes, also serious horticulturalists, in a similar climate.) The Huron lived along a major river drainage east of Lake Huron. Huronia consisted of an upland area of light arable soils surrounded by extensive wetlands, with cedar and alder swamps. The cedar lowlands were winter yarding areas for deer, and the wetlands held fish and game birds, including migratory waterfowl, and other animals such as muskrat and beaver; the amphibians that lived in them are also the basis of many food webs. Euro-American settlement with its permanent fields, roads and ditches tends to lower water tables, and Huronia now is much drier and less productive of wild beings but more productive of housing and field crops than originally. In addition to crops, 10-15% of the people's calories is thought to have come from fish, 5% from meat, the rest from gathered wild produce. Calories are not everything and that 15-20% of the Huron's calories came from protein-rich wild foods is impressive for such a large, more or less settled population. The Huron were healthy, and much larger in stature than the Frenchmen who first contacted them. Since body size is an indication of dietary quality, one can assume their diet was better. (Native Americans were also cleaner. When the six-foot tall Osage of the Missouri River who bathed twice a day met the five-foot four Frenchmen who considered bathing dangerous, the Osage were appalled by the Europeans' stink.) The Huron's population has been estimated at 20,000 before the smallpox epidemics of the 1630s, in a territory of 340 square miles (217,000 acres). About 7000 acres were in crops at any one time. Since fields were used for 10 years, about 3 times this was in long-term fallow. So about 13% of their land was "cultivated," and 3-4% was under actual cultivation at any given time. Thirty years in fallow was the time the Huron thought necessary to renew the soil, which is similar to estimates in colonial New England or in tobacco and corn growing Virginia. The numbers yield a population density of 60 per square mile for Huronia (a high number), but 'Huronia' refers to the agricultural center of their territory does not include all the area the Huron controlled. The fall hunt for deer, for instance, took place in the oakwoods of southern Ontario. Did shared resources expand the Huron's footprint? The territory under their control didn't include all the watersheds of their fisheries; or the faraway lands used for breeding and wintering by their migratory waterfowl. The Huron were surrounded by nomadic non-agricultural tribes who also used these animals and fisheries, and so different peoples fished for the same fish and hunted the same water-fowl, either nearby, or during the animals' migrations, or on their breeding and wintering grounds.

Where available, fish and shellfish were important supplemental foods for most of the more sedentary tribes. Some Pacific Coast tribes lived almost entirely on marine resources. Fish in North America were taken with dip nets, bag nets, seines, weirs, hook and line, and with spears and torches. Freshwater mussels were abundant in unspoiled midwestern and southern rivers, many of which were wide and shallow, ideal for shellfish, and marine shellfish were abundant along the coasts. The Huron had settled along a major migratory fish path. They fished the spring spawning runs of walleye, sucker, pike and sturgeon from March to mid-May, the fall spawning runs of whitefish, lake trout and cisco in November, and for lake trout, perch and smelt under the ice in winter. The Iroquois confederacy of New York State (a related Iroquoian group) straddled the junction of two great river drainages, the Saint Lawrence and the Mohawk-Hudson, both inhabited by several migratory species of fish. A Jesuit living with the Iroquois in the 1640s reported that a man could spear a thousand eels a night in the waters of Onondaga; this was during the fishes' fall migrations back to the sea. (One can find such statements about most early American rivers: a Protestant missionary in New Jersey in 1680 observed that when the fish were migrating you could catch them with your bare hands in the pool below the Great Falls of the Passaic River, and colonists in Connecticut claimed you could walk across the rivers on the backs of the migrating fish.) A petroglyph of a large shad is pecked into a granite boulder on the headwaters of Meadow Brook, a tributary of the Winnepesaukee River in New Hampshire. Meadow Brook is far from the sea. The Winnepesaukee flows into the Merrimack River, which flows into the Atlantic north of

Cape Ann. Shad made their way up to Meadow Creek to spawn. The natives of southern New England had several species of migratory herring, as well as Atlantic salmon, sturgeon and shad and the shellfish of the marine mud flats and river estuaries. A tidal fish weir 4000 years old was found during excavation for a building foundation in what was once Boston harbor (and is now landfill). The weir worked by leading fish coming in on a high tide into a complex of pickets from which it was difficult to escape when the tide fell.

Further west, on the edge of the plains, meat replaced fish as an alternative source of protein for the horticulturalists, though a few tribes netted fish from the prairie rivers. The Hidatsa supplemented the corn, beans and squash they grew along the floodplain of the Missouri with buffalo. They used their large, earth-covered pit houses in late winter, spring and summer. During the summer they grew their crops. The cool weather of fall and early winter were spent in tents on the plains hunting buffalo. When they left for the hunt, the harvested crops were hidden in pits in the ground from their enemies the Sioux, who knew they would be away on the hunt, and would raid their stores.

Many of the tribes in the lower Mississippi drainage and the southeastern United States were as agricultural as the Huron. Generally, agriculturalists modify their landscapes, especially their immediate landscapes, more than hunters and gatherers. Their effects reach further if they have grazing animals, or if populations grow rapidly or develop fuel-using technologies, such as pottery-making and metal smelting. At the time of European contact, agriculture in America north of the Rio Grande seems not to have supported large, organized, hierarchical, warlike states, like those in Mesopotamia, Mexico or Peru. Some peoples came close, such as the Huron, or the Haudenosaunee Iroquois, the chiefdoms of the lower Mississippi Valley, the Hohokum and Anasazi of the Southwest, but the effects of their environmental manipulations were limited compared with those of other "high" civilizations. Except perhaps in the Southwest, agriculture was not extensive enough and not associated with sufficient population growth or craft use to be environmentally destructive on a large scale. Throughout most of the country, native use of hilled or ridged fields tended to keep soil on the site. In the Southwest juniper and pinion pine woodlands were cut for fuel, for cooking or pottery making, and the cleared land sometimes

used to grow crops. Once the land lost sufficient nutrients (through cropping or subsequent erosion) the forest would not regenerate.

North America had no domesticated grazing animals, whose effect
on the landscape can be dramatic. Two candidates, the horse and the camel,
died out during the Pleistocene extinctions. Native agriculture had little
or no apparent effect on fisheries, and little on sedimentation in estuaries,
while Iron Age settlement in Europe had an immediate and lasting effect
on both. Studies of river bottoms in the eastern United States indicate large
episodes of erosion between 1100 and 1300, along with a disappearance
of tree pollen in the sediments, signs that river bottomlands were being
cleared to grow maize. The erosion ceased after 1300, though corn was
still grown. Native Americans modified forests and grasslands by fire, but
for the most part so as to maintain plant cover. Populations of some game
animals were suppressed, for instance, the buffalo living near the corn cultures of the Mississippi Valley. In California, some marine mammals were
hunted to extinction, and in the estuary of the lower Hudson and about
Chesapeake Bay, the oyster shells at the bottoms of 4000 year old middens
are larger, so one could argue the oysters were (somewhat) over-harvested.
(One sees such effects on harvested shellfish worldwide, from the Chesapeake to Denmark.) In areas such as coastal southern New England, the
human population had probably reached its limit, in terms of the land-use
strategies the people employed. Most hunting and gathering peoples limit
their populations to 60% or less of the maximum an environment can support, perhaps a calculation of the landscape's long-term carrying capacity;
perhaps a response to the difficulty of harvesting a very high percentage of
a landscape's resources. The land of the New England tribes was nowhere
near the state of Europe in 1650, where essentially all the cultivable land
was occupied, and the remaining rough uncultivated landscape (the half-
fearful wilderness, home of temptation and God) was subject to grazing
and timber cutting, where riverine fisheries were reduced by sedimentation
and dams, and wood shortages were common. The Native Americans still
relied extensively on gathered and hunted foods and still lived in a whole,
integrated landscape. Their burned woodlands, full of game and nut trees
(some probably planted), with rivers full of fish, were another sort of cultivated ground: a forest and riverine horticulture. (Southern New England

had walnuts, hickory nuts, butternuts, chestnuts and five kinds of edible acorns.)

The effects of agriculture vary. In the American tropics, swidden or slash and burn farming encourages the return of the forest through the clearing and burning of relatively small plots and the planting of multi-layered gardens, including tree crops, which follow the crops of annuals and the beds of cassava. The ash from the burned vegetation provides an initial burst of fertility. The mix of crop plants keeps the ground covered and protected from sun and rain. People may return to a garden plot for 20 years to harvest tree fruits as the forest returns (longer if the trees are palms). Most of the nutrients in tropical forests are in the vegetation, and are rapidly recycled by the forests' plants and animals; the soils holding up the forest are leached, acidic and poor. Weed pressure usually becomes too much to keep annual crops growing after a few years and yields also fall as nutrients are used up or erode away. Over sufficient time much of the area suitable for cultivation in the landscape may be cleared, more than once, in a very long rotation.

Native Americans in the Amazon developed a method of improving tropical soils by adding charcoal, creating so-called black earths or *terra prieta* soils. (Some modern tribes also do this, making charcoal from plant waste with cool, slow-burning fires.) Charcoal provides a site for nutrients to attach, prevents leaching and encourages the growth of bacteria that increase the soil's biomass. Black earth soils were enriched with plant wastes, human urine and manure, fish guts and bones, turtle bones, broken pots. Fields of black earth vary in size from 5-700 acres, with soil up to 7 feet deep, and remain fertile today. (The beneficial effects of charcoal may last 50,000 years.) While often found on the low bluffs that edge the Amazon floodplain, areas of black earths occur throughout the forest. Tree fruits, swidden gardens, and crops grown in black earths, along with fish, turtles and manatees, fed the large populations along the Amazon described by early explorers. Some estimate 12% of the forest cover in the lower Amazon basin has been altered, partly by repeated clearing, partly by the planting of useful trees, especially fruit-bearing palms. Others estimate that 25%

of the Amazon basin was farm or agricultural forest at European contact. Peach palms, an apparent domesticate (the fruits vary in size) introduced by Native American farmers to the Caribbean and Central America, produce an oily fruit high in proteins and vitamins that yields more calories per acre than rice or maize. The fruit can be dried to make a flour or fermented to make beer.

In Central America perhaps 40% of the forest is second growth because of clearing for gardens, and much of the rest more or less modified. The Maya essentially remade a considerable part of the Yucatan lowlands with stone tools. Much of the jungle there now is second-growth, 1000 years old. (Throughout the tropics, swidden gardening probably increased with the iron axe, which made it much easier to fell trees.) Parts of the southern Ontario woodland where the Huron went to hunt deer in the 1600s were wooded with even-aged stands of oak and white pine when the Europeans arrived. These woodlands may have been the result of fire, or they may have been cornfields in the 1400s, but they were likely human artifacts. Any evaluation of the ecological effects of the eastern or midwestern agricultural tribes is difficult, because their effects were relatively subtle, and because so many of them were so badly decimated by European diseases so early. When disease struck, 50-90% of a band might die, leaving a handful of adults. Cultures collapsed, tribes consolidated and their power relations changed. The lure of iron tools led to the indiscriminate killing of animals for the fur trade and distorted the natives' ecological relations with their landscapes.

The Onondaga of central New York State were an Iroquoian tribe like the Huron. It is thought the Onondaga comprised 8-10,000 people (a high estimate) in a territory of approximately 17,500 square kilometers (about 4.25 million acres). This comes to 1 person per 450 acres or about 1.3 people per square mile. An acre of cultivated ground yielded 15-20 bushels of shelled corn for an Iroquois cultivator, plus beans and pumpkins. Farmers growing local strains of corn in Oaxaca, Mexico, today get about 1000 pounds an acre, a comparable amount. A limit on fields is imposed by the amount of land the female members of the family (the farmers) can cultivate. If we take a middle figure of 2 people per cultivated acre (the Huron may have supported 3, and other estimates claim an acre supported from 1-5 people), the Onondaga had 4-5000 acres of land under cultivation at any one time. This is a negligible percentage of their landscape. Fields were used for 10 years, more or less, and might be used longer if they continued to bear. The Onondaga had been farming the upland areas of their landscape for about 800 years at European contact. When they started to cultivate corn, the people who became the Onondaga moved their villages out of the river valleys of central New York and into the uplands, where the soils were lighter and warmer, and where they built large, palisaded settlements near year-round creeks. Abandoned fields in the northeastern United States return to forest quite quickly (the Huron would have had a forest of 4-6 inch thick saplings after 30 years), first passing through a berry-and-browse-producing stage, which is useful to both people and game animals. Primary forest returns in 150-300 years. Such tracts are identifiable as fields for some time, because of the even age of the trees, and the dominance of so-called pioneer species, but in time windthrow and the movement of more shade-tolerant species into the canopy (beech, hemlock, and sugar maple in the Northeast) will create the more characteristic pattern of early and late successional trees in an old growth forest. Over 800 years, 80 ten-year periods, the Onondaga would have farmed 320-400,000 acres of their upland, that is, something less than 10% of their total homeland, but a much greater percentage of the land they found suitable for farming (perhaps most of it). Slope, aspect, drainage and soil quality were important considerations in locating fields. Villages had to be near water; they also had to be defensible. During 800 years of agriculture some fields were undoubtedly cleared more than once, and a considerable percentage of their woodland would have shown the effects of former cultivation.

The effect of a village was not limited to cultivated fields. Firewood was gathered daily. For some years the girdled trees in the fields provided much of it. Shortage of wood was another reason for moving a village. Some New England Indians thought the Pilgrims had come because of a shortage of wood in their country, an idea reinforced when the newcomers began splitting trees into clapboards to ship home. Iroquois villages were palisaded. A village of 2000 people occupied three acres (at a minimum), with a perimeter of 1300-1400 feet and so required 1300-1400 foot-thick logs, twice that if the palisade was doubled. Birches and elms were stripped of their bark to cover longhouses, leaving them standing dead in the forest. Such buildings, which could be 100 or more feet long, took a lot of material. One writer states that a Huron longhouse needed 450 poles 10-30 feet long, 7-8 large interior posts, and 4500 square feet of bark covering. A Huron village of 1000 people required 36 such longhouses, on 6 acres, and 600 acres of cultivated cropland, or about a square mile. Early writers report Iroquois villages in central New York surrounded by 6-8 square miles of cropland. Sometimes the whole cultivated area was enclosed by a palisade: 4 miles of posts. Birch bark was also used to make canoes and containers. A Chippewa woman (like the Onondaga, the Chippewa of the central Great Lakes made maple syrup) would store several hundred bark containers with their spouts in a bark shed in her sugar bush, near the framework of the hut her family would inhabit during sugaring season. Basswood trees were also stripped of bark, the bark soaked in water, the inner part stripped off to make twine. Black ash trees were cut down and pounded with clubs to separate the annual rings; these were cut into strips and used for basketry. A village of 2000 people also produces a considerable amount of feces and urine (about 10,000 pounds a day), which was apparently distributed haphazardly about the area. (The feces may have been scavenged by dogs.)

Hunting pressure on small game and some sorts of gathering were greater the nearer one was to the village. Many effects varied with distance. One writer postulates a hierarchical use of fire by horticulturalists like the Iroquois, the Huron and the Cherokee. Fire was used to clear fields near the village; suitable woodland within easy walking distance was burned often (every few years, sometimes every year) to promote the growth of nut trees and berries, and to produce firewood; more distant areas were burned periodically to produce forage for game animals; and the most distant areas were burned occasionally (every 15-30 years) in hunts. Fire surrounds for deer were used by the Delaware, the Huron, and also in the southeastern United States, where burning by the Florida natives in their pursuit of game converted the natural oak prairies of Florida to pinewoods, which supported fewer deer. Whether the Onondaga burned their landscape so extensively isn't certain. Some burning is likely, since burning, by speeding up the turnover of nutrients in soil litter, increases the quantity and quality of forage available to game animals by several times. But except for occasional burning and periodic gathering and hunting, the great bulk of the landscape of 4.25 million acres was left more or less alone. Lightly burned temperate forests remain forests. The Onondaga's collecting of bark, building materials and firewood resembled a light selective cut of a small part of the forest, done by hand. The standing dead trees left after their bark was stripped became part of the forest architecture. As they fell, they would be used for firewood. Indian fields were essentially large gardens. By late July the squash leaves had grown sufficiently to protect the soil from erosion. Dead growth was left on the fields over the winter and then burned off in the spring, mineralizing the nutrients in the dead vegetation for use by the new crop. The winter cover helped against erosion, and the use of permanent hills to plant into also tended to keep the soil on the fields. Any soil that eroded off the hills was hoed back into them in the spring. The effect of the hills was to increase the depth of topsoil for the crop plants; they may also have been warmer, allowing for earlier planting; whether this effect was intentional isn't known. In the upper Midwest flat lakeside soils were turned into ridged fields. The ridges would extend the growing season for up to two months, partly because of cold air drainage off their tops, partly because they warm up more on a sunny day, partly because of the heat storage capacity of the wet earth, or of the standing water, in the trenches between them. The Onondaga's overall effect on the hydrologic cycle was small, though not quite nothing. Iroquois farmers settled near Crawford Lake, Ontario, in 1268 and built a village near the lakeshore in 1325. The nutrient runoff from their fields, and probably from the village itself, was enough to make the lower levels of the lake anoxic. In 1486 the Iroquois left and the nutrient input dropped, but the water at the bottom of the lake remained anoxic for several centuries. Crawford Lake however, is extremely susceptible to small amounts of nutrient disturbance.

The effect of human material lives on the environment runs along a continuum. Aboriginal horticulture and fire modified ecosystems, but they didn't much affect their ability to conserve nutrients. Fire had some effect: one of its purposes is to speed up nutrient turnover. The level of use, as well as the cultural practices themselves, were very different from the largely cleared landscape, with its more or less permanent fields, many kept bare over large parts of the year (a good farmer plowed up his cropland in the fall, to be ready for harrowing and seeding in the spring) that European settlement would bring. By 1930, after a hundred years of Euro-American settlement, Onondaga County (the heart of the Onondaga homeland) was 90% deforested; and Onondaga Lake one of the most polluted bodies of water in the country, perhaps in the world; polluted with both industrial waste and sewage. (The industrial particulates in the air produced beautiful sunsets.) Much of the farmland in Onondaga County couldn't compete with midwestern lands, and forest cover amounted to 20% in the 1950s, when I first saw the county. But the influence of Indian settlement was not negligible. One writer speculates that the elimination of 90% of the native peoples of the Americas by European diseases after 1492 allowed sufficient forest regrowth to cause climatic cooling in Europe, that is to say, the cooler centuries of the Little Ice Age. The regrowing forests in the Americas absorbed enough carbon to bring the atmospheric concentration of carbon dioxide down several parts per million and lowered temperatures in northern Europe by 1°- 2° C. in winter, part of a degree in summer. (That forest regrowth caused the Little Ice Age is doubtful but it might have reinforced it.)

In most of the United States, fire modified the landscape more than agriculture. The mesic hardwood forests of the eastern United States are susceptible to manipulation by frequent light burning. Low intensity fires volatilize few nutrients and in general increase nutrient mobilization (making them more available to plants) and soil fertility. Fires also increase a landscape's heterogeneity—its mix of plants and forests in different stages of succession. Europeans found wooded open parkland with oaks, chestnuts, pines and much grass in southern New England, the Chesapeake region and central Mississippi; and thousands of acres of grasslands (so-called barrens) along the Rappahannock River, the Potomac, in the Shenandoah Valley, in western Kentucky, eastern New York, Alabama and Missouri. In the eastern United States grasslands imply frequent burning. Light burning can increase the yield of browse and herbaceous forage in deciduous forests by 3-7 times, the animal biomass by 4 times. At the time of European contact, the tribes in southern New England were burning their woods in spring and fall on regular rotations. Burning tended to favor oaks and chestnuts, useful for their nuts, which were also eaten by game animals, and created an open grassy landscape of large nut-bearing trees, with pines and whatever other species could tolerate the fires or sprout from stumps. Burning regenerated browse (for game) and berry plants (for game and people). It expanded areas of shrubland (created naturally by river flooding, beaver meadows, windstorms and wildfires), habitat for birds like the yellow-breasted chat, brown thrasher, indigo bunting, chestnut-sided and mourning warblers, birds which were rare or lacking in primary forest. (Audubon saw a chestnut-sided warbler, a common species of the forest edge today, once.) Only parts of any landscape were suitable for burning. Beech and maple forests tended to be too wet. The conifer forests of the higher elevations or more northern latitudes of New England and Canada were also too wet, and when they burned, burned catastrophically, setting regeneration back to zero. (The Cree occasionally burned them, sometimes by mistake. But the boreal forest was also intentionally burned. It was burned along trails and traplines to help access or improve the habitat; areas of dead trees or blowdown might be burned to clear away the trees; blueberry barrens were burned every 2-3 years to improve the crop.) The extensive cedar swamps of New England were deer yarding areas and also escaped burning.

Pollen studies of pond sediments indicate that forests in the Cumberland Mountains of Tennessee began to be burned over about 3000 years ago, some time after people in that area had begun domesticating native plants. In the Cumberlands burning both altered and partly cleared the forest, changing a forest of cedar, hemlock and mixed hardwoods to one of oaks, chestnuts, pines, and sugar maples, with many sunny openings. In the same period, 200 miles southeast of the Cumberlands, people were firing the upper slopes of the Blue Ridge mountains in spring and fall to create a similar forest of widely spaced chestnuts, oaks and hickories, whose nuts and acorns were useful as food and also attracted elk, deer, bear, turkeys, perhaps buffalo (which crossed the Mississippi heading east about a thousand years ago). Such manipulation altered the tree composition of the forest and changed its nutrient cycles. Some nitrogen and carbon would have been volatilized, and the nutrient cycles speeded up. A frequently burned landscape was a mosaic of young and old trees, with small clearings.

Populations of birds, amphibians, and soil organisms in the burned-over areas changed, some reduced, some made more abundant. Fires skip around. Much of the landscape was left unaltered and so able to take up what nutrients were released. Human settlement was concentrated in the alluvial valleys of large rivers and on lakeshores with their good soils (and thus more abundant game) and fish and shellfish. While such regions were managed relatively intensively for hunted and collected foods, and for horticulture (though much less intensively than the Europeans, whose landscapes would support 10-100 times more people), much of the landscape was still little used. In frequently burned landscapes old trees predominated, with their large crops of nuts. If Native American cropland supported 2 people per acre, then a population of 5-10 million people in what is now the eastern United States had 2.5-5 million acres of cropland at any one time, plus as much as 150-200 million acres of young successional and maturing second growth forest. These figures are probably over-estimates, since many tribes in the Mississippi drainage farmed river bottoms, where their lands were renewed by floods, and so used the same fields for long periods of time (there was no need to rotate them into forest). Cropland in the eastern United States today amounts to 278 million acres, all of it in use at once; commercially cut forestland, where old trees are rare, constitutes much of the forest (that is, much of the undeveloped landscape).

The California tribes may have modified their landscape the most. In the better watered coastal ranges, 1000 people might inhabit a village of wooden-sided pit houses. Villages were inhabited for centuries. As with the natives of the Northwest coast, there was some division of the population into elites, ritual specialists, craftspeople and ordinary people. California remained the great exception in the Americas, in that the people grew no crops in a climate in which they could have. Californians lived on salmon, acorns and marine resources. The combination had the advantage that droughts, which would reduce the acorn crop, were usually associated with upwellings of deep water along the coast, which increased the supply of fish and shellfish. California has a Mediterranean climate of winter rains and summer drought and so is more suitable for the winter grains of the Old World than for the summer crops of corn and beans of the New World. The Californians could have grown corn on the flood plains of receding rivers, as the Mohave did along the banks of the lower Colorado, or practiced irrigation like the Pima of Arizona, or ditched wetlands like the corn growing Middle Westerners. They certainly would have been aware of such practices. The Europeans began to cultivate the margins of seasonal lakes that were formed by spring snowmelt from the Sierras, as soon as they realized the possibilities. One example was Tulare Lake created by the Kern River, near present-day Bakersfield, California. The receding edges of the lake were first used to grow grain. Soon the new people dammed the Kern above its canyon, making the whole of the lake bottom cultivable, and spreading out the irrigation water over the summer. When all the land and water was used for agriculture, a wonderland for waterfowl was eliminated. In modern times, irrigated lands created by similar projects in Africa have generally been much less productive than the wetlands destroyed by the dams, where farmers once planted rice on the banks of receding lakes and rivers, herders used the new grass, fishermen caught the fish that bred there.

Perhaps the Californians didn't see the need for additional supplies of carbohydrates; they had several species of oaks with edible acorns scattered over the uplands and in the gallery forests along the rivers that crossed the Central Valley. They may have domesticated a walnut, a thin shelled form of which grows about former villages. In the high deserts, pinion pines were managed for their nuts. In the Mohave Desert mesquite trees were pruned and cleared, and desert fan palms, which produce an edible fruit, maintained by burning and clearing. The palms themselves may have been dispersed by people. California was relatively crowded, with well over a hundred tribes. It may have held 10% of the Native American population north of Mexico. Population densities along the north coast, with plentiful rainfall and a varied habitat, reached 6 people per square mile. The landscape included grassland that supported deer and elk (and antelope in the Central Valley), chaparral (more deer pasture), rivers with salmon and other fish and crayfish, and several species of oaks with edible acorns. Acorn yields went as high as 700 pounds per acre. Such yields, if fully exploited, would have supported 50 times more people than were in the area when the Europeans arrived. Acorns are laborious to process; the shelling, pounding, leaching and boiling required to process five pounds of acorn meal (enough for several days) would take a California woman 7 hours. In some tribes the right to harvest certain trees or groves was inherited. Oaks would continue to bear for centuries.

Six people per square mile is high for a gathering people. The Cree landscape supported one person per 75 square miles (0.013 people per square mile); the Onondaga 1.3 people per square mile (for the whole occupied area). The woods of Maine were supposed to have supported 0.4 people per square mile, the corn growing land of southern New England almost 3 people per square mile. Nineteenth century Americans considered one person per square mile the density of the frontier. An early version of the Homestead Act allotted 160 acres, 0.25 square mile, to a family. For a family of 5, that comes to 20 people per square mile; but for a market, not a subsistence, economy. In aboriginal California (not a market economy) perhaps the oaks, deer, wild grass seeds, marine mammals, salmon, and other fish and shellfish made crops unnecessary, from the point of view of the native population.

Fire was the major tool of landscape management in California. Light annual fires kept Douglas fir from invading the oak savannahs and increased the productivity of the native bunch grasses, whose seeds were harvested for food and which were grazed in winter by antelope, elk and deer. The Yurok and Karok burned to increase the productivity of huckleberries, hazelnuts (for nuts; and the stems for basketry), bear grass (a lily), grass seeds, acorns, and wild tobacco. The Chumash burned grasslands after the seed harvest to promote new growth that attracted rabbits. In the Willamette Valley of western Oregon burning encouraged the growth of camas, bracken and nettles (all food plants) and the grassland (the animal pasture) in which they grew. By speeding up the turnover of nutrients, burning increased the yield of oaks. Grasslands and chaparral were usually burned in the fall. Chaparral would be burned once every several years. Firing chaparral increases the number of deer it can support by a factor of 3-4, from about 30 per square mile in unburned scrub to 98 in the first year following burning, to 131 in the second year, after which the numbers slowly fall. Fire drives were also used in hunting.

An estimated 1.5 million antelope inhabited the Central Valley and its wetlands were a winter home for 60 million waterfowl. (In the early 1900s householders in the Los Angeles Basin still shot passing waterfowl from their porches.) The valley's native bunch grasses could be burned in late winter to lure grazing animals, such as antelope. Bunch grasses do not survive continuous grazing pressure (deer, elk and antelope are browsers as well as grazers) and cattle grazing has eliminated much of the native grassland of California. (It has been replaced by grasses more adapted to cattle.) Salmon, also a basis of the California diet, were, if the rivers allowed, caught with weirs. A good weir will catch many of the fish. So one could say the fishery was managed. Different tribes inhabited different reaches of a river, and the downstream tribes could have taken most of the fish. But, probably under threat of retaliation, they left fish for the upstream people; and everyone left fish to spawn. Some writers claim the Californians reduced the populations of deer, elk, some shellfish and sturgeon. They killed mostly female and young seals and sea lions (they tasted better and their skins were easier to handle) and over time eliminated the nurseries of these animals on the mainland. They hunted the large, kelp-grazing sea cows to extinction. Thirteen thousand years ago, to judge from sea urchin remains in middens, Indians on the Channel Islands, seriously depleted the local population of sea otters. On lands that are now undersea, some west coast natives built stone walls long ago to hold silt in beds, probably to encourage clams.

Would such landscapes have appeared pristine to our eyes? In many cases their vegetation was altered by burning. Since the fires were periodic and light, much of this might only have been apparent to an acute observer. Audubon however remarked on the intense smokiness of the autumn air in the Mississippi valley in the 1820s; and John Wesley Powell mentioned that in a summer spent among the foothills of the Rockies in the 1880s, months entirely among the mountains, he never saw a mountain peak because of the smoke in the air. Fires naturally burned millions of acres annually in the dry grasslands and forests of the West and more fires were set by the native inhabitants. It is now thought that the open stands of longleaf pine once characteristic of the southeastern United States were a fire subclimax of the eastern deciduous woodlands. Fire triggers a growth spurt in longleaf pine seedlings but kills most young deciduous trees. Modern longleaf forests that are no longer burned are invaded by oak-hickory forest. To keep oak from dominating, the woods must be burned every 2-5 years in the spring; both the time of the burn and its frequency are important. So these forests, which covered millions of acres at European contact (and are almost all gone, replaced by fields of corn and soybeans), were probably human artifacts. In the Northeast a more subtle effect on forests may have been caused by heavy hunting of white-tailed deer. Browsing by deer differentially effects the survival of tree seedlings, and so the future trees of the forest. Thus reducing the density of deer changes the composition of the future forest. Similar effects would have occurred wherever large herbivores were reduced in numbers. Such reductions are not necessarily negative and were not universal. It has been proposed that the no-man's lands between tribal territories functioned as sanctuaries where game animals increased in number and from where they repopulated the areas in which they were hunted more heavily.

Near villages, forests would have shown the effects of the gathering of firewood, bark and building materials. The Nootka, one of the salmon tribes of the Pacific Northwest, built timber-framed houses sided with red cedar planks. The planks were split from living trees. A set of house planks represented considerable effort on the part of the owner and were carried between summer and winter dwellings by canoe. (Several trips must have been required.) Old-growth red cedar are large trees, 6-8 feet in diameter above the root buttress. To get a plank from one, a man cut a groove at the bottom of the tree deep enough to form the bottom of the plank, and then another groove at the top, 20 to 60 feet up the trunk, depending on the length required. A split was opened at the top cut and a long pole worked into it, and the pole left in place for the movement of the tree in the wind to split off the plank. Once free, the plank would be cut loose and squared up with an adze. (Nineteenth century travelers remarked on the "nibbles" made by native adze-heads of copper, stone or bone.) A house required many planks and a village had tens of houses. So a walker in the woods in Nootka country would have come across stands of strangely deformed cedars, many of which would probably stand for centuries after such mutilation. Of course, since the trees stood, and if alive produced seeds, and if dead still constituted part of the architecture and nutritional status of the forest, such "mutilation" was much less harmful to the woods than modern logging. Such use would only have affected accessible stands of trees, and only some of those.

At the time of European contact, while human use would have affected the appearance and biological structure of much of the North American landscape, human effect on the overall ecological functioning of that landscape was still small. As a collection of nutrient recycling ecosystems, the landscape still worked. One can see this in its abundant fisheries, songbirds and game animals, all of whose populations collapsed after European settlement, as the landscapes were altered and the native beings replaced by crops and by domestic and introduced animals. This is why the extinction of the Pleistocene big game, which unquestionably coincided with the spread of people, and which was probably largely caused by them, seems such an anomaly. Perhaps big targets are tempting; and perhaps the climatic changes of the Pleistocene were more abrupt and severe in North America, whose shape (a north-south plain between mountain ranges) tends to amplify swings in temperature. Extinctions caused by people are common on islands. The Maori, and the egg eating Pacific rats they brought with them, killed off the large flightless birds of New Zealand within a few hundred years of settlement. The unwary birds were easy to kill and the rats (eaten by the Polynesians) increased in numbers quickly. With the moas went the world's largest eagle, Haast's eagle, which weighed 30 pounds, had a wingspan of 9 feet (short for its weight), and approached the size limits for powered flight. The eagle depended on the large moas, which were 15 times its size, for prey. It killed them by striking them in the head.

Chapter 7

The End of the Golden Age: the Effects of Agriculture and "High Civilizations" on the Environment

Human transformation of the landscape, like its transformation by trees, whose leaves and fruits provide food for animals, and whose shade, fire cycles, evapotranspiration rates and root secretions change the environment for everything, falls along a continuum. Its effects depend on the size of the human population and on its relative demands. One looks for a discontinuity: when did an edible landscape like that of the Cree, or even the horticultural Huron, become a marketable one, like that of sixteenth century Europe, Anasazi Arizona, Sumerian Kish? The permanently cultivated fields of agriculturalists, whether irrigated or not, mark an intensification of use in a defined habitat. Regions of early agricultural civilization include the Mesopotamian uplands of modern Syria and Jordan; the valleys of the Tigris and Euphrates; the valley of the Indus; the lower Nile Valley; the terraced uplands and riverine lowlands where wet rice cultivation developed in Asia. Somewhat later came the Americas, lost in their postglacial isolation, with the Valley of Mexico populated at the time of the Spanish conquest more densely than China; the dry uplands and swampy lowlands of Mayan Central America (whose salty ground waters made the area chemically hostile to dense settlement, requiring paved water catchments and plastered water reservoirs, the plaster underlain with crushed limestone); the grassy altiplano about Lake Titicaca, probably deforested by hunters and herders for llama pasture thousands of years before the farmers, along with the straightened rivers, terraced fields, and the ditched shore of the lake itself. Because of its tropical location and because more moisture is available at higher elevations, much of Peru's cropland lies at elevations over 9000 feet. The people of the pre-Incan Wari culture constructed terraced fields at high elevations, irrigated by water from mountain springs and glaciers, throughout the Andes. Most, like the raised fields of the Tiahuanaco people beside Lake Titicaca, are now abandoned. Some of this agricultural practice worked well. Terraces in Peru's Cola Valley, in use for 1500 years, have topsoil horizons 1-4 inches thicker than nearby uncultivated soils, and greater amounts of soil nutrients.

All early agricultural regions sustained intensive development of limited areas through the creation of fields, the diking or diversion of rivers, the construction of irrigation canals, the digging of raised fields in wetlands. Such developments were necessary to produce the agricultural surpluses that supported high civilizations, that is, those with grandiose, permanent buildings; calendars derived from astronomical observation; armies; craftspeople (weaving, pottery, metalwork, stonework); some sort of record-keeping ability. An Egyptian peasant produced 5 times the food required by his family, a Mayan farmer less than double his family's needs; which makes the Mayan achievement more surprising and explains why the Mayan states, despite their bellicosity, could not afford standing armies, or support them long in the field. Perhaps their very bellicosity was a result of their tenuous holds on their material lives, their long history of nutritional stress. (Some Mayan cities fed women—the bearers of warriors—better than men during difficult times.) The floodplain of the lower Amazon supported a population of maybe 3 million people (some writers claim more), but they for the most part lacked permanent buildings or permanent records. The area exposed by the seasonal rise and fall of the river, some 40-60 vertical feet for 1000 miles above the estuary, was used for cultivating crops, as was the forest itself, many of whose trees (especially the palms) produce edible fruit in amounts that rival or surpass the yield of grains. This population melted away upon contact with the Portuguese, leaving little trace, except in the cultures of some of the forest tribes (Levi-Strauss suspected some had descended from more complex cultures); in the structure of the floodplain forests, many of whose tree populations seem to be the result of human manipulation; and in the extensive areas of black earth (*terra prieta*) soils. (*Terra prieta* soils are manmade soils enriched with charcoal ideal for growing crops. It is now thought man-made *terra prieta* "black earths" may constitute 10% of Amazon soils. At the mouth of the Tapajos River in the Amazon is a zone of black earth soil 3 miles long and 0.5 mile wide. Accessible Amazonian black earths are mined and sold to garden centers.)

The effects of large agricultural civilizations extend far beyond their fields. Settled agricultural populations, with their ability to accumulate things, their constant growth of population, their rising wealth, their need for building material and cooking fuel, their technologies (pottery-making, brick-firing, metal smelting) that further increase their need for fuel, their domestic animals, construction programs and wars, have a tremendous effect on the surrounding landscape. Primarily this effect is through deforestation: trees cut for building material, for cooking fuel, for fuel for smelting metal, firing pottery, firing bricks (1 cubic meter of burnt brick required 150 cubic meters of wood in Roman times). Deforestation in ancient Eurasia seems to have been essentially related to demand, that is, the timber was cut as needed. To a certain extent in the ancient world, forest exploitation was limited by the technology of transport, so only more valuable woods were shipped (such as Lebanese cedars to Egypt). But forests near mines, such as in Cyprus or Spain, might be cut and recut several times to smelt the ores, and the (more valuable) metal shipped much greater distances than the wood itself could be. The plain of Cyprus was cut over several times to smelt Cyprus's copper ores and for material to build ships. Finally the land (whose trees were probably kept from regrowing by browsing goats) was granted free to farmers, in order to enrich the *polis*.

Herds of sheep, cattle, camels, goats, used for meat, milk, wool, traction power and transportation, were pastured on the areas surrounding the agricultural zone, sometimes at great distances, the animals moving from pasture to pasture up the slopes of distant mountains as summer progressed, returning to the lowlands near the villages in the winter. In Mesopotamia, pastoralists grazed their animals near the rivers in winter, moving them out onto the plain in summer, until growing droughts after 2200 BC forced them to stay near the rivers all year. Overgrazing causes erosion. Pasturing cutover forests prevents their regrowth. Greek shepherds girdled trees to encourage the growth of grass. For Greeks, cutting the forest meant the desecration of paradise, for man had sprung from an oak;

but also the birth of civilization. Cutting forests and grazing would change the local climate and hydrology, increase temperature extremes near the ground and runoff from rain or snow, raising flood levels and levels of salts and silt in rivers. Cutover forests would become scrub, or grassland; grassland, if overgrazed or if its native grasses were not adapted to grazing (as in the American Southwest), would turn into more unpalatable vegetation: the thymy uplands of Greece, the sage, creosotebush and mesquite of the Great Basin of North America. In this way over thousands of years whole landscapes were altered. By the time of the fall of Rome, the Mediterranean basin was a largely denuded landscape. The marshes at the mouths of its rivers were human artifacts, created with soil washed off the uplands, its steeper hills were eroded, often to bare rock, deforested, sometimes terraced, or used for pasture. (Terraces accumulate soil by mass wasting, and help store runoff water, passing it slowly down the hill, thus making it more available to crop plants; by slowing water movement downhill, they may help recharge ground water levels.) The vine and olive, with wheat grown on the flat patches among the trees, were crops that would grow on the eroded soils.

Humans didn't do it all: a slowly drying climate, which deforestation reinforced, modified the air circulation patterns over the Mediterranean and further reduced rainfall, especially over northwest Africa. When the rain was hard, the soil slid away: the natural erosional processes of the Mediterranean proceed by catastrophic rain events that rearrange landscapes in a stroke. But people created many of the changes. Anatolian and North African lands on the south side of the Mediterranean basin, those "fat lands" colonized by the Greeks, and used, along with the Nile Valley, as a source of grain by the Romans, were also severely eroded by deforestation and agriculture. (By the time of the Peloponnesian Wars in the early 400s BC, Egypt and Sicily were providing 30-75% of the food for Greek cities; and 900 years later the silt carried by the Nile was still feeding Rome.) It was said that in early Roman times one could walk from Egypt to Morocco in the shade. But by AD 100, the Middle East of the Roman Empire was completely deforested. Most soils in North Africa used to grow wheat were degraded by AD 200. Former Greek port cities in Turkey lie several kilometers inland from the modern coast, and tens of meters of soil and gravel cover valley bottoms in Israel and Lebanon. Deep under such spoil lie Roman bridges. Ancient Antioch, once in Syria, now in Turkey, lies under 28 feet of eroded soil. Such bare lands no longer hold much water. There are few springs; houses are built with cisterns. Similarly, the irrigated landscapes of Mesopotamia and the Indus valley turned into deserts; the Thar desert in India was a jungle at the time of Mohenjo-Daro, and Mesopotamian Iraq was a fertile land of swamps and partly wooded uplands, with tigers, date palms, and herds of gazelle. The Thar and Mesopotamia also suffered from long-term postglacial drying, but again people's civilizing activities helped things along.

Life was still possible after a collapse, if "high civilization" wasn't, and the rise and fall of human affairs gave the landscape some respite. The Maya population in the 800s fell from 2.5-5 million to 500,000 with the civilization's collapse and the tropical jungle recovered. Iron age agriculture in Britain left erosional terraces a meter or more high along rivers like the Severn; then the people disappeared and the land regrew to trees before the next invasion. Climates change and cultures come and go. Settlers of "new land" in times of population expansion in Spain would come upon the stone walls and old olive trees of their predecessors amidst the fragrant scrub. Some writers trace the rise and fall of civilizations to the 30-70 generations (800 to 2000 years) it takes a civilization to occupy and erode its soils. During the Bronze Age, Mediterranean uplands were settled, abandoned and resettled over 1000-year periods, as their soils eroded under the plow (fields were plowed 3 times a year), were abandoned to pasture or scrub, the population of the hills declining, until the topsoil recovered and the upland was settled again.

All previous "high civilizations" have collapsed, some from conquest, some from internal dissolution, often helped along by environmental problems, such as a cooling or warming climate, soil depletion, drought or new diseases. It has become the fashion to put such collapse down to abuse of the environment; that is, to exploitation of fields or forests at an unsustainable level. But some economies collapsed because demand for their products fell (the southern Arabian trading cities that depended on the collection of frankincense and myrrh are examples), or caravan routes were cut off by wars, or were no longer usable because of a naturally drying climate (both seem to apply to the overland route from China to western Asia and the Mediterranean between Roman and Renaissance times). A hundred year drought brought down the Tiahuanaco civilization (a pre-Inca culture) in the Andes. A decline in rainfall of 10-15% caused a fall of 40-45 feet in the level of Lake Titicaca, which dried up the raised fields about its shore, and slowly depleted the ground water levels that fed the rivers and the fields that were irrigated from them. Ground water reserves are large. They take a long time to dry up and a long time to recharge. In Tiahuanaco, ground water also fed fields that had been excavated down to just above the ground water level.

The climate change that caused the drought in the Andes was worldwide. At the same time a shift in rainfall patterns in Central America weakened the Maya civilization and eventually caused the collapse of the southern cities. Rainfall seems to have decreased in the uplands, terraced to grow corn, and increased near the coast, where crops were grown on ditched swampland. Severe droughts in the 800s and early 900s correspond with the abandonment of many Mayan cities. The Maya uplands were vulnerable to erosion, the ditched swampland to siltation. Both were labor intensive habitats to maintain. The canals among the swamp fields were used for transporting crops and also were a source of fish. The Yucatan is a karst landscape of eroded limestone with little surface water. Much of it was cleared during Maya times and remained cleared for centuries. (The Maya may have planted woodlots for building material and thatch.) In the southern lowlands, with 100 inches of rain a year, cities got their drinking water from plastered reservoirs fed by paved or plastered catchments, in the north from wells or natural pools. Lake cores show episodes of severe erosion, presumably from the terraced cornfields on the uplands about the lakes. Erosion would have started with forest clearance (for wood for building, cooking fires, to manufacture plaster), perhaps lessened with terracing, but worsened under a high rainfall regime. After the erosional events, rainfall seems to have dropped off. The corn crops would then have suffered from lack of water. All along soil nutrients were declining from continuous cultivation. The swamp fields were kept fertile with silt dug from the canals and with composted water lily plants, but North America had no large domesticated animal like the cow or the water buffalo to bring in nutrients from the surrounding landscape to upland fields in the form of urine and manure. Without animal manure (whether the Maya used human manure isn't certain) maintaining the fertility of their terraces must have been a problem. Erosion, which carries away soil nutrients with the soil, was a perennial problem: soil traps were constructed so that eroded soil could be carried back uphill to the terraces. The traditional *milpas* of the Maya were a mixed forest garden, planted in a patch of cleared forest, where a variety of crop plants were grown, annual crops giving way after a few years to perennial and tree crops, as the land slowly returned to forest. Thus the soil was almost always covered by vegetation. New land has to be periodically cleared, and in modern times in the face of population pressure the fallow may be reduced to as little as four years. (Modern *milpas* in the Yucutan, if used continually, erode to bedrock in two decades.) Higher populations require constant production and permanent fields; thus the terraces, with their exposed erodible soils and successive crops of nutrient-demanding corn. (Central Mexico also shows signs of extensive pre-Columbian soil erosion. Old corn-growing soils there—still cropped, now with plows—consist of thin mantles of broken rock while nearby soils with little evidence of cultivation have 18 inches of topsoil.)

Disaster in the Americas was associated with a warming trend (1° C., 1.75° F.) in Europe: the Medieval Climate Optimum. The temperature rise of 1° C. increased growing seasons in northern Europe by a month and extended the range of many crops. The Mediterranean climate, with its mild wet winters and hot dry summers, shifted north to include most of Europe, replacing the cooler damper Atlantic climate. Grapes were grown on south-facing terraces in England, and human settlement and the tree line moved further up in the Alps, and into the uplands of Scandinavia. The warmth corresponded with a burst of population growth and forest clearance. The wheeled plow was invented in the sixth century. It wouldn't come into common use until the eleventh century, when, pulled by a yoke of 8 oxen, it allowed the heavy clay soils of the European oakwoods to be broken. During the warmth, population and settlement increased throughout Europe.

While food production was the purpose of both horticulturists like the Onondaga and of "high civilizations" like the Inca, the "high civilizations" required more per farmer. This meant more land under cultivation and more work per person. Generally speaking, peasants under Inca rule were supposed to produce a third for themselves, a third for the gods (the priesthood and their temples), and a third for the Inca, the god-head of the state. Besides cultivating crops, peasants had to provide labor in construction, in transportation of crops and goods and in mining; they also had to provide military service. Food was hard to transport without wheeled vehicles, and heavy foods like potatoes were usually stored in stone warehouses near fields. Domesticated llamas brought warm season crops like maize (eaten and used for brewing beer) and cotton, as well as beans, cocoa, squash and chilis from terraced fields lower down on the slopes of the warm and humid Amazon basin to the east. The llamas also carried forest products such as woods, resins, honey, feathers, animal skins, wild fruits, medicinal and hallucinogenic plants. Such luxury goods were grown, collected, and transported by free peasant labor. Under the Incas, more people created a more powerful state and more wealth, and put more stress per person on the landscape. The Inca state, like many of the precontact American states, was not a trading state like those of Europe, but a theocratic one, whose power was based on the success of its agriculture and thus on its successful connection with the mountain gods. The size and impressiveness of its monuments, and the wealth of its upper classes, celebrated this. (A massive Tiahuanaco fountain building seems to show the mountains yielding their water to the valleys below.) The Inca economy depended on traditional labor-sharing agreements and on the ceremonial exchange of goods among kin groups (of lowland crops, for instance, for upland wool and potatoes). Such exchanges were not monetized (their value calculated in coins) and thus were less likely to be influenced by economic developments and less capable of including other groups.

But Inca agricultural systems were capitalist systems in that their cultivation produced a profit. Roughly speaking, it would seem that part-time adult labor produced a surplus double or more that of the subsistence needs of the farmer and his family. The situation is a little cloudy because other labor was required of the peasantry and the participants in the labor brigades were probably fed with food they themselves had grown. The profits from agricultural activity accrued to the organizers of the state: the Inca, his nobles, the priesthood. Profits were re-invested in roads and bridges, in monumental public buildings—all built with conscripted labor—and in maintaining the lifestyle of the priests and nobles. All this was made possible by an organized, labor-intensive manipulation of the local landscape.

The Tiahuanaco people that preceded the Inca constructed extensive raised fields on the shores of Lake Titicaca. These fields were not reconstructed by the Inca state and are now an undulating upland used for cattle pasture. Originally the fields constituted a sustainable agricultural landscape of some size. Lake Titicaca lies 12,500 feet above sea level. It is a large, deep lake and its great mass of water (it never freezes despite its altitude) moderates the local climate. It fluctuates unpredictably in level, by as much as 15 feet over 2 years in historic times and by much more over the last 12,000 years. The lake is home to substantial populations of fish and waterfowl (ducks and flamingos) and has extensive shallows of totora reed, parts of which are eaten, and whose stems and leaves are used for thatch and for making textile boats.

Raised planting beds are common in peasant agricultures. Their primary purpose is drainage, but those about Lake Titicaca also functioned as solar collectors. These raised fields were 15-30 feet wide and up to 600 feet long, their planting surface raised about 5 feet above the bottoms of the canals, from where the earth was excavated to build them. The water in the canals, warmed by the strong tropical sun, irrigated the growing plants with warm water from below, drawn upward by capillary action. The heat stored in the water, and transferred to the beds, improved plant growth and protected the plants from summer frosts, which are not uncommon at this altitude. (Yields in raised beds nowadays are double those of dryland crops nearby.) The depth of the water in the canals was regulated, so as to provide sufficient heat storage and also water for irrigation. Water came from the lake, from groundwater, and from rivers that flowed into the lake. The canals thus constituted a sort of extended delta in the lake. (The rivers were also used for other sorts of irrigation.) This is certainly a modified landscape, though not necessarily a degraded one. The modifications of the lakeshore and river deltas may have improved things for the fish and waterfowl of Lake Titicaca. These raised-bed fields are in some ways similar to rice paddies, except that the plants are cultivated above the water table. The crops included quinoa (a grain), potatoes, and various other Andean roots. The blue-green algae that colonized the canals fixed nitrogen and provided one base of the food chain that lived in the canals, itself a part of the food web that connected the waterfowl, fish, birds, turtles, amphibians, and higher plants of the lake. The silt and bacterial scum dug from the canals made the fertility of the fields self-maintaining; they needed no fallow and could be planted to crops every year. The Tiahuanaco people, somewhat unusual in the Americas, also used human manure on their fields.

(They may also have pastured their large flocks of llamas and alpacas on them after harvest, manuring them further.) The nutrients added to the soil and produced in the canals were recycled in the canal-field ecosystem, or incorporated into the larger food chain; they didn't leach out significantly as mineral elements into the lake water downstream and cause algal blooms, as modern agricultural nutrients do.

Similar agricultural systems include the Maya's planting beds in lowland swamps; and the raised beds dug in seasonal wet prairies in the American Middle West. The disadvantage of such systems to a modern farmer is that probably half the area is lost to waterways. This is not a disadvantage in a biologically friendly system, where agriculture is considered part of the whole environment; the water is not only necessary as a heat trap, but for fertilization and irrigation; and the ditches extend the habitat of the lake or swamp, whose inhabitants may be useful as a source of food. It is a disadvantage in a modern capitalist agricultural system where land in crops must be maximized, cultivation intensified, machines are available to install drainage tile, fertilization comes from artificial fertilizer, irrigation water from pumps, and the polluted runoff is exported downstream. In central Florida large lakes provide heat storage for the orange groves planted among them, at the northern edge of the frost free zone. Natural wetlands (several thousand acres) about the inlet to Florida's Lake Apopka were drained to grow oranges: this maximized use of the available landscape. Natural wetlands in Florida are about 2° C. warmer on frosty mornings than drained farmland, probably because of the saturated soil. The cold temperatures also last less long. John Bartram pointed out the sheltering effect of the palm canopy over wild orange groves in Florida on cold mornings in the 1770s. An agro-ecosystem that grew oranges under a canopy of palms (also useful for various fruits), the trees planted on raised beds that allowed the wetland to function, and to maintain a flow of clean water downstream, would provide more frost protection and filter the water entering the lake. In this case Lake Apopka would still be a clear swimmable lake, with a sandy bottom and a large natural bass fishery, and not a murky and eutrophicated one, its bottom covered with a layer of silt and decomposed algae.

Inca landscapes were modified much more than those of the Onondaga. Rivers were straightened, fields leveled, or excavated down to ground water level, irrigation canals dug. Mountain springs were led into stone-lined trenches to irrigate descending terraces, changing (probably not eliminating), their watery contribution to rivers. Much terraced land was inherited from the Wari, a people who built high altitude terraces over much of the Andes. A mountainside with terraces constitutes a different ecosystem but not necessarily a degraded one.

The deforestation of the Andean uplands, perhaps partly for fuel, but mostly to create better pasture for deer and the semi-domesticated camelids (llama, alpaca, vicuna), happened beyond the time of folk memory. The llama and alpaca have been domesticated for 7000 years. Like the Tibetan plateau, which was once wooded with cypress, and which was also cleared several thousand years ago, to grow barley and graze animals, the Andean highlands were once wooded. So the natural vegetation of the altiplano disappeared a long time ago. A change in vegetation, without a concurrent increase in erosion and in nutrient losses, is not necessarily degradation, especially if the new system turns into an adequately functioning ecosystem. (Some habitat will be eliminated; and fire clearing, which was probably used here, certainly involves an initial loss of nutrients.) Ecosystems change under natural conditions all the time. The new landscape may be a little more leaky of nutrients and have a different hydrology than the former one. The relative abundance of its plants and animals will change, more or less, depending on the extent of the change. The modern landscape, dryfarmed where possible, and otherwise pastured by sheep, is clearly degraded. A similar upland in Andean Columbia, planted by an utopian community to forest, is reverting to a natural forest. Trees and shrubs, with their associated birds and animals, have appeared that were not planted, planted from the droppings of those animals and birds. Is this better, or only different? It is certainly better for the hydrological environment than modern grazing and farming practices. Whether turning the Andean uplands into pasture for the native cameliids significantly degraded the altiplano depends on many things, including the rate of stocking and the response of the vegetation to the animals. The buffalo and elk barrens in the forests of the eastern United States were not degraded landscapes. They were human artifacts, maintained by the animals and by man-made fire, but not degraded. Nor were forests burned to maintain their crops of nuts, at the expense of other, later successional trees degraded; such systems were capable of indefinite survival and fitted seamlessly into the surrounding landscape, including the hydrological landscape; though the forests that resulted were not those that would have been there without human intervention.

What is interesting is the difference, so much as one can make it out, between the agricultural (proto-capitalist, marketable: these words probably work better with Eurasian cultures than with American ones) and the edible landscape. One thinks of the edible landscape as subsistence and sustainable, but this is not necessarily so. Subsistence landscapes, even gathering and hunting ones, can be over-exploited and thus become unsustainable in the long run. Modern agricultural landscapes are for the most part unsustainable, but older marketable landscapes were sometimes sustainable and sometimes not. The landscape of Tiahuanaco was proto-capitalist and some of it was sustainable. Fish remained an important part of the diet of people about Lake Titicaca. The Maya terraces were not sustainable because of soil erosion and nutrient depletion from the continuous crops of corn, but their swamplands were. The continued productivity of waterways is a good indication of the health of terrestrial environments.

Some writers argue that the large-scale adoption of agriculture 8000 years ago helped create our present climate. During previous interglacial periods, levels of carbon dioxide and methane (two natural greenhouse gases) slowly fell, as plants and bacteria stored carbon away, while the intensity of summer sunshine decreased due to cyclical changes in the earth's orbit. These changes reinforced each other and would have set the earth into another cycle of cooling. (The reason for the falls in the levels of the greenhouse gases aren't completely known, but have also occurred in past glacial cycles; during interglacials much more land is available, especially in the north, for carbon storage by plants and bacteria.) In the current interglacial (our time), carbon dioxide fell until 8000 years ago, then began to increase. It should have continued slowly to fall. The increase corresponds with the extensive clearing of forests in India, northern China and southern Europe for agriculture, and the corresponding release of carbon dioxide from burning trees and cultivated soils. Estimates of the amount of cropland and modified forest needed to feed a person with Neolithic agriculture are quite high (about 7 acres), so the amount of carbon dioxide released to the atmosphere by a relatively small farming population (a few tens of millions by 7000 to 8000 years ago; perhaps 200 million by 2000 years ago) would be considerable. Methane, another greenhouse gas, began to increase about 5000 years ago, with the cultivation of paddy rice. (Like natural wetlands, rice paddies produce methane.) The growing abundance of livestock, which produce methane in their digestive tracts, also added to the amount in the atmosphere. Cattle were domesticated in Greece or the Balkans about 8000 years ago and spread to the Middle East, sheep and goats at least a thousand years earlier. Looked at this way, agriculture may have contributed 0.8° C. of warming globally before 1700, 2° C. at higher latitudes. The temperature increase at higher latitudes prevented the re-glaciation that should have started 4000-5000 years ago. In this case agriculture created a new golden age. (Other writers argue that the current orbital configuration implies a long current interglacial, an unheard of 60,000 years; in either case we're lucky.)

Chapter 8

Limits of Sustainability in the More or Less Edible Landscape: Varieties of Desert Agriculture

Some landscapes, like the north European plains, or the humid prairies of North America, are more tolerant of human agricultural occupation; some are less so. (Both the former are part of the loess belts of thick glacial silts deposited by the wind, which make excellent, if easily erodible, agricultural soils.) Limits often appear more dramatically in arid landscapes. The Papago Indians (the 'Bean Eaters': they call themselves the Tohono O'odham) inhabited the Sonoran Desert of Arizona and northern Mexico. They were flood-water farmers; or partly so: it is thought 20-25% of their calories came from cultivated foods. Among the 75% of their diet that came from hunted and gathered foods, 4 times as many calories came from plants as from animals. The Papago farmed the mouths of desert washes, where the flood waters from the July rains spread out. Low embankments, shallow ditches and brush weirs were used to divert and distribute the flow. A brush weir constructed across the flat slowed the rush of water, made it spread out and drop its load of leaves, small sticks, rodent droppings (all excellent fertilizer); slowing its flow helped the water sink into the ground. Corn, squash, beans and pumpkins were seeded in the damp ground. Fields used for many years developed their own associations of wild plants. Some of these were allowed to grow and were in a sense cultivated. (So-called cultivated devil's claw, a plant used for basketry, may have been among them, along with goosefoot, a weed harvested for greens; and cross-pollination with wild chilis growing around the edge of the garden added heat to next year's planted chilis.) To mature well, the crops usually needed a second and even a third watering. This was provided by the later, gentler, female rains, though tepary beans needed less water than corn, and corn less than pumpkins. Yuma corn, grown on the floodplain of the Colorado River to the west, matured in 60 days in ground left saturated by the receding floodwaters. So this was a system dependent on natural regularity, in a very variable environment; that is, the amount of rain and its timing varied from year to year. The environment's variability helps explain the low level of the Papago's dependence on cultivated foods.

In all ways water is the desert dwellers' problem. The Papago dug shallow reservoirs near their fields to catch additional runoff (which they might use for hand irrigation as well as for household water) and also dug wells in the beds of washes. These usually held water until October or November. By then the crops had been harvested, the agricultural year was over, and the deer dance was performed to mark the start of the winter season. The tribe moved from the drier mid-altitudes where its fields were located and where the desert fruits, the foods of spring and summer, grew, to its higher, better-watered perennial grasslands: the animal pastures.

Flood-water fields were partly created; they might be leveled with earth dug from the reservoirs, and enriched with nitrogen-rich earth collected under mesquite trees or other nitrogen-fixing desert shrubs, and carried in baskets to the site. Their layers of silt were increased by the brush dams. The settling silt helped level them and also made them fertile; the yearly flood of silt and other organic matter renewed the soil's fertility. Older fields had over a hundred characteristic plants, besides crop plants, and a rich insect and invertebrate life. The man-made concentration of water and nutrients made them local centers of biological diversity. (One hundred and ten species of birds are found about Papago fields and they attract many species of mammals, reptiles and amphibians). Some of this life, especially the competitors for the crops, like rabbits, would be eaten.

Many of the collected foods of the Papago were strongly seasonal; abundant for two weeks or a month, and then gone. Such foods included the fruit of the sahuaro and organ-pipe cactus, which ripened in June; cholla buds and fruit; mesquite beans (mesquite yields something like 150 pounds of pods per acre, and the beans don't require cooking to be edible); the seeds of ironweed and paloverde, and of several desert grasses, out of which a nutritious mush was made, higher in protein and fat than wheat; various greens; and sandfood, a parasitic fungus that grows on the roots of a desert shrub. A family might also consume 12-15 black-tailed deer per hunter, as well as other game. If these figures are accurate, it puts Papago meat consumption from large animals far above that of Northeastern tribes like the Iroquois, whose take of deer and bear in the winter hunt is estimated at 40 pounds of dressed meat per person, or less than one animal each. (The Creeks of the Southeast reportedly used 25-30 deer per household annually for meat and skins; in the 1700s during the high years of the trade in deerskins for European goods, Creek hunters killed 100 deer per household per year.) With these numbers, each member of a family of 5 Papago is consuming 2-3 deer. The desert also provided black-tailed jackrabbits, an animal of cyclical, sometimes enormous abundance, which the Pueblo Indians killed in drives; white-tailed deer; desert bighorned sheep; Merriam's elk (now extinct); pronghorn antelope; doves and quail and their eggs; and wild turkeys. Grasshoppers were gathered for a protein rich mush, and the larvae of the lined sphinx moth, extremely abundant after the summer rains, were dried and stored.

The neighbors of the Papago, the Pima (Akimel O'odham), who lived along the Gila, were floodplain irrigators. Where the Pima lived, the Gila floodplain was four miles wide, much of it floodplain forest. The oval fields of the Pima were surrounded by planted hedgerows of mesquite, willow and aspen. Sticks were woven among the trunks; the fence helped slow and control the irrigation waters. Mesquite was common on the floodplain, and its beans were harvested for food. Pima canals were up to 10 feet deep and 4 to 6 feet wide. The water was forced into the canals by diversion dams. Pima crops would mature with one heavy, pre-planting irrigation. The muddy floodwaters, a product of the winter's snowmelt, also fertilized and helped level the fields. The spring flood of the Gila was caused by mountain snows, which were more reliable than summer rains, but would fail on average once every 5 years. Additional irrigation during the growing season produced more abundant crops and the Pima also constructed temporary embankments and dikes on the flats beyond the floodplain to divert sheet and arroyo runoff from the summer rains onto their fields. Such structures may have helped them garden in years when the spring flood failed. It is thought the Pima got 50-60% of their calories from cultivated crops. Their additional manipulations of the landscape somewhat reduced their dependence on climate variability. Instead of being dependent on a cloudburst flooding the drainage of a wash, they were dependent on a season's snowfall over a large mountain drainage, and, to a lesser extent, on the summer rains that fed their other water diversions. Since they cast a wider net, the likelihood of failure was reduced, and they could be more dependent on agriculture. In a good year the Pima could grow 2 crops (one with the spring flood, one with the July rains), and so have harvests in both July and October. Many of the Pima's other foods (such as cactus fruit and mesquite beans) were similar to those of the Papago. In bad years they ate more collected foods. For this reason they may have encouraged the growth of mesquite trees on the floodplain. They also ate fish. Before Euro-American settlement, the Gila, like many southwestern rivers, was a slow-flowing, low-banked river, with canopy forests of cottonwood, willow, mesquite and walnut, grassy meadows, and rich marshlands with waterfowl and amphibians (the *cienagas*). Fish from the Gila were a staple (the Spanish claimed the Pima supplemented their diet of fish with corn and beans), especially the humpbacked sucker, which was netted in commercial quantities from the Salt River until the 1940s. The humpbacked sucker is now extinct in the Gila. Beaver were also eaten, in moderation. Beaver were plentiful along the tributaries of the Gila until trapped out by the Americans, another indication that much or most of the near-river floodplain remained densely wooded.

Like many New World tribes, the Pima practiced no soil fertilization, except through the yearly flows of muddy water. North America had no large domesticable wild animals, whose keeping would encourage the use of pasture crops, which could be rotated with grain, and whose fertilizing manure could be spread on the fields. A Hidatsa woman, whose people gardened in the Missouri floodplain in the mid-nineteenth century, remembered carrying horse droppings out of her gardens: they were a source of weeds. Pima soils remained productive, though it was said new land produced better, and the wheat and barley introduced by the Spanish, and adopted by the Pima as winter crops, tended to be soil depleting. The three main native crops (the Pima also grew cotton and tobacco, both notoriously hard on soils) were corn, beans and squash, planted together in the same field. Up to a point, these were soil maintaining. Corn, the depleting crop, formed a standard for the climbing, and nitrogen fixing, beans; squash, another nitrogen user, spread its trailing vines with their large leaves out between the corn hills and sheltered the ground from sun and erosive rain. Where floodwaters did not fertilize the ground, as with the light upland soils favored by Native Americans in New England, New York State, and southern Canada, fields were only cultivated for 10 years or so, then the field, and often the village, moved. Free-living, nitrogen-fixing bacteria will provide a yearly pulse of nitrogen to continuously cultivated fields, but growing cultivated crops year after year, without a rotation into sod or forest (the poplar plantations among the cornfields of Italy and France, the plantations of pine pulpwood on the Minnesota prairie), tremendously slows the mobilization of nutrients from the soil. So productivity falls, to a low, constant level. The fields of the Narragansetts of Rhode Island were unusual in that they were apparently capable of bearing continuous crops of corn, though some accounts indicate the fields were left fallow in alternate years. (There is also now doubt that the New England Indians used spring run herring for fertilizer: it has been suggested that the use of fish was an English innovation; one writer pointed out that the Narragansetts would have been better off eating the fish.)

Landscapes like those of the Pima are easily over-exploited. All rivers are subject to cycles of downcutting and channel migration, as their floodplains, built up by layers of springtime silt, mature and grow above the reach of the spreading floodwaters. This cycle is more rapid in the young, rapidly eroding landscape of the American Southwest. For some combination of reasons (a decrease in vegetation, perhaps from drought; an increase in rainfall; tree cutting or grazing that lets storms produce more runoff and more silt) a flood begins cutting a deeper channel. This leaves the old floodplain out of reach of the floodwaters. More water is confined to the channel, which increases its erosive power. The deepening channel migrates inexorably upstream, sometimes dramatically so, in the form of a receding waterfall. Floods on the Colorado River in 1905 and 1906 led to its taking over an irrigation canal leading to the Salton Sea, a natural depression in the desert, as its main channel. According to Indian legends, this was not the first time the river had abandoned its delta in the Gulf of California. The rush of water created a headwall recession in the form of a waterfall 40-80 feet high and 1000 feet wide, that moved upstream at a rate of several thousand feet a day. People came out from Los Angeles to see it. After an episode of downcutting, parts of the old floodplain are undercut and washed away by the river, which finally adopts a shallow and braided (rather than a deep, single) channel. Trees, or other perennial vegetation, which might stabilize the channel, have trouble establishing themselves because of the annual floods. Few fish survive in the shallow, warmer water and beaver cannot colonize the river, and help stabilize it with their dams, without trees. In time, given a sequence of favorable events (two or three good years for aspen seedling survival, for instance; or colonization by the invasive tamarisk), a single channel will form again, and a floodplain re-form.

Something like this seems to have happened to Chaco Canyon in north-central New Mexico in the 1100s. The canyon had been occupied by the Anasazi people for 600 years and was densely settled. Wide ceremonial highways connected massive public buildings, and carefully laid out roads, for transporting pottery and grain in backpacks, connected Chaco with other Anasazi settlements. The Anasazi of Chaco were a state trading culture, exchanging pottery and other goods for grain among their various settlements. Thus they redistributed grain in a region of variable climate. The Chaco River was a shallow seasonal stream with a floodplain several hundred feet wide. Chaco Canyon caught runoff from a wide area and had a high rate of soil renewal and a high alluvial water table. Its relatively low elevation gave it a long growing season. The early Anasazi grew crops in the damp alluvial soils and diverted sheet runoff over the canyon bottom into irrigation channels for further irrigation. They hunted deer and collected wild grass seeds. Later, as populations expanded, they also grew crops on the mesa tops, under a mulch of stones. The stone mulch broke the force of a hard rain, letting water seep in; reduced evaporative losses from the soil; reduced erosion; reduced the temperature variation between day and night; and condensed dew. (In modern reconstructions, rock-mulched soils have double the soil moisture and four times the average yield of unmulched soils. Rock-mulched gardens were especially common in the Southwest during the droughts of the 1200s and 1300s.)

The nearby watershed of the canyon, including the mesa tops, was originally occupied by a pinion-juniper woodland. The Anasazi collected pinion nuts for food and cut the trees for building material and for fuel for cooking and firing pottery. Regrowth of trees is slow in the Southwest and as the population grew, the nut and deer harvest declined, people ate more rabbits and mice, as well as domestic turkeys raised on corn, took more wood to fire pottery, and cutting overtook the rate of regrowth. Southwestern pinyon-juniper woodlands are a mix of bare ground, grasses and trees. Approximately 20% canopy cover by the trees constitutes dominance (the roots go into the openings and compete with grasses for moisture); herbaceous cover amounts to 15-20%; the rest is bare ground. When too many trees are removed by cutting, or when the forest occupies too much of the ground because of fire suppression, or when the grasses are overgrazed by cattle or sheep, the patches of bare ground connect and erosion is inevitable; net soil loss has been measured at 0.5 inch per decade, or 5 inches per century. While the stone mulches on their upland gardens should have prevented erosion, erosion probably explains why 1000 years later pinion pine has not recolonized the mesas above Chaco.

The buildings of Chaco Canyon also took tens of thousands of ponderosa pine timbers, carried from up to 30 miles away. In the century that preceded abandonment, 75,000-100,000 large pines may have been cut in the Chaco River's watershed (some writers say 200,000 trees; not a lot of trees, really, for a population of some tens of thousands of people over a century). The tree cutting may or may not have contributed to the downcutting of Chaco Wash but it illustrates the Anasazi's use of wood. Trading pottery (and to a lesser extent worked turquoise) for grain let the Anasazi maintain large populations in an uncertain climate. Firing pottery requires fuel (the large pines were used for building material); as does cooking corn. Yields fell under continuous corn, so new fields had to be constantly opened up, sometimes in less fertile or more distant territories; rainfall varied from place to place and year to year, which affected yields; and the Anasazi population kept growing. Trade was a means of redistributing grain, and thus of evening out agricultural production over a large area. Grain distribution was controlled by the people in the Big House, and depended on a ready source of fuel to manufacture pottery. This was a hierarchical society. The Anasazi elite were about 2 inches taller than the farmers who lived in more simple dwellings, whose teeth show signs of episodic starvation. Child mortality among the elite was 9.5%, among the farmers 26-45%. The Anasazi farmers (like the Maya at a time of high population) showed signs of overwork and nutritional stress from their restricted diets: severe dental caries, tooth erosion from the stone grit in cornmeal, periodontal disease, arthritic diseases, osteoporosis.

The main flourishing of the Chaco people occurred from AD 950 to AD 1150, during a period of abnormally high rainfall. About AD 900 the irrigation channels on the canyon floor began to downcut below the level of the fields. Clearing of vegetation, in this case the pinion-juniper woodland as well as the upland Ponderosa forests, increases the speed and amount of overland flow from rainfall. Downcutting could have begun with a heavy rain; or with a drought, which would further reduce the vegetation about the canyon, followed by lighter than normal rainfall. Under these conditions a light rain would produce heavier then normal runoff. But rain was still relatively abundant and to compensate for the downcutting, the Anasazi dammed streams in the side canyons and used the stored water for irrigation. They also captured the water that came off the cliff on the north side of the canyon and built rock dams across the river itself, to raise its level. The population of the canyon remained high until a long-term drought began about AD 1140. By then, the population in the canyon had been a net importer of corn for 200 years. A more general drought meant crops would fail all over the area and the trading system would have too little grain to distribute. Corn can be stored for 2-3 years, but no culture stores more than 2 years of grain. (Thus the saying that 3 years of drought kills everyone.) The natural foods of the area (the nuts, animals, wild grasses) were no longer sufficient to feed the population and were reduced by the drought. (Such supplemental foods had helped when population densities were relatively low.) Some people dispersed to areas of higher elevation, which had more rain, but a drought late in the 1200s destroyed the Anasazi as a culture. Imports of food, and of luxury items such as pottery, stone, turquoise, macaws, shell jewelry and copper bells, into Chaco Canyon continued until near the end, which was rather sudden.

Six hundred years later, in 1849, though its uplands had not recovered, an explorer found Chaco stream clear and continuous with no signs of gullying. The stream had healed itself. By 1871 (after Anglo-American settlement had begun) the stream had cut an arroyo 16 feet deep and 40-

60 feet wide; by 1930 the arroyo was 25-30 feet deep and 300 feet wide and the stream had become intermittent. Once it begins, downcutting is difficult to reverse. Heavy machinery can be used to stabilize braided floodplains, although at considerable expense (hundreds of thousands of dollars per mile).

At European contact, many Southwestern rivers flowed through wide floodplains with meadows, cottonwoods, willows and connected marshlands. The floodwaters distributed seeds, soil, organic matter and other nutrients; their slow recession let cottonwoods and other trees germinate. (The seedlings' roots followed the water down.) Throughout the Southwest and Great Basin, beaver, with their frequent dams on smaller watercourses, and their maintenance of a dense riparian forest of poplar and willow, are thought to have been a major factor in stabilizing floodplains. The streamside vegetation slows the movement of water in floods, making it spread out and drop its fertilizing sediment on the floodplain terrace. With less sediment and less water, the flow descending from the tributary streams is less erosive. The Pima ate beaver but apparently, like the Cree, they took a limited number. The Pima also farmed only a part of the floodplain. Why is uncertain, but that is a conundrum only for people like us, inhabitants of a capitalist, market society, for whom growth is life. The river was important to them for fish and the uncultivated floodplain for other foods. Unlike the Anasazi, they never developed a state culture that demanded a surplus for its support. Perhaps only parts of the floodplain were within reach of their irrigation techniques. Most likely, they also saw the limitations of their environment and controlled their population.

Anglo-American settlement of the Southwest (the final and by far the most environmentally demanding settlement to date) was preceded by the trapping out of the beaver. Timber, mostly mesquite, which then was restricted to river floodplains, was cut for mine timbers, building material, fuel to process ores, and for cooking and heating. Cattle were introduced from Texas. The cattle, by depositing the seeds from the nutritious mesquite pods in fertile pats of manure, helped mesquite make its move from stream bottoms to the uplands, where its takeover of former grasslands has tormented cattlemen ever since. A railroad in 1881 made it possible to ship cattle and cattle products, and cattle in the Arizona Territory increased from 5000 in 1870 to more than a million in 1890. The grazing pressure on the river floodplains was tremendous. The 1700s and 1800s were mostly dry in the Southwest. It was wet with floods from 1881-1884. A flood in 1891 altered the channels of the Salt and the Gila rivers, destroying the irrigation systems and canals along the rivers. A drought from 1891-1893 reduced the cattle herds by half. Then a period of somewhat above normal rainfall continued the downcutting begun by the 1891 floods. Most of the major river floodplains were converted to shallow, braided channels. Streams downcut deeply. Native floodplain forests, deprived of moisture, declined, and were invaded by the deep-rooted saltcedars (*Tamarix ramosissima* or *T. chinensis*), alien plants introduced as ornamentals to the Southwest (some were planted by government crews in the 1930s to control erosion), which now cover 1.2 million acres of southwestern river bottoms.

The bunch grasses of the American Southwest and Great Basin are not adapted to continuous grazing and under grazing by cattle and sheep tend to be replaced by less palatable plants (the thorny scrub of the modern Southwest). The native herbivores (black-tailed deer, bighorn sheep, Merriam's elk, pronghorn antelope) are browsers that sometimes eat grass. Elk and sheep eat more of it, elk largely in the winter when it does no harm to the plants. While it is thought that over-grazing by cattle helped initiate the downcutting of the rivers, cycles of downcutting and floodplain aggradation are endemic in the Southwest. The area had been grazed (relatively lightly) by domestic sheep for two centuries before the Texans came. One can see the effects of the new settlers not only in the rivers but in the levels of sediment deposited by the wind in southwestern alpine lakes, which rose by 5 times in the middle of the nineteenth century and are still 3-4 times above pre-nineteenth century levels. The hooves of the cattle broke the crusts on the arid soils (crusts of lichen or wind blown gravel) and let the soil blow away. In the case of the events of the 1890s, a flood, followed by a period of light rainfall, which further reduced the vegetative cover, followed by a period of heavy rain started a cycle of downcutting. Heavy grazing by cattle, tree cutting, the removal of the beaver made the downcutting more likely, more extensive, and more severe. Such practices provide more sediment, which increase the erosive power of rivers.

Much remaking of the desert landscape was deliberate. The valley of the Salt River, where the Hohokum had grown irrigated crops for 1500 years (they had been replaced about 1500 by the less state-like Pima culture who practiced a simple flood irrigation) was converted into a major producer of irrigated grapefruit, dates, cotton, lettuce, melons and salt tolerant winter grains. The river was dammed to provide water. The dams dried up much of the channel and the floodplain forests (when not removed for fields) collapsed from the falling water tables. The Hohokum had dealt with the salty groundwater of the Salt Valley by rotating their fields every ten years; only 9-10,000 of the 100,000 hectares they cultivated were under cultivation at any one time. The Hohokum also grew agave on the uplands and ate fish from the river. Unlike the Anasazi, who were dispersed by drought and replaced as a culture by the less environmentally demanding Pueblo, the Hohokum seemed to deal with the environmental stresses they faced. They were probably eliminated by European diseases before actual contact with Europeans. The modern irrigators dealt with the salty groundwater by digging drains, and sending the salty runoff back into the river channel. Irrigation works along the Salt were destroyed by the floods of 1891 but were rebuilt.

Continued heavy grazing in the twentieth century has speeded up a shift in the vegetation of the Southwest from grasses to shrubs. This change in vegetation results in a change in the distribution of soil nutrients from even to patchy, as nutrients are concentrated under the shrubs. Like the downcutting of streams, this shift in vegetation seems irreversible. Why is unclear. Many desert shrubs use carbon more efficiently in photosynthesis than desert grasses. The shrubs thus need to allow less carbon dioxide to filter into their leaves from the atmosphere (through openings called stomata) and correspondingly to let less water out; so they are also more efficient users of water. A shift from plants less efficient at using carbon dioxide to plants more efficient at using carbon dioxide has been going on under limited rainfall regimes worldwide for several million years. It is probably part of the long-term adaptation of terrestrial vegetation to the tremendous decrease in the carbon dioxide content in the atmosphere since the Carboniferous, when carbon began to be locked up in fossil fuels. Since carbon dioxide uptake by plants is related to water use, drier conditions favor plants that can take in more carbon dioxide with less transpiration of water. Shrubs are also deeper rooted; the individual plant draws water from a deeper soil profile and essentially depletes the water available for any grasses growing in between them. Nutrients from fallen leaves, from fine root turnover, from small animal droppings then accumulate under the shrubs, creating the so-called patchy distribution of nutrients, and making it harder for grasses to establish themselves on the bare ground in between.

Agricultural systems like those of the Papago, Pima, and Anasazi were used in dry environments worldwide. The Nabataens of the Negev Desert of modern Israel and Jordan, whose capital, Petra, was carved out of the walls of a canyon, lived on what was 2000 years ago a caravan route from the Red Sea to the Mediterranean. From the first century BC to the third century AD the route connected India, Africa and China with Egypt and Greece through ports in Gaza, handling trade in cinnamon, cassia, myrrh, frankincense, cotton, silk and ivory. The Nabataens fed a population of 60,000 people on grain grown with runoff irrigation. Rainfall then is thought to have been very similar to that now: from 2-15 inches a year (that is, highly variable on an annual basis), coming in showers over a 2 week period. The Nabataens dammed some wadis (canyons) in their upper reaches and led the captured water through canals to fields below. They dammed the whole courses of other wadis with low stone dams (so-called rubble masonry, stone dams with an earth core, the dams up to 2 meters high), the crest of one dam level with the base of the next. These dams converted the wadi bottom into a series of stepped fields, that were filled with silt and partly leveled by the runoff of silt and water that flowed with the rains down the wadi floor. Crops had to mature with this single, pre-planting irrigation. The density of dams was great: near Ovdat, another ancient city of the Negev, 17,000 have been found within 50 square miles, a square a little over 7 miles on a side. Such reconstructed wadis were probably also oases for desert animals.

At the mouths of the wadis, natural terraces at the bottom of moderate slopes were used for another form of runoff agriculture. These fields used runoff funneled down from the slopes above them. In modern reconstructions, the farms require about 20 acres of water catchment for 1 acre of cultivated ground. The degree of slope and the soil cover of the catchment are important. Steeper slopes tend to lose more water to rock fissures and depression storage. Moderate slopes yield about 20% of a light rainfall to the fields. The yield rises with a heavier rain. All slopes in the Negev are essentially devegetated. The system works precisely because the Negev is such a destroyed landscape (probably destroyed several thousand years before the Nabataens by people cutting wood and grazing goats, together with a slowly drying climate). The impervious soils, moderate slopes, and low rainfall lead to a rate of runoff sufficient to irrigate the terraces. In some cases, stones on the slopes of the ancient catchments have been collected into regularly spaced piles. Stone clearance increases runoff under very light rainfall regimes. (While it seems unlikely the stone piles create runoff from dew, such structures can also function as catchers of dew during cool desert nights. Stone towers 30 feet high and 100 feet across at the base condensed water from the air to supply the ancient Crimean city of Feodosia; the average measured modern yield is 35 gallons per tower per day.)

The Negev receives about 100,000 gallons of water per acre per year. In modern times, experiments have been made with individual water catchments on the desert floor. This modern innovation involves much soil movement, but these catchments also work. The ideal size of the catchment varies with the location, but in general pomegranates will grow in 250-500 square meter catchments, which corresponds to a planting of 8-16 trees per acre (20-40 trees per hectare). By comparison, standard apple trees in humid regions are planted at 100-200 trees per acre. Grapes can be planted at 32-40 vines per acre (80-100 vines per hectare). The creation of such microcatchments also increases the density of forage plants like saltbush 20-30 times (they grow around the rim of the catchment and are used as forage by goats), and over time the water filtering down will clear the accumulated salts from the catchment's soil. In such a ruined or severely altered landscape, such manipulation seems a definite improvement from a human point of view; and perhaps biologically. Catchments, colonized on their edges by forage shrubs (which help prevent soil movement), last more than a century, so the cost of construction has a long pay-off period. Regenerative agriculture in the Negev may mean encouraging the nitrogen fixing cryptogamic crust typical of deserts (which foot traffic and bulldozing will destroy); and the rock-eating snails that are now the chief agents of soil formation in the Negev. The snails rasp into rocks to get at the lichens that grow under their surfaces; the rock comes out in the snail's feces converted to soil. Such biological weathering amounts to about a ton of material per hectare annually, and makes the snails critical agents in soil formation and nitrogen cycling, processes necessary for the successful growth of higher plants.

Vineyards and pomegranate orchards would accumulate carbon in the soils under them, and so help remove carbon dioxide from the atmosphere. Millions of square miles of land world-wide that have been ruined by agriculture, grazing, logging or mining, much of it dryland, including perhaps 30% of former agricultural lands, are candidates for such innovative, marginally profitable, carbon storing schemes. Unlike modern commodity agriculture (much of it also marginally profitable and heavily government supported), such schemes usually involve much hand work; that is, they also produce employment. One could not in modern times grow grain in the wadis of the Negev, or on the terraces of runoff farms, as the Nabataens did (the relative price of grain is now much, much lower than then) but one could grow other crops. Pomegranates or grapes in the Negev; salicornia, a halophyte that can be irrigated with seawater, whose yields are similar to alfalfa, and which makes a good livestock feed, in seaside Middle Eastern or Indian deserts, alongside greenhouses that use cold seawater to condense moisture out of the desert air, at a gallon a day per square foot of glass (the water also cools the greenhouse, making the plants more efficient users of water); jojoba, a desert shrub whose seeds yield a useful oil, in Arizonan or Mexican fields that now grow irrigated alfalfa or cotton; Christmas trees, pulpwood or horticultural evergreens on the strip mined hills of humid West Virginia, Pennsylvania and Ohio; native perennial grasses and large herbivores underneath the towers of windmills, on the newly shaped, strip mined hills of Wyoming or the Navaho Reservation in Arizona. Destroyed sunny landscapes can hold photovoltaic collectors; or solar concentrators to make electricity from steam heated by the sun. Regenerating such economically worthless landscapes makes them economically and ecologically useful.

Chapter 9
What was Wrong with the Edible Landscape? Why Agriculture?

The reasons for the adoption of agriculture are a mystery. A writer has summed up the effects of agriculture on human populations as malnutrition, periodic starvation, recurring epidemic diseases and class division; but along with these came the invention of writing, metallurgy, higher mathematics, astronomy, the ability to dominate one's neighbors, 'civilization.' Population pressure is the usual explanation for the adoption of agriculture. During the 90,000 years of the last continental glaciation (most of human existence) the climate was very variable. The two periods of most stable climate were during the deepest cold at the peak of the ice accumulation about 20,000 years ago, and the last 10,000 years. Glacial times are cold, windy and dry. Productivity in the oceans, limited by a lack of iron, was high as winds carried iron rich dusts from the Sahara and the central Asian deserts about the earth; and as wind driven upwelling brought nutrients to surface waters. (The productivity of tropical forests is also thought to be limited by iron arriving on the wind but their extent was limited by the dryness.) About 30% of the earth's surface was covered by ice at the last glacial maximum. Sea levels were 360-400 feet lower. When the ice began to melt about 17,000 years ago, sea levels rose, sometimes rapidly. They stabilized (that is, the rise slowed to a level manageable by marine organisms such as oysters and coral reefs) about 6000 years ago. As with the 90,000 years during which the ice grew, the 10,000 years during which it melted was marked by abrupt climate changes, the most dramatic of which was the Younger Dryas (there is also an Older and Oldest Dryas) from 12,800 to 11,500 years ago. During the Younger Dryas cold windy conditions returned to much of the earth. At the end of the Younger Dryas temperatures in central Greenland went up 15° F. in a few decades. The Younger Dryas was followed by another cooling event about 8200 years ago, with changes perhaps half as great for a tenth as long, and by a very mild cooling event during the late Middle Ages (the Little Ice Age), at the end of which the Norse colonists abandoned Greenland (or perished there).

In most parts of the world horticulture closely followed the extinction of the Pleistocene large mammals at the end of the Younger Dryas. Rising seas flooded continental shelves and ocean banks. Some of these habitats, such as the steppe connecting Alaska and Siberia, and the continental shelves off the north European coast, were very large. Animals and people were driven inland as their habitats shrank. The warmer weather resulted in deciduous forest covering much of temperate Asia, Europe and eastern North America. Much of this landscape had been steppe, or shrubby grassland with clumps of trees, inhabited by herds of horses, bison, hairy elephants and reindeer, and by bands of human hunters. While some habitats shrank, others opened up. The Sahara was very dry during the late ice age from about 20,000 to 15,000 years ago. It grew wetter about 11,000 years ago during the glacial meltdown, and remained so for several thousand years, then began drying about 5500 years ago. Rainfall in the Sahara depends on the northward movement of the summer monsoon from the Indian Ocean; very small changes in rainfall make the desert expand or contract. The changing climate dried up the grassy Saharan savannahs and their rivers with crocodiles and hippopotami. The rock paintings on desert walls changed from buffalo, crocodiles and elephants to small Saharan cattle, apparently newly domesticated. The changes stranded a population of African elephants along the North African coast. This population of animals declined as the climate dried further. The last of them may have become Hannibal's elephants, led over the Alps to Rome.

So agriculture began about 10,000 years ago during a time of changing climate. It didn't begin in the Middle East. It began worldwide, with people growing plants in many different places. Animal herding is supposed to have preceded agriculture, but grains were being grown 12,000 years ago in the Near East, 1500 years before the usual dates given for the domestication of sheep. Signs of domestication are not always easy to read and the herding of semi-domesticated sheep and goats, as the Lapps and the Siberian tribes herd reindeer, may have preceded all this, while the first domesticated plant may have been figs (which grow on trees). We Westerners, living in an era of unlimited fossil fuels, and with an extremely progressive view of history (a product of our constantly improving material lives), see no mystery in this. Agriculture pointed the way out of the cave. Agriculture gave people more apparent control over their environment, through a more intensive manipulation of it, and thus more control over their lives; it made possible a sedentary life with permanent dwellings, easily storable foods, tools, clothes, beer, cooking utensils, decorative objects; more children. Agriculture can support many more people than hunting and gathering. Modern calculations indicate that solar-powered agriculture, with work performed by men and animals, can support 64 to 256 people per square kilometer; in warmer, wetter climates multicropping is possible and the potential population doubles. Gathering and hunting can support up to 4 people per square kilometer (this is a low estimate). The worldwide average density of hunter-gathers is 1 person per 26 square kilometers (about 0.04 people per square kilometer). Most hunter-gatherers live at 20-60% of the calculated maximum carrying capacity of their environments; this may indicate their environments' actual ability to provide continual subsistence. Similarly, wolves in Yellowstone keep elk at 20-30% below what the weather, through its effect on vegetation, would support; and thus create a more stable elk population.

With dense populations of sedentary agriculturalists, the arts bloomed: pottery making, metal working, tile making, wall decoration. Many of these arts were heavy users of fuel. Mathematics, architecture, astronomy and writing advanced along with larger, more complex and more hierarchical societies; societies with hereditary classes of rulers, priests, warriors, artisans, farmers. Growing agricultural surpluses, that is, more excess production per farmer (as in Egypt, where a peasant produced 5 times the food required by his family), made possible the support of the non-farming classes. War was a sort of investment policy; a means of adding to the society's wealth. It brought wealth through tribute and booty: in most premodern societies it is thought wealth came more from increasing a group's lands or from war than from technological innovation. Behind all this was population growth. One of the effects of settled life is more frequent childbirth. A mother who moves camp every few days or weeks cannot deal with more than one non-ambulatory child. Some anthropologists think that nursing, which helps inhibit ovulation, went on for 4-6 years in a normal, that is, a gathering and hunting, society. (Gathering and hunting was by far the longest human adaptation.) At any rate, births tended to be spaced every 4-6 years. In an agricultural population births are limited mostly by the food supply and tend to come every 2 years. So populations began to rise. And as long as people were willing to work more, and more land was available, as long as the rains came, or the rivers rose and fell in a predictable way, agriculture would continue to feed many more people (10-100 times more) per unit of land. In Neolithic Europe an agriculture of cattle, goats and hoe-cultivated wheat is thought to have supported 1 person per 120 hectares (50 acres), or about 8 per square kilometer (13 per square mile), twice that of the acorns and salmon of northern California. Most of the land was needed for browse for the animals. In Mesopotamia, cultivated land supported 100 or more people per square kilometer. The population may have reached 20 million, with two-thirds of its 35,000 square miles of arable land irrigated, before the societies collapsed. A larger population is always the aim of biological evolution; and in human society, as in wolf society, numbers matter: the larger group wins the competition for resources. More resources allow the population to grow further. (Of course, technology can compound or confound this.) Some anthropologists think the advantage of being able to mobilize more people for war explains the development of agricultural chiefdoms and early states.

But none of this would have been apparent to people who harvested a few baskets of wild wheat, spilling some as they cut it down and carried it home (thus fulfilling the dispersal strategy of the wheat plant); or planted a hill of gourds (useful as containers) at a campsite to which they were likely to return in three months. Gathered wild grains may have been special foods at first, used for feasts, or to make beer. (Animals that eat fruit like elephants, American robins and people have a long acquaintance with fermentation, which is one way fruits attract attention—and explains those robins flying into autumn windows—but standard histories put the fermentation of beer several thousand years after the cultivation of grains.) The advantages of agriculture come at a later stage when people have, voluntarily or not, gathered into villages, when populations are denser, and agriculture is providing 90% or more of dietary calories. With the advantages of agriculture comes a down side: starvation, disease, more hierarchical societies, poorer health. The poorer diet, mostly of grains, expresses itself in shorter stature (the height of men may have fallen 6 inches, of women 5); in nutritional diseases like anemia and osteoporosis; in vitamin deficiencies; in overall poor health. Death comes earlier than among many of the gatherers, perhaps at 19 years in early agricultural societies, compared with 26 in groups of hunter-gatherers. (Such figures are disputed and vary. Cistercian monks in the 1300s died at an average age of 35, but it isn't clear this figure takes into account a childhood mortality of 25-30%.) The heavy manual labor of agriculture causes degenerative skeletal diseases. Infectious diseases become more of a problem, partly because of the crowding, partly because of close contact with domestic animals, which are reservoirs of diseases that directly infect or adapt to humans. About 50 diseases are shared with cattle (including smallpox, measles, tuberculosis, and diphtheria), 65 with dogs (our oldest companion), 46 with sheep and goats, 42 with pigs (including flu), some with birds (other flus), a few with horses (perhaps the common cold). Drinking water contaminated with human feces spreads cholera, dysentery and intestinal parasites. The parasitical diseases schistosomiasis and malaria are associated with irrigation in warm climates.

The idea that population growth led to the development of agriculture conflicts with the idea that hunter gathers kept their populations below a landscape's carrying capacity; but not if a changing climate makes the landscape less habitable. One scenario for the development of agriculture in the Middle East involves the Natufian people of today's Israel and Jordan. About 15,000 years ago the Natufians lived on wild acorns and pistachios and hunted gazelles and wild sheep, in a landscape managed by burning. Their permanent villages were made possible by the abundant acorn crop, itself a consequence of more rainfall in the eastern Mediterranean with the retreat of the glaciers north. They also collected wild grains (wheat and two kinds of rye); these large-seeded grains returned a lot of food energy for the effort exerted in their harvest.

The shutting down of the Gulf Stream between 12,800 and 11,500 years ago during the Younger Dryas Event (the result of the draining of a glacial lake in the center of North America into the Labrador Sea), caused temperatures to fall in northern Europe and a thousand year drought in southwest Asia. The acorn and pistachio crops lessened, then failed. The Natufians, already living in villages, began cultivating cereals and keeping domestic animals. Large-seeded annual grasses such as wheat, emmer wheat and rye already formed large natural stands. (In the 1960s, botanists found wild grains growing in more or less pure stands over hundreds of hectares in the same area.) The domestication of grains like wheat involves selection for plants that don't drop their seeds when harvested, and whose seeds will sprout simultaneously when planted. It is thought the domestication of emmer wheat would have been rapid, taking 20-100 years, not several thousand, as with maize. (Emmer wheat is a natural hybrid of two grasses that occurred in the Near East 30,000 years ago. Bread wheat is a hybrid of emmer wheat with another grass that occurred 10,000 years later, at least 8,000 years before bread wheat was cultivated.) When warming resumed in Europe and rainfall improved in the eastern Mediterranean, farming and the keeping of domestic animals had taken hold; the villages had become larger. A later drying of the climate in the Near East, which made the growing of rainfed winter grains difficult, is thought to have led to the development of irrigation agriculture.

At some point an agricultural population outgrows its supply of wild game and gathered foods. Then return to an earlier way of life is not possible. (When their populations were still small and harvests were bad, the Anasazi, like the Pueblo people that survived them, made up the difference with wild grass seeds and pinion nuts; later, their population was too large and their landscape too altered for such foods to help.) As populations grow, domestic animals, kept for meat, milk, fat, hides and traction power, are pastured on the stubble fields and in the surrounding wasteland, increasing the pressure on the local environment. Crafts such as pottery and metal working require fuel, buildings need timber, grains must be cooked, and so trees become scarce and distant. Grazing prevents the re-establishment of forests cut for timber and fuel. As forests disappear, small streams dry up, floods increase, the land erodes, rivers grow more salty. Farmlands yield less, under continuous crops of grain. Crops also sometimes fail, and as the population grows, the threat of starvation becomes more constant. Unforeseen or unmanageable environmental problems appear, sometimes the consequence of greater environmental manipulation, sometimes the result of natural disasters or of natural shifts in climate. Populations near the carrying capacity of their environments make collapse of the civilization more likely. Past problems include soil erosion about the Mediterranean basin and in upland Mesopotamia; too great or too low floods in Egypt; the slow saltation of irrigated lands in lowland Mesopotamia (these lands, like those in the American Southwest, require a long fallow to avoid saltation and waterlogging); droughts in the Andes, and in the Mayan highlands; a tectonically mediated rise in the water table along the lower Indus River that flooded the foundations of the city of Mohenjo-Daro, and required the firing of more and more brick to build the buildings up, with more and more wood cut from the drying jungle; the "dry fog" in 535 AD (from a volcanic eruption? an exploding comet?) that cooled the climate and caused crop failure and famine in Europe, southwest Asia and China. Trade, perhaps allied with weather conditions, such as the warm, wet spring weather that increases the presence of bubonic plague in the marmots of the Asian steppe, introduces new diseases, to which populations are not adapted. Rome suffered from several plagues in the years after 144 AD, at a time when a warmer, dryer climate in the southern and eastern Mediterranean and in North Africa was disrupting its agricultural base. Conquest had made the early empire wealthy but the lightly settled northern lands last occupied lacked the wealth of the older Mediterranean cultures. From the first century on, the largely agricultural tax base of the empire, though heavy, fell short of paying the costs of administration and defense. The population of the empire never recovered from the plagues of the second century. The plague of Justinian that started in 542 AD (probably bubonic plague originating in China—several years after the famine caused by the sulfurous fog) may have killed half the people in the empire.

In an overpopulated society disasters and epidemics could be an economic blessing. The Black Death, which is thought to have killed a third to a half of the population of Europe from 1347-1349, was followed by a century of recurrent epidemics of diseases with severe mortalities (plague, smallpox, flu, dysentery), along with some episodes of starvation, so that in 1450 the population of Europe was probably half that of 1347. (Without the additional epidemics, the population would have recovered from the Black Death in 30 years.) The population of China, which was affected by the Black Death earlier, also fell by half from 1200 to 1400. The depopulation of Europe gave a considerable push to the economic expansion of the Renaissance. Poor farmland was abandoned, so crop yields rose. Food became more abundant, labor scarcer, wages higher (50% higher in Florence, compared with 50 years before); the poor gained wealth and bargaining power, which led to the peasant revolts of the late 1300s and the gradual erosion of serfdom. Technological innovation grew. Much wealth was transferred from the dying, some of it to the new European universities. In England, land without heirs went to the local nobility, who grew richer, and financed development schemes, such as mines and the reclamation of wasteland for agriculture. Writers claim that in general, the landed gentry lost wealth and power compared to the laboring classes. Depopulation resulted in a more diversified, more capital intensive, more technologically advanced economy, and a more prosperous population.

Chapter 10

Were Ancient Marketable Landscapes Sustainable?

We Westerners are taught that civilization began in the Fertile Crescent of Mesopotamia. This is a matter of cultural focus, for while agricultural civilization developed earlier there, similar civilizations elsewhere began not much later and eventually reached comparable levels. But agricultural civilizations in the Americas, southeast Asia or highland New Guinea, did not lead in a straight line to the modern West. Rather they were all taken over by that bellicose, capitalist, Christian culture that rose in Europe over some 7000 years out of its Mediterranean and Near Eastern roots.

In Mesopotamia, the adoption of agriculture seems more obvious than in many places. Wild annuals like emmer wheat and barley grew in large natural stands; for wild crops, they had exceptionally large seeds. These plants formed a natural climax in the wet winters and long dry summers of the eastern Mediterranean, in microclimates where the length of the summer drought made the support of woody vegetation difficult. Especially on fertile sites, these tall annuals grew quickly and used up the available soil moisture, out-competing trees and shrubs. Their seeds were adapted to survive the long summers and sprout and grow quickly in the returning winter rains. A ton of seeds per hectare could be gathered with a hand sickle from such wild stands. The warming and drying climate at the end of the Pleistocene greatly expanded the range of these plants altitudinally. People could harvest them from month to month as summer proceeded up the mountain terraces, and store the surplus. The Middle East also had many other wild precursors of modern crops, especially the pulses (the protein-rich lentils, peas, chickpeas and bitter vetch); and also flax, whose stems produced a fiber and whose seeds produced an edible oil. Combining grains and pulses in a dish produces so-called complementary proteins, a mix of proteins that corresponds more closely to what the human body needs—not as good as animal protein, but much better than grains alone: so one has wheat and chickpeas, rice and beans, beans and corn. (Chickpeas also contain tryptofan, a precursor of serotonin, which improves performance under stress and promotes ovulation.)

Wild cows, sheep, pigs and goats also lived in the area; animals that were domesticable because of their tractable natures and herding or flocking behaviors. (Other crops soon included barley, oats, grapes, olives, dates, figs, apples, pears and cherries—apples and cherries from Turkey and Asia Minor, pears from the Tien Shan mountains.) So here one can more easily envision how an increased birth rate, thanks to a more sedentary life based on acorns and pistachios, and also on gathered cereals and pulses, carried down to the village in woven woolen bags lashed over the backs of goats and sheep, led to an upward population spiral that then led to more settled agricultural communities: permanent fields, domestic animals (pastured after harvest on the fields), stone or mud brick dwellings, pottery making. Such villages, based on hoe culture of grain and on domestic animals, existed in upland Mesopotamia 9000 years ago in areas with sufficient rainfall. By 8000 years ago, in what is now central Jordan, such communities were being abandoned, partly because of soil exhaustion and deforestation, and partly because of a drought associated with another breakdown in the North Atlantic circulation, this one caused by the final meltdown of the Laurentide Glacier in North America. With the end of that drought about 7000 years ago, Europe and western Asia entered a climatic optimum that lasted 2000 years.

The progression to agriculture is less obvious elsewhere. In eastern North America the Hopewell mound builders of the Ohio and Illinois valleys were cultivating 7 crops about 2500 years ago. Four of them were grains with tiny seeds but good protein and fat content; the other crops were a squash and a sunflower grown for their seeds, and sumpweed, a large oilseed, whose pollen and leaves are often irritating to people. These crops were grown on the floodplains of the rivers after the flood had receded. While seed crops provided a considerable part of the diet, the people in the Hopewell cities also ate fish, shellfish, deer, migratory waterfowl, turtles, passenger pigeons and wild nuts. Their cities have been called cities of hunters and gatherers. Later mound builder cities, such as Cahokia, near today's St. Louis, were based on maize. Cahokia was a city of several thousand densely packed small farms—the quintessential "garden city"?—apparently without significant trading relations with other places.

Shellfish were very abundant in Middle Western rivers. Nuts and acorns had been important foods for people dwelling in the deciduous forests of eastern North America for thousands of years. Most acorns have to be leached of their tannins to be edible. Hickory nuts were pounded shell and all in a mortar, the mass boiled in water, and skimmed of froth and particles of shell. Further boiling reduced the mass to a nutritious, storable paste: hickory milk. In the late 1700s John Bartram watched Creek families in North Carolina store hundreds of bushels of hickory nuts. While corn would transform life in aboriginal North America, the domestication of corn in highland Mexico took thousands of years. (Domestication may date from 9000 years ago.) The plant that became corn was grown for several thousand years before its seeds became usable as a grain. Corn may have been first grown for its sweet stalks, fermented to make beer. (Modern maize stalks are 16% sugar.) An entire ear of *Teosinte*, the plant that was turned into corn by generations of Native American women, has less nutritional value than one kernel of modern maize. For corn to move north to the temperate zone took many more years, since it had to adapt its daylength characteristics from the even days of the tropics to the long summer days of the north. The advantages of corn over the native domesticates were immediately apparent, and when corn appeared, the tribes adopted it. Corn changed social structures; many societies became more hierarchical and their villages larger.

In the Asian subtropics, the cultivation of paddy rice involved the invention of a method of cultivation. The method is not obvious, though wild rice gathered for its grain grew in wetlands also hunted for their wild animals and fish; one could argue the rice paddy was an elaboration (with hand planted rice plants and opportunistic fish) of the swamp. In the case of corn and rice, the immediate advantages of agriculture are more difficult to see, and the process that led to a settled agricultural life more difficult to imagine.

In the Near East, irrigation societies and large cities (50,000 people and up), with their organized hierarchies of nobles, priests, courtiers, warriors, artisans and farmers; their massive public buildings of brick, mud brick or stone; their calendars based on astronomical observation; their use of writing; and war for the purpose of gaining land and tribute, developed after some thousands of years of upland village agriculture. Early cities were storehouses for grain: a means of distributing food and storing it against a risk of crop failure. (Similarly, one can regard the cities of modern societies as means of storing and distributing wealth.) These societies exploited a new environment: the flat and arid Mesopotamian plain, with its high summer temperatures (40° C., 104° F.), high evaporation rates and relatively impermeable soils. Initially agriculture here consisted of the cultivation of rainfed winter grains, but the shift of the Indian monsoon south about 5800 years ago reduced winter rainfall so that winter crops would no longer grow. The timing of high water in the rivers meant that irrigated crops would have to be grown in the hot summer. So more water was needed and soils accumulated salts more quickly. Irrigation produced a more apparently controllable supply of water than rain (it depended on the reliability of the river's flow patterns), and a much larger harvest per unit of land. The agricultural work was seasonal. In the off seasons, farmers could extend and maintain the irrigation system. Thus the surplus per farmer was considerably more and the other levels of the society considerably richer than in the upstream societies that still depended on rainfed agriculture. Irrigation also made large areas of new land cultivable.

The Tigris and Euphrates took parallel courses across the Mesopotamian plain, but at slightly different elevations. The plain was essentially flat; the rivers dropped 30 meters over a distance of 700 kilometers. Overflow channels connected the rivers during spring floods, and the first irrigation schemes exploited these channels. The land between the rivers was a mix of swamps, stands of date palms, forest, shrubs and grassland. Large herds of gazelle, along with other herbivores and their feline and canine predators, originally occupied the plain. The Euphrates carried a lot of silt; its delta moved out into the Persian Gulf at 15 miles per 1000 years, so ancient trading cities built near the gulf are now 50-70 miles inland. Irrigation agriculture would have had all the disadvantages of rainfed agriculture: a poorer diet (with sufficient calories, but less balanced, with less protein, and less calcium, potassium and other minerals and vitamins, and thus leading to poorer overall health, more skeletal disease and shorter stature); the diseases and parasites that come from contact with water containing human excreta (such as cholera; and various worms, including those that cause schistosomiasis), and those that come from close contact with domestic animals (many of the latter are also crowd diseases that require a sufficient density of human population to maintain themselves and so flourish in cities); the skeletal problems caused by hard physical labor; and earlier death. But at least at first, some of these irrigation societies were rich and grew a wide variety of crops, including pulses and some vegetables and fruits (such as onions and figs, sources of sugars, minerals and vitamins). Animal protein came from herds of goats and sheep that were pastured out on the plains in summer, along the rivers in winter. Wooden rafts supported by inflatable goatskins brought commodities down the Tigris in summer. Timber from the rafts and the commodities, including semi-precious stones and copper, were sold, and the goatskins packed back upstream. The actual extent of nutritional diseases would have depended on how well food was distributed (the extent to which the needs of the people, compared with those of the elite, were taken into account); of waterborne diseases on the city's organization of things like water supply and sanitation (whether drinking water and human excreta were separated).

There were the usual advantages of agricultural societies: a higher birth rate and higher population; permanent dwellings; fine and varied craft work; the development of the sciences of mathematics, engineering and astronomy; the construction of large public buildings; improved methods of war; and writing. In the several times writing has been invented (Sumer; Mexico; perhaps Egypt; probably China; perhaps Peru) it always flowed from irrigation, or quasi-irrigation, societies. Sumerians seem to have invented writing first, part of the great head start of Eurasia in cultural evolution. This was a gift of geography and biology, as Jared Diamond has explained. The easily domesticated plants and animals of southwest Asia could be grown without much adaptation north and west through Europe and east through much of Asia, an immense area, which led to societies influencing each other with new ideas, new crops, new inventions; thus cultural development was rapid. Writing in Sumer began, like the knotted strings of the Inca, from the need to keep agricultural accounts. By making possible the recording of knowledge (as well as that of wealth) writing made possible the accumulation of knowledge in mathematics, biology, astronomy, physics, law, natural history, human history: that is, it made possible the modern world.

Irrigation societies in Mesopotamia would all eventually collapse because of the character of the landscape's soils and the climate. Failure through saltation, accumulation of toxic minerals, or waterlogging is the fate of virtually all large-scale irrigation schemes in hot, dry climates. (Today, accumulating salts remove 1% of all irrigated lands from cultivation yearly.) The muddy waters of the Tigris and Euphrates had a relatively high salt content. This was partly natural, partly the result of some thousands of years of agricultural erosion and deforestation upstream. The high summer temperatures of the Mesopotamian Plain meant a high rate of evaporation and of water use. The time of spring high water meant that crops, unlike the rainfed grains of the Mediterranean uplands or those in Egypt, where the Nile flood allowed for a late autumn sowing, were grown in the summer heat. Thus they required a lot of water. The high salt content of the water and the relatively impermeable soils meant that the soils would inevitably begin to accumulate salt. Salt interferes with the ability of plants to take up water and thus reduces their growth. This soon became apparent and new land was brought into cultivation as the fertility of the old fields fell. (Modern irrigation schemes in impermeable soils help solve this problem by installing drains to carry away the excess water. The salt leaves with the water.) The development of new lands maintained total agricultural productivity and accommodated the rising population for a long time. But the climatic drying that began 5800 years ago continued. As new land ran out, and population pressure prevented the use of long-term fallows, crops in Sumer shifted from wheat to the more salt-tolerant barley.

The cities also needed wood, which had to be within an economical hauling distance (some came down the Tigris in rafts), to smelt metal, fire pottery, brew beer, burn brick, cook the pulses and grains, and for building material. Deforestation added to the salts and silt carried by the river water. The irrigators kept cows, pigs, chickens and goats, and so the cutover forests were grazed, which prevented their regrowth. Such use slowly transformed the whole landscape within the reach of the Sumerian cities and intensified the load of silts and salts that fed the rivers. Rising silt loads made keeping the canals open a constant problem, and sent the Euphrates migrating across its plain. Such problems take time to develop on a large scale, but unless dealt with, become (like our rising curve of carbon dioxide) more and more inexorable in their effects. About 5500 years ago equal amounts of wheat and barley were grown in Sumer, by 4500 years ago wheat amounted to 15% of the crop, by 4100 years ago 2%, by 3700 years ago no wheat was grown. Crop yields remained high until 4400 years ago—that is, for 1100 years, far longer than we have been growing industrial grain in the American Middle West. After 4400 years ago, no new land was available and crop yields fell by 40% over the next 300 years. From 4200 to 3900 years ago there was a region-wide drought. (Egypt's Old Kingdom collapsed from drought about 4200 years ago.) Water levels in the Euphrates may have fallen below the beds of the canals. The drought was so severe that in fields near the Syrian border the earthworms died. The first external conquest of the region occurred almost concurrently with the end of new land, 4375 years ago. By 3800 years ago yields were less than 20% those of earlier times and the society had effectively collapsed.

Sumerian agriculture might have been sustainable if it had been handled differently, using fields over a long rotation, like the Hohokum of the American Southwest. (Iraqi fields are still cultivated though 70% of them are degraded by salt.) Some ancient irrigation agricultures were sustainable: Egypt's; the Mayas' ditched swamps; the raised beds of the Tiahuanaco; paddy rice in Asia; floodplain agriculture along the lower Mississippi or Amazon. None of these are irrigation agricultures in the classic sense. In the Nile Valley the flood was usually high enough, and the soils beneath were more permeable, so the water washed the salts from the flooded land. Agriculture along the Nile was a sort of improved floodplain agriculture, much like that along the Mississippi and the Amazon. In the swamps of the Maya and the raised fields of Tiahuanaco, the water percolated up from below. Paddy rice involves a constant slow flow over a more or less impenetrable substrate.

Until the building of the dam at Aswan, Egypt's Nile Valley was probably the most naturally productive agricultural landscape in the world. When invaded by Napoleon in the 1790s, Egypt's wheat yields were twice those of France. By then the valley had been cultivated continuously for 7000 years. Half or more of Egypt's cultivable land is in the delta. The geology of the lower Nile Valley and the use of its natural overflow basins for agriculture removed the problems of saltation and waterlogging. After the flood, the water table would drop 10 feet below the surface, letting the flood waters drain away. The timing of the overflow also helped. Water from the spring rains in the uplands of east Africa (today's Ethiopia and Uganda) reached the Nile Delta in September, so crops were sown in late fall and matured in the cooler temperatures of winter. Spring and summer irrigation was restricted by the available technology to areas near the river, which the floods would leach clean of salts every year. Fertility was provided annually by the silt, and also by human and animal manure. Fertilizing minerals came with the river water from the highlands of Abyssinia, humus from the jungles of central Africa. As in Mesopotamia, the volume of silt may have been increased by deforestation for agriculture in Ethiopia and for metal smelting in the Ugandan uplands. But the river was also cleaned and regulated by its passage through the swamps of Nubia, before it fell to the lower valley.

The Nile also provided fish, both in the river and off the delta. The nutrients and fresh water brought down by the Nile supported fisheries throughout the eastern Mediterranean. The problem with the Nile flood was its unpredictability. Low floods occurred about twice a decade. The heights of Nile floods are connected to El Nino Events, which influence the northward reach of the Indian Ocean monsoon, and thus the extent of spring rains in Ethiopia. Two low floods in a row were a disaster. A flood can also be too great, remaining on the land too long and preventing the sowing of the winter crops. So, loosely speaking, the agricultural area would have supported a population that could store sufficient grain for 2 years. But population control is not the point of high civilizations, the timing of low or high floods wasn't predictable, and since Early Dynastic times the population kept rising above the food supply. Periodic starvation was common. Of course over such a long period of time Egypt also had more severe climatic disasters. The end of the Old Kingdom was caused by 300 years of low floods. A drought in the eastern Mediterranean 1000 years later (3200 years ago) stressed all the civilizations of the area. It helped bring about the collapse of Mycenaean Crete, already weakened by a tsunami from the volcanic explosion on Santorini that washed over west-facing coastlines throughout the eastern Mediterranean. The earthquakes that accompanied the eruption apparently compromised Crete's aquifers (so wells went dry and streams stopped flowing). The Cretans had traded olive oil and wine for grain from the mainland; they also grew some wheat themselves. The drought reduced their crops and meant less grain was available from the mainland. So even in their lucky landscape, the Egyptians were caught between the constant increase in the human population and the behavior of the river.

Paddy rice constitutes another sustainable agricultural system. Tropical soils in humid climates, once cleared of their natural vegetation, are in general impervious and nutrient poor. The soils are old, and have been depleted by rainfall, plant respiration, and internal erosion. Most of their nutrients are concentrated in the vegetation that is removed. In paddy rice cultivation, the nitrogen that feeds the rice plant comes largely from blue-green algae that colonize the warm, slowly moving water that covers the paddy. The water itself, descending from the uplands, provides some nutrients (phosphorus and other minerals). Trampling and working of the paddy bottom makes the soil more or less watertight, so water and nutrients are retained. Fish and freshwater invertebrates grow in the paddies and the feeder canals. In the Lake Biwa basin of Japan, very old paddy fields are used by the lake's catfish as a spawning area and nursery: that is, as an extension of the lake itself. This use is a benefit both to the farmers, who harvest some of the fish and whose rice is more productive, and to the fish. In some cases the nutrient-rich overflow from the paddies is led to duck ponds, whose algal and insect life (food for the ducks) is supplemented by grain. The ponds, their bottom muds fertilized by the algal and bacterial growth and by that cycled through the ducks, are periodically drained and planted to vegetables. Fertility in the paddies is also maintained by adding manure from the animals that plow them, human manure, ashes, the walls of demolished mud brick houses impregnated with soot and grease from cooking.

The water buffalo whose manure ends up on the paddy and increases its yield of rice, is fed from forage cut in the nearby forest. Irrigation water also comes from the forest. So the forest is part of the paddy system. It also provides firewood, small game, material for building and basketry. Lowland rice growing systems, irrigated by large rivers, also depend on a predictable supply of good quality water. These systems can be overwhelmed by timber cutting in the watershed and the consequent change in the quality, timing and amount of water delivered to the paddies.

So some agricultural systems of "high" civilizations were sustainable, some not; some landscapes could stand more human development, some less. Any human civilization is an expression of an existing landscape and climate, that is, of levels of rainfall and temperature, of sea levels, flood levels, the likelihood of extreme weather events. All agricultural systems can be pushed over the edge by too much development about them or by small climatic changes. Increase in population is often the problem, though technology also matters, in that technology magnifies the effects of population. In the 1800s, perennial irrigation to grow cotton (three crops a year) began to turn Egyptian agriculture into a modern, unsustainable agricultural ecosystem. Salinity in the fields became a problem. The British built a low dam at Aswan in the early 1900s to provide more irrigation water. The Aswan High Dam of 1958-1976 ended the Nile flood, with its natural flow of irrigation waters and fertilizing silt. The water stored by the dam increased irrigable land in Egypt by five times; and (at first) provided half of Egypt's electricity (now 10-15%). The salinization and waterlogging that accompanied artificial irrigation were mostly corrected by installing subsurface drainage. The increased amount of standing water in the irrigation canals greatly increased the incidence of bilharzia (or schistosomiasis, a debilitating parasitical disease spread by snails). The end of the flood ended a sardine fishery off the delta; and the loss of silt a brick making industry. It made necessary the use of artificial fertilizer, which pollutes the river. The ongoing erosion of the delta (from the lack of silt) will lead to the loss of a brackish lake fishery that is the greatest source of fish in Egypt; and to a continuing loss of agricultural land. (Lake Nassar has a fishery for Nile perch but is far from major population centers.) Aswan High Dam changed a working agricultural ecosystem capable of feeding a large but limited number of people with little harm to the natural world into a modern, energy intensive, polluting agricultural system capable of feeding many more (but still a limited number) of people, with many negative effects on the natural world.

Chapter 11
Europe until 1800: Limits of a Fully Settled Agricultural World

Using the natural production of the forest or waste to increase the fertility of cropland is a common strategy of folk agricultures. It usually depends on domestic animals that eat leaves or grass from the surrounding uncultivated land and whose manure is then used on the fields. In a modern African variant of this system, branches from leguminous trees, grown in hedgerows between the fields, are used to mulch, and thus also to fertilize, crops; no animals are involved. The rapid decay rates of the tropics make this system possible.

Overexploitation of systems that depend on the surrounding forest or wasteland for a portion of their fertility is easy. While usually caused by over-population, over-exploitation can also be caused by an increase in market demand. For example, trekkers in Nepal seek local food and shelter. Villagers build small hotels of native lumber to take them in. Trees are cut for this. They also grow more food to feed them, and cut more fuel to cook it and to warm their guests, thus increasing their income at the cost of over-cutting the forest for timber, fuel and fodder. Cutting the woodland increases erosion and the risk of landslides in steep areas. (Much of upland Nepal is steep, its slopes held in place by shrubs and trees; under traditional management, firewood was taken from dead vegetation; live ones are lopped for browse.) The rise in the number of ski areas in Swiss mountain valleys is a more high tech example of market based over exploitation of steep forestlands. Two hundred and fifty years ago in the French Alps mountain slopes were cleared for farmland, with catastrophic erosion, remarked by travelers; now steep slopes are cleared not for food or fuel but (similarly to the Nepalese) for business income. Erosion, avalanches and changes in streamflow result. In such cases the demand for wealth magnifies the effect of population.

In the grain and cow culture of the Near-Eastern agriculturists who settled Europe 7000 years ago, using products of the wasteland to fertilize fields already had a long history. Fertility of upland fields is maintained on the one hand by in-situ weathering. This is the release of mineral elements from the soil by bacterial and fungal action and the erosive effect of natural rainfall, which is slightly acidic (root secretions make it more so and increase the release of minerals). Fields have much simpler plant, animal, fungal and microbial populations than forests or grasslands, and lack their nutrient cycling ability. Leaving them bare for much of the year exposes them to extensive leaching and erosion. Nitrogen in fields is provided by free-living nitrogen fixing bacteria in the soil, and those living in nodules on the roots of leguminous plants. The annual pulse of nitrogen may be greater in fields than in grasslands or forest, because of the warmer temperatures of the cleared ground. However a good part of this nitrogen is leached out by rainfall, even from hayfields, that is, cultivated grassland. The problem is that in the simple system of a field, the bacterial activity that releases nutrients, and nutrient uptake by the plants, do not always coincide; then nutrients escape. For instance, bacteria may mobilize nutrients before crop plants have been seeded in the spring, or after they are done growing, or have been plowed under and are starting to decay, in the fall.

While cereal crops use nutrients equivalent to the forests or grasslands they replace, much of their growth, along with the nutrients that created it (such as calcium, phosphorus, nitrogen), is removed in the annual harvest. The fertility of fields then depends on the balance between what is removed—by crops, by leaching, by soil erosion—and on what nutrients are produced within the field or added to it. If the soil is inherently fertile, and soil erosion is not too great, the crops not too demanding, and leaching of nutrients by rainfall remains within bounds, a field will retain a low level of fertility indefinitely. (Temperate loess soils are an example.) But the steady fall in fertility after the clearing of the natural vegetation is the reason for rotating fields back into forest. The earliest agriculturalists in Europe (the devotees of the horned earth goddess) cultivated river floodplains with hoes. Such soils were good to begin with and are renewed by floods and by soil washing down from the hillsides. But the slash and burn agriculturalists of upland Europe apparently moved on, with their cattle and stores of grain, when the fertility of their fields fell. Denser populations require permanent fields, however, and with the manure from domestic animals the fertility of fields can be maintained. Animals are pastured in the woods and kept nights on the fallow; fed cut branches, hay and grain straw; and put to graze on grain stubble (where they also deposit their urine and manure). The biological productivity of uncultivated lands is a major support of such continuous grain-growing systems.

The introduction of cattle was a tremendous innovation in Neolithic agriculture. Milk provides 4-5 times the protein and energy for the same amount of feed as meat; cattle provide traction power; and manure for grain crops. The development of lactose tolerance in adults (the ability to digest milk usually disappears after childhood in humans) is thought to have increased the number of a person's descendants several times (perhaps 10), so it would have been selected for in cattle keeping cultures. In Neolithic Europe each person needed 20 hectares (48 acres), some cleared, for cropland, fallowland, pasture, hay meadows; and some forested, for firewood, building material and forest browse. (Branches were lopped and brought to the animals, sometimes stored in piles by the trees for the winter. This is still done in parts of the Mediterranean.) A village of 30 people needed a herd of 40 cattle, 40 sheep or goats, 13 hectares of wheat or other grain, and about 5 square kilometers of forest for firewood, timber and animal pasture. As in medieval times, the cropland (hoeland, plowland: the plow was invented about 6500 years ago on the Sumerian plain) was probably communal and divided into two: one field was used for winter grain, the other rested, its stubble and weeds grazed by the domestic stock, which were also kept there at night. Since cattle and sheep do most of their eating during daylight but continue to excrete during the night, they deposit approximately half of their daily feed as urine and manure on the fallow. The fallow period allowed for the build-up of nutrients from bacteria, decayed plants, animal urine and manureA variation on this system in medieval and renaissance Europe was performed by the so-called transhumance pastoralists who took their flocks of sheep up into the mountain pastures of the Alps or Pyrenees during spring and summer, returning in fall and winter to the grain-growing lowlands: Spanish wheat growers paid for the privilege of having such flocks kept on their fields for a night or a week (as long as the stubble and the roadside grass would support the animals); manuring by the sheep is thought to have doubled wheat yields. While the manure produced by the flocks came from the immediate surroundings (the wasteland, steep banks, roadside ditches, and grain stubble), the animals themselves were at least partly supported by the mountain pastures and the landscapes in between. That is, their total biomass was much greater than the local landscape would have supported. Sheep are good at converting biomass to dung, producing 10 times their weight in dung annually. So this was a way of bringing the biological productivity of the mountains to the plains and making the mountains useful to people at lower elevations. (All the same, overgrazing during the medieval period by huge flocks of sheep in La Mancha and Estremadura—those impoverished lands that produced the shrewd and pitiless *conquistadores*—converted large parts of central Spain to poor quality grass and scrub; and the general decline in Mediterranean forests after the Middle Ages is thought to have been caused by overgrazing by sheep.) Whether such grain growing systems were sustainable over the long run depended on the underlying fertility of the soil (a gift of nature) and the rate of erosion (a matter of climate, soils and management), but they supported (or helped support) many of the Mediterranean and Near Eastern high civilizations.

The organized settlement systems (an early state capitalism?) of the Greeks, with their colonies in Turkey, Sicily, Egypt, the Black Sea, the Mediterranean coast of France, provided surplus grain for mainland Greece. By 400 B.C. perhaps half the food eaten in Greek cities was imported. Were such colonies a sign of erosion in Greek agricultural soils? Many Greek sites show thousand year cycles of use and abandonment. Cycles of expansion and contraction of agriculture and population during the Neolithic and Bronze ages occurred throughout the Mediterranean basin and central and western Europe, especially on upland sites. As people filled the better soils of the river valleys and lower slopes, the population continued moving up to the poorer soils of the surrounding hills. Such settlement was followed by massive erosion, visible in cores from lakes or swamps, followed by the abandonment of land, depopulation, the regrowth of scrub or forest, until some centuries later, when the soils had rebuilt themselves, settlement began again. Many European and Mediterranean landscapes were thus deforested and cleared several times over 7000 years. Such cycles continued into the classical and medieval periods.

Greek colonies were followed by those of Rome, the citizens of whose capital were entitled to a daily ration of grain, and whose grain-shed included most of the Mediterranean basin. Egypt was called the granary of Rome; there were also the more or less new lands along the North African littoral, in Turkey, and in southern France. Under a law of 111 B.C., any Roman citizen could claim up to 20 acres of public land to cultivate; by bringing it under cultivation he established ownership. This was 10 times the size of the individual holdings Romulus passed out during the settlement of Rome 600 years before, an indication either that agriculture had become more commercial or soils had become less productive. In 750 BC a man with a hoe could cultivate two acres of olives, grapes, vegetables, cereals and fodder crops. The multistory canopy saved labor, prevented erosion and took half the land to feed the same number of people as plowing with an ox; but for large landlords plowing with an ox was more profitable. At any rate cultivation in Italy expanded; the land near Rome, once full of orchards, became large estates of grain, and then, as the soil declined or eroded away, uncultivated wasteland. Wood use in Rome has been estimated at 1-1.5 cubic meters per person per year, in total about the wood in 30 square kilometers of forest. Is such a number high or low? Per capita wood use in North America before the Revolution was 17 cubic meters a year, about 4.5 full cords, that is 11-17 times as much. (Two hundred years later modern people in the northeastern United States use 4-5 cords per winter to heat their houses; or less than 1 cord if their houses are super insulated.) Deforestation for metal smelting, pottery making, brick and lime burning, building material, for new agricultural land, for pastureland, meant the continuous exposure of bare and overgrazed soils to the elements; and led to slow, massive, cumulative soil erosion. Composting, crop rotation, and the use of manures in maintaining soil fertility were known to the Romans (and probably to earlier peoples: the slow charring of vegetation, along with composting, began producing black earth soils in the Amazon Basin 2500 years ago), but such practices were not widely followed, and soil exhaustion and erosion imposed long cycles of settlement, abandonment, and resettlement on river valleys throughout the Mediterranean. One assumes soil erosion was among the reasons for the colonies and the importation of grain, though expansion is the natural result of any successful political system, or way of exploiting nature, no matter how brief.

Many former Greek and Roman port cities now lie several kilometers from the ocean, behind soil washed down from the hills. Some of this erosion would have occurred without human intervention, as the Mediterranean climate became drier, and the landscape more susceptible to erosion from its intense rainstorms, but human manipulation of the landscape speeded things up. Some Mediterranean uplands have little soil left to erode. The vine and olive, with winter wheat on flat ground, the tree fruits that date from Roman times (many brought from Persia), and the sheep pastured on the aromatic but not very palatable herbs of the once forested mountains of Crete or Lebanon constitute the modern and beautiful Mediterranean landscape. Springs and streams dry up in summer; the total runoff from the landscape is greater. On hilly land near Rome, farmers plant hazelnuts by blasting small holes in the light volcanic rock, then plant the shrubs, and water them until they take or die. Those that die (perhaps half of the first planting) are replanted. No natural topsoil is left. Such persistence constitutes land rehabilitation: the re-creation of soil with dynamite, tree roots, tree litter, hope and hard labor.

The agricultural remaking of Europe has left various signs, some of which we can read. Pollen cores from English ponds show pollen of oaks replaced by that of weeds, wheat, rye, hazel and birch. Hazel and birch are early successional species that appear after clearing the forest; hazel was often coppiced for fuelwood, that is, cut at short intervals from stands that sprout from stumps. Agriculture, by baring and stirring the soil, mobilizes the soil's lead in airborne dust. Airborne lead from Roman silver smelting shows up in cores from the Greenland icecap. (The lead—*plumbum*—went into plumbing.) Cores from peat bogs in the Jura Mountains of France show variations in airborne lead in the surrounding landscape over the last several thousand years. An initial rise 8000 years ago corresponds with a volcanic eruption in France. Soon afterward Neolithic agricultural activity tripled the relatively constant, postglacial background level. A further rise in lead 3000 years ago corresponds with smelting at Phoenician lead mines in Spain. There were rises corresponding with Roman and Greek metallurgical activity (a layer of lead from Roman silver smelting is found in lake muds all over Europe), a decline from those heights in the Middle Ages, and a rise with the Industrial Revolution that peaked in 1905, the rise in this case caused by coal burning as well as metal smelting. Total airborne lead peaked again in 1967, from lead in gasoline, on top of all the other sources, when it reached 85 parts per million. The postglacial background concentration was 0.28 parts per million. So anthropogenic lead in the modern atmosphere was 250-300 times that of the hunting and gathering background, that to which one assumes modern people and animals are adapted. (Lead is a neurotoxin. Lead levels in the atmosphere have since fallen, so much so that one can call the control of atmospheric lead levels one of the success stories of modern times.)

The two field system supported Greece and Rome. The fields, fallow or cropland, were plowed in spring, summer and fall to control weeds (thus the erosion); they were planted in the fall. Together with Egypt's Nile Valley, and some irrigated lands, the two field system supported the Islamic civilizations of Turkey, the Middle East and North Africa. ("Some irrigated lands" includes lands watered by qanats, underground tunnels that collect groundwater from mountain slopes. Qanats are found in the Middle East, Cyprus, Iran, Central Asia, and in parts of North Africa. Their design makes them self-regulating, though they must affect surface waters. Their flow in Iran in 1960, to provide urban water and to irrigate farmland, has been estimated at that of 12 Nile Rivers. For the most part qanats have been replaced by pumps, that is, water taken from deep wells and rivers, but the cities of Bam and Irbil still use water from qanats dug by the slaves of Sennacherib 2700 years ago.) The two field system supported the Christian civilization that followed Rome in Europe north of the Alps: its surpluses (wheat yielded only twice the seed sown) built Romanesque churches, fortified castles, early walled towns. About 800 A.D., when Charlemagne was crowned king of a united Europe in Aachen, a three field system had come into use in some villages in northeastern France. The common plowland was divided in three parts. One field was planted in autumn with a winter grain (wheat or rye); this is the traditional method of Mediterranean or Near-Eastern agriculture. Another field was planted in spring with a summer crop of oats, barley or peas (the last a nitrogen fixing legume); this was new. The third section was left fallow. This system increased total crop yields, putting two-thirds of the plowland into crops yearly. It increased the land in crops by a sixth. It also increased crop variety, provided more fodder for the animals and spread work more evenly over the year. More fodder meant more manure and greater yields, a positive feedback. In the 1600s, planting the fallow to clover, which fixed nitrogen in the soil and which the animals ate, further increased the production of meat, milk and manure, as did the planting, in about the same period, of leafy winter fodder crops (rutabagas, turnips) for the animals. Later, stall feeding let the farmer use virtually all the manure, and much of the urine, from his animals, at the expense of hauling and spreading it.

The resettlement of Europe that followed the crowning of Charlemagne was intended to remake the European landscape into a cultivated and holy earth. Charlemagne renamed the months (then, as now, named for Roman gods and goddesses) for their agricultural activities (the month to plow, the month to plant, the month to cut wood). Around 1000, the wheeled iron plow, pulled by a yoke of 8 oxen, came into common use. This implement, invented several centuries earlier, made possible the conversion of Europe's heavier soils (the clays on which the oaks grew) to agriculture. The invention of the shoulder harness and the nailed horseshoe led to the growing use of the horse for traction power. Horses are several times more expensive to maintain than oxen and must be shod to protect their hooves from the northern European damp, but can exert more force and work faster for a longer time. Such developments in agriculture, along with a warming climate, opened up new lands in Europe and by 1100 led to prosperity across the continent from the Atlantic to the Dnieper. The 1100s brought the first European manufacturing age, powered by wind and water mills. Europe had abundant resources of wood, flowing water and minerals. Water mills were used to mill grain, full cloth, process hemp, for tanning, laundering, milling logs, crushing and grinding ores, sieving, turning, polishing, stamping, for iron-making (operating bellows, puddling and beating iron, drawing wire). Watermills averaged one per 50 families in England.

So the Dark and Middle Ages that followed the death of Charlemagne were a time of boom: in population, in land clearance (sometimes of land abandoned after the collapse of Rome), iron manufacture, stock raising, the founding of new towns. Fields were 5% of Europe in the sixth century, 30-40% in the later Middle Ages. Religious orders established monasteries in the wilderness and granted colonists their forestland to clear and cultivate. Interested in increasing their income, the nobility also established colonizing settlements and began the reclamation of marshland and heath. Forest cover in Europe was reduced from 80% (95% originally) in AD 500 to 50% or less in 1300. (Some writers claim only 20% of the forest was left by 1400, that in France perhaps 25% of the forest remained.) That forest was heavily exploited for fuel and timber. Land use had fallen to 2 hectares per capita from 16-20 in Neolithic times. The climate was also good. During the so-called Medieval Climate Optimum (from about AD 1000-1400; some writers now put it a century earlier) temperatures in Europe were about 1° C. warmer. This lengthened the growing season by a month. The climate change was worldwide. The warm period in Europe coincided with a warm period in the Arctic (southwestern Greenland was settled by the Norse and grain was grown in Iceland), while most of the earth was slightly cooler and civilizations in Central America, the Andes and the American Southwest collapsed from droughts.

Population in Europe doubled from 1000 to 1220, from 38.5 million to 75.5 million (from a base of 18 million in 600). New land was gone by 1300, and the population was reaching the limits of its renewable resources. Overall yields fell as more and more marginal lands were brought into cultivation; wages also fell. By 1300 Europe's population had overwhelmed its productive capacities. Trade was still a small part of the economy, which was largely agricultural. Roads were poor and travel unsafe. Religious views discouraged many kinds of economic activity (for instance, lending money at interest was usury, a sin). Land in the medieval economy was held by right of occupation and was difficult to sell. Labor could be hired but work was governed by a customary web of rights and obligations. Many of the so-called prerequisites for economic growth did not exist. These include secure property rights, the rule of law, more or less working markets, some social mobility, a desire by the individual for financial improvement. A famine from 1315-1322 was followed by the beginning of the Hundred Year's War, a general European war in 1337. Catastrophe arrived 11 years later as a plague, the Black Death, which turned out to be a blessing in disguise.

Europe in 1300 still depended on renewable resources. The primary limit was food. Production per acre would rise 2.5 times during the succeeding centuries, with better forage crops and legumes, animal breeding (which produced more milk or flesh from the same amount of feed), more complex rotations, new crops and animals from Asia, Africa, India and the Americas; but yields never kept up with population, and periodic starvation in Europe continued until the Industrial Revolution. Death from famine and cold were common in Europe in the 1700s. Many episodes of starvation were local, a matter of food distribution rather than absolute shortage (as is still the case in Africa now). The last famine in Europe caused by an absolute shortage of food was in the early 1800s, when the eruption of Krakatoa in Indonesia injected enough dust into the stratosphere to cause two years of climate cooling worldwide. In New England in the year following the eruption, frost occurred in every month. The Irish potato famine, which followed this, was not caused by an absolute shortage of food (Ireland exported food throughout the famine), but by the failure of the British government to distribute food to a starving population. The Irish famine was an unexpected problem of industrialization: that of the introduction of new organisms to new environments. The blight that destroyed the potato crop was brought to Belgium on American seed potatoes imported by steam ship. The cheapness of steam transport made shipping the potatoes profitable, and the rapidity of the trip across the Atlantic allowed the fungus to arrive on the potatoes alive. The damp summer that followed spread fungal spores all over northern Europe.

Another renewable limit was fuel. Until the 1700s (earlier in England) wood was the primary fuel used for industries and crafts, as well as for cooking and heating. Brick burning, glassmaking, iron smelting, salt evaporation, lime burning, sugar refining, soap making, brewing all required fuel. Heating and cooking probably required the most. Shortages made wood expensive; in 1600 the average city dweller in France spent 10% of his income to keep a fire burning in one room for part of the day. In general, from 1500-1700, 7.5% of an ordinary budget went for light and heat. More efficient brick heating stoves in central and northeastern Europe made it more comfortable in 1700 to winter in Warsaw than Toulouse. Timber was necessary for buildings, tools, ships and furniture. Hazel and oak, species that sprout well from stumps, were cut on short-term rotations to provide fuel and also materials like tanbark. Some oak stems (the standards) were allowed to mature for timber. Producing the beer and bread for London's 80,000 people in 1300 required the wood from 500,000 acres of coppice. But supplies were limited to what growth of the forest provided.

In the European wars of the eighteenth century, English blast furnaces, needed to forge cannon shot, could only operate intermittently; when they ran out of charcoal, they had to shut down. (Eight tons of wood made two tons of charcoal, which would smelt just under a ton of pig iron.) Similar shortages occurred all over Europe. Since iron making depended on a renewable resource, production of iron had to remain at or below what the wood supply could handle. If production were increased to meet an increase in demand, the supply of wood in the future would be reduced. Future production of iron would have to be lowered, or stopped altogether. Everything depended on the growth of the trees, which put an inexorable limit on production. Where available, coal could be used for cooking, heating, and processes such as brewing that simply required a source of heat. Coal was commonly used for such purposes in England, where wood shortages developed early and coal was abundant. Coal had largely replaced wood for household heating and cooking, and most industrial processes except iron making, in England by 1700. The adaptation of essentially unlimited European coal supplies to the smelting of iron during the eighteenth century was a fundamental factor in the rise of the Industrial Revolution.

There were other problems related to an overexploitation of a renewable environment. One of the more serious was soil erosion. This reduced soil fertility (nutrients were lost with the soil and the depth of topsoil was also reduced); caused the siltation of streams, which increased flooding; and ruined freshwater fisheries. (The gravels in which fish laid their eggs silted over, rivers became shallower and warmer.) Mill dams also destroyed fisheries. Partly because of overfishing riverine fisheries were failing in Europe by 1000 and were replaced by the cultivation of fish in ponds. Many fishponds in the 1100s and 1200s were the dammed sections of rivers and streams, but with different species of fish, especially carp. Eroded soil also filled waterways and harbors. Harborworks of cities in the Rhine Delta suffered severe siltation. Bruges in present-day Belgium, the commercial center of northern Europe in the 1300s, watched its harbor on the Zwin disappear, silted past the ability of the city to clear it. Rivers were also polluted by metal works, dye works, tanneries and sewage. Wells were polluted with seepage from cesspits and from rotting bodies in churchyards. The smell of cellars about the cemetery of *Les Innocents* in Paris was notorious.

As the population grew and freshwater resources declined, marine fish began to replace freshwater fish in the European diet. The trawl was invented in the 1300s, with (eventually) disastrous effects on fish stocks and the life of the seafloor. With the trawl, cod off the English coast became so easy to catch that the surplus was fed to pigs. Diking to reclaim land in the Rhine Delta caused major losses in sturgeon, once a key item in the European diet, by destroying its spawning habitat. The disappearance of the commercial herring fishery in the southern Baltic in the late 1200s-1300s corresponded with increased runoff from land clearance in its basin. By 1500 stocks of herring in the Baltic and North Seas and of cod all about Europe were failing. (Except for the Danish herring fishery, which collapsed with finality in the 1300s, the cod and herring fisheries would recover to fail again.) The Baltic and North Sea herring fisheries originally amounted to billions of fish annually. Fish was one of the sources of wealth of the cities of the Hanseatic League. (Even in the late 1600s, when fish were fewer, work in the herring fishery constituted 20% of the Dutch economy.) In the late 1400s fishermen from Bristol sailing west of Iceland discovered the Newfoundland cod fisheries. Did they stop to take on some of the Norse colonists in southern Greenland, whose sheep and cattle culture was failing in the cooling climate of the Little Ice Age? By 1600, 20,000 European fishermen were salting and drying cod off Newfoundland. Dried cod from the North American banks would provide a cheap source of protein for Europe for 500 years, helping support a world in which the Church declared 40% of the days meatless.

Until the Industrial Revolution, the European world was Malthusian. That is, its population tended to increase faster than its food supply. In a downward spiral, more people meant more demand for grain, less crop rotation, less fodder for the animals, not enough manure to raise grain yields, the animals dying from parasites and malnutrition in late winter and spring. People died from hunger and exposure. Infant mortality was 25-30% until the late eighteenth century, life expectancy was 30-40 years, epidemics were common, most of the population was undernourished and depended on vegetable foods (bread, gruel, potatoes). Much of the population was very poor. At the time of the French Revolution, 80% of the population of France is thought to have been poor or destitute; that is, they owned the clothes on their backs. Many were more or less homeless, not being able to afford both food and lodging. In medieval and early modern Europe four-fifths of disposable income went for food. The nobles and the members of the middle class ate well. These distinctions lasted in England long into the period of industrialization: in 1800, boys taken into the Royal Navy from the slums of London or Liverpool were 8 inches shorter than boys from the upper classes; in 1940, working class draftees were still 4 inches shorter than boys who had gone to elite schools. So one could recognize an officer by his stature. Until about 1700, periods of growth in the standard of living of the common people were followed by periods of reversal (the long waves in the economy were cyclical), so the standard of living of an agricultural worker in Europe in 1500 or 1600 was only slightly higher than in Roman times. When Europeans came to North America, where food was plentiful, their numbers, instead of doubling every 150 years, began to double every 23 years. Most of the first born children in Puritan families were illegitimate. (When introduced to North America in the late 1800s, English sparrows underwent a similar population explosion. The sparrows raised five broods a year, the young of one clutch brooding the eggs of the next, as they expanded into the empty niche of the American barnyard.)

This is not a portrait that leads to a progressive view of history (constant progress onward and upward). That view would come out of the cascading improvements in material life with the Industrial Revolution (railroads, steam power, gas lights) and their consolidation and elaboration in the twentieth century (electricity, cars, radio, penicillin, TV). During the 1780s, as the English were losing their American colonies, industrial growth in England rose from 1% to 4% a year. It remained near 4% (some writers claim 2%) for a century. The United States was more egalitarian and less bound by social traditions than Europe, but remained agrarian longer. After the economic boost of the Civil War, markets expanded there with the population and economic productivity grew at 2% a year from 1870 to 1970.

But this is also not the whole picture. From 1200-1800, during a period of worsening climate, agricultural yields in Europe rose 2.5 times, while population rose 10 times. Land under cultivation increased several fold. During the so-called Little Ice Age, from 1430-1850, the climate was 1- 2° C. lower than during the Medieval Optimum. A severe famine from 1315-1322 killed 10-15% of the population and set the stage for the Black Death of the 1340s and 1350s. (Fetal malnutrition interferes with the development of the immune system.) The Black Death killed a third to a half of the population. Recurrent epidemics followed for the next century and there was another famine in the 1430s. For the most part crops near their natural temperature limits (at high altitudes or latitudes) failed. During the Little Ice Age cereals could no longer be grown on the hills of northern and western Britain, glaciers and the tree line descended in the Alps, and the northern limits of vineyards in France and Germany retreated 300 kilometers south. Yields of grain on newly cleared land during the warm summers of the Optimum had been twice those of late Roman times but inevitably fell as soils eroded and fell further as the weather worsened. Most of the cooling took place in winter (this was a time of ice-skating and ice festivals) but summers were also 0.8° C. cooler.

After the calamitous 1300s came a period of recovery. Europeans rebuilt their water mills for water-powered industry. In return for paying rent, tenants gained heritable rights to their land. Some common lands began to be enclosed to create rentable tracts. (In the late 1500s, it was said a living was three acres and a cow. People were still saying that in upstate New York in 1850.) Towns, supported by agricultural surpluses, grew into small industrial centers. Industry in late medieval Europe was for the most part cloth, wool or linen, woven on hand looms at home. By establishing chartered political units, surrounding themselves with walls, and mobilizing their citizens into a defensive force, towns established a degree of political independence, both from the church (many towns were founded as the seats of bishops) and from the countryside and the nobility and bandits that ruled it. Intellectual life bloomed as paper replaced animal skin parchment in books (paper was 13 times cheaper than parchment at the time) and printing (also much cheaper) replaced hand copying. Books became more affordable; before the printing press, a professional man's annual salary bought two cows or four books (and nothing else). After the printing press, a middle class person could afford a couple of books a year. As land became more valuable, property rights became more exclusive, and began to extinguish the traditional rights of the nobility, such as that to ride or hunt where they pleased, and those of the peasantry, which included the right to common grazing land and to fuelwood gathered from the forest. Trade spread once again (memories lingered of the fairs of the 1300s), its bankers and merchants operating under the protection of the towns. Some of this trade was for commodities like Baltic timber and grain and some for luxury products like the silks and spices of Asia, to which the caravan routes, closed for several centuries by drought and the expansion of Islam, were once more open. Imitating the nobility, families of the middle class tried to keep their wealth through such devices as late marriage (thus limiting their fertility); by advantageous marriages between families; or by restricting inheritance to first born sons. A writer has said that among the wealthy, population control was positive, while among the poor, population was controlled by starvation. In Paris at the time of the Revolution one-quarter of all children are thought to have been abandoned for adoption. Such infants were taken in by church orphanages, where the great majority of them died.

The medieval boom had ended with the famine of 1315-1322. Every season of 1315 was wet. Crop yields were half of normal. Hay was put up wet and rotted in the barns. During the winter and spring of 1316 people ate their seed grain. The year 1316 continued wet with another crop failure. The price of wheat tripled. Cannibalism was widespread. The famine continued until 1322. Twenty-five years later, after several warm, wet springs favorable to the spread of plague among the rodents of Central Asia, the Black Death arrived with people on ships from trading posts on the Black Sea and killed a third or more of the population of Europe. The people from the Black Sea were refugees fleeing the Mongol invasion. Parts of Europe would not see the population reached during the Climate Optimum for another 450 years. But the decline in population was followed by a period of development and prosperity: the Renaissance. The average age of death for adults was still something like 35. Still hemmed in by the Islamic world, Europe began probing its limits. Genoese bankers (who had financed the trading cities on the Black Sea) now financed Portuguese explorations around Africa to the Spice Islands of the East and out into the Atlantic, where large semi-tropical islands were discovered. New crops had been appearing in Europe thanks to trade with Asia and contact with the Moors in Spain; these included rice and sugar cane and the new livestock of silkworms. After 1492 came American crops. The yields of Mexican corn and Andean potatoes would dwarf those of European grains. A writer has speculated that the introduction of maize, peanuts and manioc from the Americas to Africa, which substantially increased the human population there, made the trans-Atlantic slave trade possible. New diseases (such as yellow fever) came with trade and with African slaves to Europe and the New World.

The high point of European row crop agriculture was reached in England in the eighteenth and nineteenth centuries. The development of rotations between cereals and crops of legume hays (which raised the yield of grain per acre), the raising of animals on legume hays (which increased the numbers of animals that could be kept), and the use of animal manures on the fields (which further raised the yields of grain), together with plant and animal breeding that resulted in higher yielding varieties of plants and animals (more meat, wool, or grain for the same nutritional input)— all raised European upland agriculture toward the heights of successful overflow agriculture (such as in the Nile Delta), the ditched fields of Tia-huanaco, or paddy rice. That is, it became a high yielding, self sustaining agricultural system, less dependent on inputs from the wasteland, capable of supporting many more people per acre. Enclosure laws in England, en-acted partly to ensure an adequate supply of manure for cropland (the ratio of pastureland to cropland is important in manure dependent cropping systems), better equipment, and better capitalized farms put many agricul-tural laborers and small tenant farmers out of work; some went to poor-houses, some to Australia or the Americas, some to work in the shops of the Industrial Revolution. That horses replaced oxen in the 1800s as Europe's main source of agricultural power showed Europe's agricultural prosper-ity. Able to work faster for a longer time, horses are several times more expensive to maintain. Environmentally speaking, such mixed agricultures were a high point of upland agriculture; they were sustainable as long as they occupied a more or less limited place in the larger ecosystem. Similar agricultures flourished briefly on the American prairies, and in the German settlements of the Shenandoah Valley of Pennsylvania and Virginia. One finds them still among the American Amish.

What ended the European dependence on the renewable world was the use of fossil fuels. England was the European nation most short of wood. It was largely deforested, probably not for the first time, when the Domesday Book was compiled in 1089 AD. (Deforestation of the English downlands had begun in Neolithic times 5500 years before and had con-tinued through the Roman invasion of AD 43. Deforestation, erosion and grazing converted the original downland woodland of oak, alder, willow, hazel, birch and rowan into the thin-soiled grasslands of today.) England was also the country most in need of wood for its imperial ambitions and it was here the Industrial Revolution began. While many technical developments came from the continent, the English most thoroughly exploited their economic possibilities. A writer has summarized the reasons for the development of the Industrial Revolution in Britain: the greater size and efficiency of British markets (many of the export markets created under duress and maintained by Britain's sea power); Britain's commercially minded society (the desire to make money); Britain's openness to innovation; and its accumulation of natural resources from around the globe (also a function of its sea power). England had coal and by the 1700s most industrial processes that required heat (baking bread, brewing beer, firing brick) had shifted from wood to coal. In the late 1700s machines were developed that, powered by water, would spin cotton thread; other machines wove the thread into cloth. The mechanical advantage of the machinery was so great, the multiplication of the value of the labor and investment so enormous, that the major limits on cloth manufacture became the availability of waterpower, of raw cotton, and of the ability to market the cloth. The manufacturing process was so cheap, compared to hand spinning and weaving, that the manufacturers were able to lower the price until demand met supply. Once cloth fell within reach of the poor, an enormous market was created and demand exploded. Profits were still enormous. The invention of the cotton gin, which mechanically cleaned cotton (previously slaves had picked out the seeds by hand) had a similar advantage: one didn't just double or triple the value of labor, one increased it by orders of magnitude, powers of ten.

It was the use of fossil fuels to smelt iron ore and power steam engines that finally changed everything. In England in the late 1700s processed coal (coke), rather than wood, was first used successfully to smelt iron ore. Unprocessed coal had too many impurities compared with charcoal and the iron smelted with it was too brittle. The coke making process eliminated these. Coke was much cheaper than charcoal. And coal to make it was available for the mining, as was iron ore. Coal fired the steam engines, made of iron smelted with coke, that pumped the water from English mines and let the miners produce more coal. Coal powered locomotives, built of iron and running on iron rails, hauled coal and iron ore to mills. Coal powered spinning machines took over from water powered ones, removing another constraint from the manufacture of cotton cloth. Coal a century or so later was joined by oil and natural gas, useful because they were fluids rather than solids, and so flow under gravity or pressure. Such fossil fuels became unlimited sources of energy in an otherwise renewable and limited world. Fossil fuels will likely remain available for a long time. (For coal the current guess is another 200 to 400 years, depending on the rate of use.)

Coal powered steam engines eliminated the limits set by low water or frozen waterways in water transportation, by replacing canal and river traffic with rail. In the 1860s coal powered steamships, built of coal smelted steel, carried three times the cargo twice as fast as sail. With transportation faster and cheaper, Europe reached out to the rest of the world for its resources. From 1860-1920 one billion acres of new land was converted to agriculture, 40% in the United States (much of it in the Corn Belt), 20% in the Russian Black Earths, 20% in Asia. Grain was imported from the "new lands" in the Ukraine, North and South America, and Australia. (Land clearing would continue, with another billion acres added to agricultural lands from 1920-1980, mostly in Latin America.) Refrigeration and pasteurization of milk made milk more transportable and its production rose enormously, with beneficial effects on the European diet. Milk now constitutes 20% of the value of agricultural production in Europe and the United States. Imported fertilizers such as rock phosphate and Peruvian guano increased the fertility of European soils. But it was primarily cheap imported food, its cheapness made possible by the exploitation of fossil fuels, that ended starvation in Europe. The European diet in 1800 was worse than that of the European hunter-gatherers of 12,000 years before. The diet worsened further during the early years of industrialization—the average height of men in both North America and Europe fell in the 1830s—but improved after about three generations. By the 1920s the shortfall in height was gone. By 1900 England was importing 80% of its grain, 75% of its dairy products, and 50% of its meat.

Coal also made the chemical industry possible. Coal provided the energy to run the reactions; while coal, oil and natural gas replaced wood as a feedstock for the chemicals. The Haber process, which synthesizes ammonia from atmospheric nitrogen, led to the manufacture of synthetic nitrogen fertilizer, which further increased the yield of soils. Artificial fertilizers and further developments in crops and animals would make Europe nearly self-sufficient in food by the mid-twentieth century. However the excess nitrogen used on crops would cause tremendous pollution problems. Anthropogenic nitrogen lies behind the biological degradation of marine estuaries and (along with phosphorus) of fresh waters. Nitrate pollution in the Thames and Rhine are now two orders of magnitude (100 times) above the mean values of unpolluted streams. (Not all of this is due to agriculture; some nitrogen and phosphorus come from sewage effluent and some nitrogen from the combustion of fossil fuels.)

By the end of the nineteenth century coal was being used to generate electricity. By two decades into the twentieth century coal and oil had produced the modern world, where the problem is not that of producing enough, but of creating demand for all that can be produced. Through gaslight and electricity, fossil fuels had eliminated night. They ameliorated the seasons through heating in cold climates, cooling in hot ones. Fast, cheap transportation eliminated distance and with it the agricultural seasons, so that now any modern expects fresh fish, fresh lettuce and fresh grapes to be available anytime, whenever he or she wants them. The telephone eliminated distance between people, public education ignorance, public health measures and medicines disease. The human habitat in developed countries, urban or rural, is a built one, of roads, telephones, powerlines, fields, heated and powered houses, internet communication. The cost of food has kept falling until it is now less than 10% of income in developed countries. Such development comes at a cost. In 1850 every Englishman used the equivalent of 1.7 tons of coal a year; this rose to 4 tons by 1919, where it remained until 1950, despite considerable economic growth, when the number began rising once again. (A year's fossil fuel use in the twentieth century consumed approximately 400 times the global net primary productivity—net plant growth in the world—of that year.) Such energy use brings us other problems. But the notion of man's independence from nature and its constraints is a hallmark of the modern.

Chapter 12

The Land Trip Revisited: the Capitalist Landscape in the Settlement of "New Lands"

Madeira entered European history in 1419 when it was sighted by Portuguese sailors blown out into the Atlantic, though the genetics of its mice indicate it may have been visited earlier by Scandinavians. Madeira was an uninhabited island 350 miles off the North African coast, on the latitude of Morocco. Legend has it that the island's thick woods were set alight by the discoverers (the first step in settlement), who then sailed away. For 7 years the pillar of smoke was a beacon for ships. Madeira is among the first examples of the modern capitalist settlement of "new lands," that is, lands newly settled by Europeans. A few of these lands were uninhabited (most weren't); some had inhabitants who were subdued by force of arms, or by arms and diplomatic pressure. (The 80,000 Guanches on the Canary Islands, another island group in the subtropical Atlantic, nearer to the North African coast, took a century to subdue.) The Americas, whose inhabitants had been isolated from the Eurasian landmass for the last 10-30 millennia, were emptied by European diseases that flew miles and years ahead of the Europeans themselves. Ideally, such lands were adaptable to European animals and plants; or to tropical plantations.

Some newly discovered lands were not really conquerable; some were not possible to settle for other reasons. Tropical Africa was difficult because its climate and its long occupation by humans made it a home for diseases and parasites to which the Europeans had no resistance. (In some colonies half the Europeans died the first year. The survival rate of colonial administrators in West Africa was calculated in so many years, so that sufficient replacements could be trained.) Asia could be dominated, but its dense population made much of it difficult to settle. Settlement of abandoned or waste lands, and the turning of marshland into farmland through drainage schemes, were still taking place in Europe; but by 1400 Europe was already largely settled, and traditional rights and attitudes also tended to limit what could be done there. Madeira on the other hand was empty. It was a mountainous island, and so work there had to be done by hand. The island's near-tropical location made it suitable for sugarcane, which had been introduced to Europe by the Arabs, who had got it from India. The Crusaders had adopted the crop for the areas they conquered. They invented the plantation system using slave labor to grow it, but in the Mediterranean a shortage of fuelwood to refine the sugar was a problem. (It took 100 cubic meters of wood to crystallize a metric ton of sugar.) Sugar was a delicacy in Europe, whose only previous sweetener was honey. Candied fruits in Renaissance Europe had been sweetened by boiling in lead vessels; the reaction of the acidic fruit with lead from the vessel sweetened the fruit with lead acetate. This is a process that dates at least to Roman times, and some writers have implicated the fondness of Romans for such sweets, and the subsequent level of lead poisoning among Roman elites, in the fall of Rome. On Madeira sugar cane was grown in narrow terraced fields, irrigated with water brought down from the top of the island in channels cut out of the rock. Its mountains gave Madeira many microclimates and its high mountaintop plateau caught a constant supply of water from rain and clouds. This, plus 700 kilometers of irrigation channels, made reservoirs not necessary. Most of the water in the system was stored in the channels and canals. Over time the island was more or less remade into an agricultural landscape. Some of the development was financed by outsiders (again the Genoese) and much of the work was done by African slaves. Sugar was already being grown by the Portuguese with African slave labor on the Cape Verde Islands. The Cape Verdes are located off the African coast, but are hotter and drier. They were used primarily as a staging ground for the export of slaves.

Madeira's production of sugar showed a typical curve for the capitalist exploitation of a new resource. The beginning was slow. Production reached 640,000 pounds in 1470 (about 45 years after construction of the irrigation system began), rose to 2,560,000 pounds in 1490, 3,360,000 pounds in 1494, and peaked at 4,508,000 pounds in 1506. By 1500, the population of Madeira numbered 20,000, including several thousand slaves, and Madeira was the world's largest producer of sugar. Madeiran sugar production had caused the price of sugar in Europe to fall by 50% and sugar had begun its journey downward through the European class structure. (Sugarcane is now the world's largest crop and sugar from cane, beets and corn a major source of calories for the poor.) The move from use by a select few to use by the whole population, the so-called democratization of the market, is typical of modern capitalist enterprise: consider cars, electricity, computer use, internet access, or, more hopefully, the market for organic food (currently 1-2% of food production in the United States, but growing at 10-20% per year). What then happened to Madeiran sugar production was also typically capitalist: production stabilized at about 2,240,000 pounds a year after 1500; and other places began producing sugar. Madeira's dominance of the market would be taken over by Brazil after 1570. Brazil, with its endless flat areas suitable for sugar production, its limitless timber resources, and its huge importation of African slaves, would eventually make sugar production uneconomic on Madeira.

Madeira had additional expenses compared with Brazil: the irrigation system, the difficult terrain that required more manpower per unit of production. But Madeiran yields were also falling from soil erosion, and from losses from rats, insects and introduced diseases of the cane plant. The tremendous yields of earlier years were probably partly the result of new land being brought into cultivation. Because of its different elevational micro-climates, Madeira had always grown some wheat, wine grapes and tree fruits, and had exported furniture made from the trees that had escaped the initial conflagration. (Many trees must have been left. Not only does sugar production requires a great deal of fuelwood, but sawn boards 15 inches wide and double the length of those available in Europe were exported to Portugal; it is said these planks made possible the large ships the Portuguese sent around Africa to the East.) Agricultural competition began to change Madeira into a wine island: vines and trees are less demanding crops, require less labor, and are better at maintaining their soils, though with the proper mycorrhizal inoculants sugar cane can be grown without fertilizer, and with the return of the composted bagasse (the crushed stalks, often used to fire the mills), or their ashes, to the soil, sugar cane may not be bad at maintaining its soils either. In the rest of the world, sugar production would continue its upward trajectory for some centuries.

As the price of sugar continued to fall, and it became a staple food, profits moved more and more from the tropical producers to the trading houses of Europe. In the seventeenth century, Caribbean islands under the control of the English and the Dutch (such as Barbados and Antigua) were cleared to grow cane (on Antigua almost no natural vegetation remains), and the Caribbean began to compete with Brazil in sugar production.

The settlement of Madeira shows the extent to which Europeans of the fifteenth century had the idea of landscape as something marketable. The idea of commodifying nature was not new. The Romans saw wilderness as something to be domesticated and used and so presumably did earlier agriculturalists. The Greeks saw the conflict between wilderness and civilization as poignant and unavoidable. In an agricultural or a capitalist world, developing a landscape greatly increases the value of the land itself. In North America relative real estate values would become prime considerations in how landscapes were used. Sugar was much more valuable than the productions of the natural landscape of Madeira, whatever they would have been, and sugar's high value per unit volume also made it profitable to ship in the fifteenth century. Logs (one product of the natural landscape) were too bulky to ship profitably; timber from Madeira was shipped as sawn lumber or chairs. (But things change: by the late 1700s hand squared white pine timbers 60 feet long were being loaded onto timber ships in the Port of Montreal for shipment to England, where they would be resawed into planks in English pits.) So under capitalist management the natural landscapes of Madeira (its associations of trees, herbs, insects, animals, natural watercourses) were restricted to areas too cold or steep to develop. The ecological functions of the landscape did not enter into the economic calculation. Of course capitalist development runs along a continuum, and some natural landscapes had economic value for some people. The Atlantic salmon that ran up Lake Ontario streams to spawn were considered part of the value of a farm in early nineteenth century western New York State. The salmon didn't last long: they were overfished; clearing of the landscape for fields made the streams too warm in summer; and silt eroding from the farmland covered the river gravels in which they laid their eggs. Farm woodlots, though forced more and more into poorer, steeper, uncultivable land, were useful in the northeastern United States for a long time as sources of fencing material, logs and fuelwood. The woodlots protected the headwaters of streams and provided some natural habitat among the cultivated ground.

The search for profit helps explain why the meeting of peoples like the Huron and the French was such a collision. The material lives of a seventeenth century Abnaki in New England or a Huron in Ontario were not that different from those of a seventeenth century European farmer; and better than the material life of a landless agricultural worker. The well-off European peasant was entirely dependant on his fields for food, owned some metal tools, a table, chairs and a bed, a more or less permanent house, chickens, a cow, perhaps a plow. The North American had better clothes, of animal skins rather than cloth, a safer water supply, was less liable to chronic injury from heavy agricultural labor, suffered from fewer diseases, had a less permanent house, a less laborious daily round and a better diet. He was taller, stronger and probably lived longer. The Abnaki's winter wickiup of bark or skins was small, dark, warm, and smoky (the indoor air of the communal long houses of the Iroquois may have been better). European houses in North America were larger, drafty, cold, also dark (because of the expense of window glass), and smoky (fireplaces with chimneys were just coming into use). Native bows were more accurate, more powerful and carried further than blunderbusses; and canoes were faster and more maneuverable than wooden dories. Moccasins were drier and more comfortable than European boots.

It was the direction of such lives that was different. The colonies in Massachusetts were financed by investors who expected to be paid back. The Pilgrims, trading with the native women for their used beaver cloaks (furs 2-3 years old, with the guard hairs worn off, were preferred by the London hatters); gathering casks of sassafras and sumac; splitting pine logs into clapboards, or oak logs into barrel staves; filling a ship with this material, must have looked very odd to the Abnaki, who thought the Europeans came from a land short of wood. The Pilgrims were soon replaced by the tall-hatted, less tolerant Puritans. What drove these people? In the 1920s, the chiefs at Taos Pueblo in New Mexico told the psychologist C.G. Jung that all white men looked mad. In the 1620s, trade was what made Europe work. The goal of all this activity for a Massachusetts Bay Puritan was a big house in town, a life free of the daily round of a farmer or a retail businessman: not the life of a shopkeeper but that of a capitalist, an investor who lived off holdings in real estate or ventures in timber or shipping. The idea was to approach the status of the nobility (always unattainable, since that was a matter of blood), for whom physical labor was anathema. For the Indian gatherer and horticulturalist the chief value of the landscape was its edibility and the daily round with its stones, trees, sweat, nuts and wild animals was life. Life existed in the physical landscape, which was encompassed in myth and in the structure of language. The thousands of petroglyphs tapped into stones below the three volcanoes on the western edge of Albuquerque, New Mexico, commemorate the entrance to the underworld from which, in mythic times, the tribes came out onto the earth. Direction in the language of one Californian tribe was indicated by one's position with respect to a sacred mountain. The Navaho homeland is delineated by the four sacred mountains. This is not to say that a landscape might not be improved, say by an annual burning, the building of buffalo jumps, the digging of pitfalls along deer trails, the clearing of fields (which often lost nutrients and eroded); such matters always fall along a continuum. But the notion of a marketable landscape opened another door. A marketable landscape is no longer important in itself, or in what it can produce by itself in biological perpetuity, or in how it works together with other landscapes up or down the watershed. (These anyway are ideas of modern biology, that sometimes correspond with edible or tribal notions.) A marketable landscape's value is in what it can produce for sale, as quickly as market demand allows (beaver skins, freshwater pearls, gold), and in its later potential for transformation into a landscape of greater value (a farm, inland waterway, factory site, house lot), with its natural productions reduced or removed.

Agricultural use moves a landscape further along the continuum towards the marketable. Most agriculturalists have some variety of property rights. They may not own the land they cultivate in our sense (that is, be able to sell it) but the right to farm it cannot be taken from them easily, they control what it yields, they can let others use it. The development of modern western ideas of private property rights can be traced back to the European commune: the walled town that managed to separate itself by economic leverage and force of arms from the nobility that ruled the countryside. Such towns were comparatively rich by the thirteenth and fourteenth centuries, especially in Italy. Merchants were financing trade in grain, silks, and spices; land reclamation schemes; public works; and (under duress) wars. In the medieval countryside, the nobility rode and hunted where they wished, and took what they wanted by force of arms; the Church also had a growing amount of land and various rights. (By the 1400s, the Church had more land than the nobility, left it by pious parishioners.) Property rights had to be regulated if a person was to invest in farmland or other development projects, such as mines. The right of unobstructed use of one's property, the right to use it without interference, was a necessary precursor to capitalist investment. This was not always a simple matter. Some land uses were objectionable. Iron mines and foundries, for instance, were almost as unpopular among the peasantry as among the nobility. Their noise was appalling; their logging and runoff polluted rivers; and their metal rich smoke poisoned nearby farmland.

For an Inuit or a Huron, private property in our sense was more or less limited to what one held in one's hand. Cornfields among the Iroquois descended in the female line, but were temporary, and were not sold, though they might be lent or given away. One could say (following Locke) that they were created by their user's labor. A Chippewa matron would have the use of a particular sugar bush, also in a sense created by labor. In some California tribes, the right to the acorns from certain oaks or groves was passed on by families. The heraldic grapevines of the coastal New England villages remarked on by the early explorers, that were located in favored spots, trellised on trees, kept clear of shading by shrubs or other trees, may have been individually owned. Groups of families had the right to certain hunting territories but this was also a limited right, under the supervision of the tribe. All this wasn't very different from medieval, rural Europe, with a village's common plowland, divided into individual allotments, and common waste and woodlot, pastured and cut by communal agreement. But common lands in Europe after 1200 would slowly become private lands: freehold property, to be farmed, enclosed, mined, sold to another party, that is, developed, as the owner wished, without interference from other landowners or from a state power. Advances in agriculture, along with the constant increase in human population, made such lands valuable. The motive behind the enclosure of common lands in England was the desire of large landowners to profit from the increased value of their land's agricultural production. So the traditional residents were removed. The development of private property rights made the landscape marketable, and the lack of traditional restraints in the "new lands" across the Atlantic meant capitalist land development would reach its greatest extent there.

Chapter 13

Some Ecological Implications of the Land Trip; a Short and Anecdotal Essay on the Environmental History of the United States, with a Focus on Streams, Especially Those in the Middle of the Country

---

Now, long after it all happened, a large literature is appearing on the ecological consequences of European settlement of "new lands" in North and South America, Australia, New Zealand, the South Pacific islands. Exploitation of the lands' resources began with settlement. (After all, it was the reason for settlement.) By the 1670s, Boston merchants were buying timberland in Maine in order to log it and abandon it. (Fernando Gorges' settlement in Maine preceded that of the Pilgrims in the 1620s, but the natives, not yet affected by European diseases, were too numerous and too unfriendly, and after attempting to set up a sawmill, the English left.)

Beaver insured the success of the colony at Plymouth. The fur trade had begun in the 1520s as a sideline of the cod fishery off Newfoundland, then expanded with the Dutch settlements in New York, and with the growing popularity of felt hats, made of beaver fur, in Europe after the 1550s. Two million skins were shipped out of New York during the 1600s, and by 1650 beaver were more or less gone from New England and New

York State. Most were gone from streams east of the Mississippi by 1700. By 1680 the French had a trading network from Quebec to the mouth of the Mississippi and west to the Rockies, where they came up against Spanish traders; the British were trading in most of northwestern Canada from an outpost on Hudson's Bay. Recurrent epidemics, collapsing villages, the desire for trade goods, and the movement of European and Native populations into others' lands demoralized the Native Americans and removed their cultural constraints on hunting. In the 1650s the tribes who hunted the immense swamplands between Lake Erie and the Mississippi (wooded swamps of elm and red maple, now drained to grow corn) were forced by the Iroquois to provide skins for the British as a sort of tribute: acting as middlemen for the British, the Iroquois reaped the profits. The Iroquois part in the fur trade led to Iroquois domination from Hudson's Bay to the Carolinas, and from the Atlantic to the Mississippi.

The land between the Great Lakes and the upper drainages of the Mississippi and Ohio was one of the centers of the North American beaver population. Some writers put the aboriginal population of beaver in the United States at 200-400 million animals. (Seton estimated 50 million, other estimates go as low as 10 million.) If 200 million is correct, at an acre of pond per beaver there were then 200 million acres of beaver wetlands, about 10% of the continental United States. (Is 5% a more reasonable number?) Beaver ponds occupied about 20% of the area around precontact Detroit. In good habitat, each 1000 meters of stream contains an active beaver lodge. Beaver wetlands supported waterfowl such as wood ducks, sandpipers and teal; small animals like fish, turtles, frogs and salamanders, and the mink, otters and raccoons that ate all these; willows and moose; populations of invertebrates and microbes; and the wolves that ate the beaver and the moose. They increased the width of the land-water ecotone, where microbes recycle nutrients, and regulated stream flow. In pre-European North America, beaver shaped the colder, wetter parts of the continent, from the Atlantic to the Rockies, north to the margins of the Arctic. Beaver wetlands produce more methane than other boreal wetlands, and so affected global atmospheric chemistry. Settlers cut hay in abandoned beaver wetlands ("beaver meadows") and drained former beaver wetlands for farms. Draining wetlands affects water quality downstream: the filtering effect of the wetland is eliminated, so most of the nutrients and sediment coming from upstream continue on downstream; and the oxidation of wetland soils releases stored nutrients and sediment that also flow downstream. In the 1600s and 1700s, as their population fell, and their cultures collapsed around them, Native Americans trapped out the beaver to trade their skins for European goods and alcohol. (The trap-out by the mountain men of the early 1800s in the streams draining the Rockies, using steel traps and castoreum bait, occurred after the Louisiana Purchase.) Managing beaver sustainably, as the Indians had done for millennia, removes a quarter or less of the population annually (50 million animals out of a population of 200 million, not a small number).

On the frontier, the settlers fished, shellfished, fowled, cut timber, hunted and trapped at will. In the mid-1700s New Jersey farmers complained to Swedish botanist Peter Kalm about the growing shortage of ducks. As Kalm pointed out, since the farmers shot ducks whenever they pleased, including in spring and summer, when the ducks were raising their young, they had only themselves to blame. New England coastal Indians, who had depended on an endless supply of shellfish for thousands of years, complained that the Englishmen's pigs rooted up their clam beds; but they were liable to punishment if they killed the pigs. Arguments over dams that prevented the movement of migratory fish would finally be lost by riverbank landowners in New England in the 1800s, when Boston capitalists started to develop New England's water resources to power textile mills. Dams that shut off fish runs changed the character of whole drainages. Dams flooded riverside wetlands permanently (turning them into shallow lakes) and changed the vegetation along the lengths of regulated streams. (Erosion from farmland might then fill the wetlands in.) Trees on a floodplain grow where the period of inundation is limited. When the flood lasts too long, the tree roots, deprived of oxygen in the wet ground, stop transporting water to the leaves, which die of drought. Dams that raised water tables along the Mississippi, together with cutting of the streamside forests for fuel and timber, changed a mixed forest of oaks, hickory, willow, sycamore, pecans and elm to a monoculture of silver maple, a pioneer species that could stand the higher water levels and shaded everything else out—with huge implications for riverine wildlife. Rates of land clearing before the Revolution were half or less the rates of land clearing that came afterwards, but the effects on streams were severe. Development spiraled upwards after wars, which pushed technological development forward; rates of western settlement increased after the French and Indian War, the American Revolution, and the Civil War. The Second World War gave the industrialization of agriculture a major push.

Early forest clearance was for farms. Logging overtook agriculture as a reason for cutting the forest only in the 1880s, with railroad transportation. Forest clearance in New England reached a rate of 0.4% to 0.5% of the land area per year by the late 1600s, a rate that would take about 200 years to bring the whole landscape under cultivation. Despite a speeding up of this process after 1750, cleared land in New York and New England reached its maximum extent in the 1880s, with something like 80% of the landscape cleared. (Some writers say 65%.) New York State, in a typical reversal, is now about 70% forestland. Before freight railroads, transportation constituted 50-75% of the cost of sawn lumber. Without transportation to a market, probably 75% of the timber cut to clear farmland was burned. If wagon roads and water transportation were available, ashes, turned into potash, were the first crop on a farm. An acre of hardwood produced 60-150 bushels of ashes, worth $0.06-$0.08 each; this would pay for the labor in clearing the land. Sometimes the yield of ashes would pay for the land and clearing it. After being "proved" by harvesting a crop, the land could be sold for 10 times its original cost. Some people established several farms during a lifetime. In the years from, say, 1650 until 1935, when agriculture was the country's dominant business, the notion of the family farm was something of an illusion. Making farms was a business. In 1910, more than half of all American farmers had been on their land less than 5 years.

All this land clearing sent tremendous amounts of sediment into rivers, raising their beds and filling streamside wetlands. Much of it ended up in the ocean, where it silted up harbors and shorelines. Many of the mud flats in small New England bays date from early land clearance. One sees the same forces at work now. Logging upland forests about previously unlogged watersheds in New Zealand sends sufficient silt downstream to allow mangroves to take root in the intertidal zone (a habitat formerly too sandy for this to happen). This is not necessarily a bad development (it would happen if the erosion were for 'natural' reasons and is a way of turning the silt into biomass—that is, vegetation, fish and birds) but one not favored by the admirers of once pristine beaches.

Woods that were not cleared for farmland were changed by cutting for timber. First growth forests in the eastern United States have 3-6 times the woody biomass of second growth forests (125-250 tons of wood per acre for good first growth, 20-80 tons for second growth). Early cutting, for prime sawlogs, fencing, and building material, was light, but changed the character of the forest. American chestnut, which sprouts from stumps, was 4-15% of woodland in New Jersey and Connecticut in early surveys, 60% by 1900. Chestnut, which resists rot, was popular as a building material and for fencing and was cut extensively and recovered quickly. Softwoods like spruce and pine (which do not sprout) were also favored, largely because they could be floated down rivers to market. Old trees of long-lived species, such as eastern hemlock, cut for tanbark, and white pine and red spruce, cut for dimension lumber, disappeared.

Logging, like burning, increases deer populations. White-tailed deer are 4-5 times more abundant in a logged forest, than in mature deciduous forest. (Deer are found at a density of 2 per square kilometer in mature deciduous forest, 8-10 per square kilometer in a logged forest. Densities as high as 20-40 per square kilometer may be carried for a limited time.) Deer suppress the regeneration of hemlock, yew and white cedar, and browse on many more species. Their winter browse lines are a characteristic of the shorelines of northern lakes. A density of 4 animals per square kilometer may eliminate species sensitive to browsing. An old estimate of the animals in precontact New England forests lists, for every 10 square miles (25 square kilometers, a square a little over 3 miles on a side), 5 black bears, 2-3 mountain lions, 2-3 wolves, 10,000-20,000 gray squirrels, 400 deer, and 200 turkeys. The number of deer is very high (about 16 per square kilometer). The woodlands of southern New England were not mature deciduous forest in an ecological sense, but managed by fire for game. Perhaps this explains the high numbers. (The numbers may also be wrong.) The numbers for squirrels look high but only come to 2-3 per acre. By 1750 most of the predators were gone from the settled parts of New England and by 1900 hunting had made deer extinct in most of New England and New York State, removing both the threat of Lyme disease (colonial records report a disease something like it) and the influence of deer on forests.

The Middle West, settled in the boom after the Revolution, was plowed and cleared at about 1% a year, twice as fast as New England. River transportation, canals and railroads, rapidly succeeding and supplementing each other, made this a more commercial proposition from the start. Ohio had 600,000 people by the 1820s, the six other states of the Old Northwest—the modern Middle West—200,000. By 1860, Ohio had more than 2 million people, the rest of the Old Northwest 5.5 million. In the first period of settlement of the Middle West, only river bottoms and wooded slopes were considered farmable. The extensive flat upland prairies were used as cattle range. Successful grain farming on the wet upland prairies of Illinois and Indiana, landscapes which we now consider quintessentially Corn Belt, had to wait for the practice of tile drainage. Drainage was expensive. It finally took off in the 1880s. One can trace its progress in the records of local manufactories for drainage tile (clay tile is heavy and expensive to ship). Draining wetlands (beaver wetlands or prairies) sends water and nutrients downstream and would have major effects on middle western streams and rivers. Drainage continued through the 1920s in the wet prairies of Illinois, Indiana, Iowa and Ohio. Drained lands now constitute 27% of Illinois, 22% of Iowa, 29% of Indiana: large numbers that imply major changes in the terrestrial and riverine habitat. Drainage created the Corn Belt.

Until recently, the turning of wild land into land that produced something of marketable value was considered a public good in the United States. (For most people, it still is.) The reclamation of seasonally wet "waste" land was held to be a public benefit by the courts in suits that gave drainage ditches the right of eminent domain to cross other peoples' land. Under some state laws, if a majority of the landowners of 60% of an area to be drained favored a drainage district, the other landowners were forced to join it, and pay for the improvements. (This potentially let a minority of landowners rule. Such laws may have been a way to force capital poor farmers off their land.) The expense of drainage was probably a factor in the fact that over half the farmers on drained land were tenants in 1914. If one could afford it, drainage paid, covering its costs in about 5 years, through increased yields. Tile drainage, the lines 40-100 feet apart, 4 feet deep, with tiles 2-5 inches in diameter, shunted the prairie's water and nutrients into the headwaters of the Mississippi. The nutrient load was high partly because the plant cover had been removed, partly because drainage let the nutrients in these formerly wet soils oxidize and flow away with the water.

By 1964 drainage in southern Minnesota, by reducing the area of surface water and changing the character of the remaining ponds and marshes, had reduced the area's once immense waterfowl population (tens of millions of birds) to insignificance. (High soil fertility also means high wildlife populations.) Probably 45 million acres of duck habitat were drained in the upper Midwest, reducing populations by at least 90 million birds. Other animals suffered similar declines.

Grain farming in the late 1800s was a mixed agriculture, with grain, animals and hay or legume-hay mixtures raised on the same farm; the hay and some of the grain was fed to the animals. Hay was a major crop. Grass and clover were grown for seed, and hay was grown to feed the farm's cattle and work horses and exported to feed the cab horses of the eastern cities. As late as 1939, Jasper County, Illinois had 47% of its cultivated area in grasses for hay and seed crops; this would fall to 1% by 1974, as soybeans, a row crop like corn, rose from 9% to 69% of the county's agricultural lands. In Illinois and Iowa in the 1990s, 30% of agricultural land was in continuous corn, 60% in two years of corn followed by one of soybeans. Virtually all the crops were heavily fertilized, the fields treated with herbicides. Farm animals were gone. Only one in six farms spread manure. Such changes were part of the general boom in the farm economy after World War II, the shift from horses to tractors, and from mixed farming with its rotations and manures, to specialized cropping with its commercial fertilizers and pesticides. In general, traditional crop rotations kept 25% of cropland in sod (hays or legume hays; small grains like wheat also form sods and hold soil better then row crops). The land in sod reduced soil erosion, which was enormous on the lighter, hillier prairie lands. Hayfields, especially if used for seed, and so harvested late in the year (or if the first cutting was delayed until the third week of June), provided habitat for grassland birds, which hayfields no longer do under the intensive cutting practices of modern management. The fields must look inviting to the returning migrants, but the time between cuttings is not sufficient to raise a brood, so the fields become a trap for breeding birds and after a few years have none.

Such changes in the landscape affected streams. Forest clearance, and cultivation and drainage of the prairie, increased the rate and amount of runoff from rain and melting snow. Soil carried with the water running off agricultural lands widened and shallowed the headwaters of middle western rivers and raised their beds. Cultivation of rich alluvial floodplains in the Midwest and South turned the bottomland forests that were originally sinks for floodborn nutrients and silt (accumulating silt at a rate of 10-20 tons per acre per year) into—as fields of corn or soybeans—donors of silt to streams (15-60 tons of sediment per acre per year). The growing Mississippi floods in the early twentieth century were caused by agriculture and forest clearance upstream. Agricultural erosion peaked in the 1920s and 1930s along the tributaries of the upper Mississippi. The 1927 flood (after which the Corps of Engineers took over management of the Mississippi) destroyed 160,000 structures in the Mississippi Valley, leaving tens of thousands of people homeless, and all the bridges for 1000 miles upriver of Cairo, Illinois. Soil conserving practices such as strip cropping, contour plowing, and the use of winter cover crops, adopted after the droughts of the 1930s, helped reduce erosion, but erosion would rise again after the Second World War, as rotation into sod crops ceased. During the War prices for grain were high and farmers brought more land under cultivation. With the use of artificial fertilizers, rotations into sod were unnecessary and considered a "waste of good land." Agricultural Extension Agents encouraged farmers to grow the crops "best adapted" to the soils and region (not a bad idea but we are talking of economic adaptation here). The crop rotations of the mixed farm had reduced soil erosion and helped control water runoff. Crop rotations also maintained soil fertility and helped control weeds and pests. On farms that grew only grain, the land lay open two-thirds of the year, erosion and runoff increased, and fertility was maintained with manufactured fertilizers. The weed and pest populations that built up were controlled with herbicides and pesticides. These chemicals ran off into streams and sank into ground water. Under traditional crop rotations before World War II, levels of nitrates in Iowa's Des Moines River were already high: rates of soil and water runoff from bare fields were also high; and cultivating corn for weed control during the summer, up to 7 times, led to high rates of erosion from cornland. Now such rates would rise. Modern agricultural runoff contains suspended solids and nutrients on the level of sewage wastewater.

After World War II, as some farms specialized in row crops, others began to specialize in cattle and hog raising; that is, they became essentially feedlots. Raising thousands of animals in one place meant that the animals' manure, rather than being a useful soil amendment, became a disposal problem; there was not enough cropland to spread it on. There are approximately 9 billion domestic animals in the United States at any one time, compared with about 300 million humans; the animals produce several times the urine and feces of the humans. Most of it is handled badly. Manure becomes a pollutant if it is applied too heavily or at the wrong time of year. Large-scale operations that raise confined animals now handle it as a liquid, since the capital costs of dealing with it this way are less. This lets much of its nitrogen volatilize as ammonia (a waste of the nutrient), makes it difficult to store for long because of the expense of the facilities, and creates horrendous pollution problems when storage ponds fail, as they do regularly during heavy rains. (During Hurricane Floyd in the 1990s, hog manure lagoons failed along 23 of the 26 river systems in North Carolina and flooded Albemarle Sound with a layer of nutrient-rich muck 6 inches to several feet thick.) Composting the manure, while somewhat more expensive, would be a far better solution. Aerobic composting locks up much of the ammonia as usable nitrite and nitrate, and releases considerably less of the greenhouse gases methane, nitrous oxide and carbon dioxide. Composting reduces the bulk of the material by half, so there is less to store or spread. After composting, the material is easily stored under a shed or a plastic tarp and can be spread as needed. Composting makes the nutrients in manure less soluble, and careful composting, say with fly ash, a waste product of coal burning power stations, greatly reduces its load of pathogens, such as the intestinal bacteria *E. coli*, and *Cryptosporidium*, a diarrhea-causing organism whose oocytes, probably derived from manure running off dairy farms, are extremely common in shellfish in river estuaries of the eastern United States. A more capital intensive solution for farms with large numbers of animals (feedlots; or dairy farms with 1000 cows or more) is generating electricity from the methane in the manure. The electricity can be used on the farm or sold and the material left over is useful as a soil amendment (it is essentially compost). Neither composting nor electricity generation will happen without some sort of enticement; the returns on capital aren't great and methane generators have to be overseen by someone, creating an additional cost. However currently the systems for handling liquid manure are largely paid for by the federal government, which could require more environmentally appropriate solutions. (All this of course ignores the question of the health, well-being and tastiness of animals held in the miseries of close confinement.)

Before European contact, the rivers of the upper Mississippi Valley ran clear. Their valleys were forested. Along the Wabash, a tributary of the Ohio, sycamores grew 200 feet tall between the river bluffs (50-100 feet taller than eastern deciduous trees), and 6-8 feet in diameter. Illinois as a whole was about a third forested, Wisconsin largely forested, Ohio more or less completely so. Farm families went out with wagons to gather walnuts from groves in the fall. The bottoms of the rivers were covered with mussels. The wide, shallow Ohio was known for its mussel beds. As with the shellfish of Chesapeake Bay, their filtering capacity (the time it took for the volume of river water above them to pass through them) is thought to have been measured in days. (Perhaps as long as a week. Within the limitations of their habitat, and of losses caused by their predators and parasites, mussel populations would have expanded to the limits of their food supply, which was plankton and detritus from the water.) The mussels filtered out the primary producers in the river (the bacteria and algae) and kept the water clearer, letting sunlight penetrate further. They thus benefited the rooted underwater plants, which grew where the bottom was less firm (mussels need a firm bottom to attach) and which grow less vigorously, or die, if algae cut off the light. The underwater plants anchor the bottom against disturbance by bottom feeding fish (like the introduced European carp); their stems and leaves slow currents, damp waves, and help silt settle out, which helps mussels, which are sensitive to siltation. By reducing turbidity the plants let light penetrate further into the water column and thus maintain a more favorable environment for themselves and for clear water species of fish, such as bass, perch and pike. Their roots oxygenate bottom muds. Their lower stems and roots provide a large oxygenated surface in the anaerobic zone, where microbes living on them convert ammonia (toxic to many aquatic animals) to usable nitrate (which the plants convert to plant tissue) and oxidize toxic metals into harmless forms. Aquatic plants generate dissolved organic matter and detrital particles that support the invertebrates, bacteria, and plankton at the base of the food chain (including the mussels' food chain); they shelter snails, aquatic insects and juvenile fish; and are eaten directly by waterfowl and muskrats. Enormous flocks of migratory waterfowl (100-200 million birds) once visited the rivers of the

Mississippi Valley and were probably important in the natural nutrient regimes of the rivers and their floodplain lakes. Waterfowl and fish ate the algae and plants, zooplankton, small fish, invertebrates, fingernail clams and various aquatic insects in the rivers and floodplain lakes. They ate mast from the floodplain forests. They processed all this into fertilizer that was released into the water.

The mussels that covered the firmer bottoms were thus part of various mammalian, avian and fishy food chains. Mussels were extremely abundant in all Middle Western rivers including the larger ones (the Illinois, the Ohio, the Tennessee, the Wisconsin, the upper Mississippi).

The discovery of fresh water pearls in mussels led to their systematic elimination. Since mussels were abundant and the work was done by hand, this took several decades. (Times to economic extinction under capitalist systems vary: during the 1820s sealers took only 5 years to eliminate the southern fur seal and elephant seal as economic resources; and modern ocean fisheries take about 15 years to fall to 10% of their former abundance and lose their commercial importance.) During this time the shells were also used for making buttons, an industry that employed 20,000 people in 1920. From 1914-1920 the upper Mississippi produced an average of 35,000 tons of shells a year. Button making might have led to a stable, renewable use of the mussels, and perhaps use of the meat as well as of the pearls and shells, but the pearl fishery was uncontrollable. One year 10,000 tons of shells were harvested from a mile and a half of the Mississippi near Muscadine, Iowa, constituting perhaps 100 million mussels. As wild animals, the mussels were free for the taking. Rivers were public. The shells were steamed open on the riverbank, the pearls picked, the marketable shells sold, the meat and unusable shells abandoned. The stench left by professional mussel-fishers made them disliked.

Other factors were also in operation to doom the mussel fishery. The eastern tributaries of the Mississippi flowed down from the pineries of Wisconsin and Minnesota. From the mid-nineteenth century on, logs were floated down the rivers to sawmills, and the sawn lumber shipped west by wagon or railroad to the growing prairie towns. Michigan's timber resources were exported via Lake Michigan to Chicago, where depending on demand, much of it might rot in the yards. The pineries of Michigan, Wisconsin and Minnesota comprised about 300 billion board feet of timber, which was cut off in less than 50 years. About 100 billion board feet had already been cut in New York, Pennsylvania and New England. (Because of continued cutting, and because catastrophic fires on the logged-over ground converted much of the pineland of the upper Middle West to aspen and scrubland, the current stand of white pine in the northern states is about 10 billion board feet.) No dam was built on the Mississippi or its tributaries while the timber trade was underway. But the end of the trade was a disaster for the timber towns. Most American towns have arisen for an economic reason: to service newly opened agricultural lands; near a mill dam site; near a water source suitable for brewing beer, or tanning hides; near mines; near good harbors. Most of these functions would sooner or later end: the timber would be used up; the hemlock or oak tanbark gone; mines or fisheries depleted; coal would replace waterpower; farms consolidate; farm products fall in value. Cheap transportation of materials and goods would make cheaper labor available, that is, more economic, elsewhere, and so (for instance) textile and shoe manufacturers moved from water powered New England to the cheaper labor of the American South, then to southern and eastern Europe and Brazil, then to Mexico, the Caribbean, Vietnam and China; while cheap, fast transportation made poorer land, in, say, New England (but land close to the market), uneconomic to farm, compared to better land in the Middle West, or to government financed irrigated land in California. Thus American cities and towns have always had to re-invent themselves. The main motive for this was to protect land values; when the businesses left, land values collapsed, and buildings were worth little or nothing. Peoples' savings (and banks' assets, most of them in real estate) were lost.

So when the cut was over in the upper Middle West, a clamor arose from the Mississippi towns for dams, and the water powered electricity and river transportation they would bring. River transport was becoming an anachronism by then; railroads into Chicago had already largely eliminated St. Louis, an old river port, as a transportation center. Railroads ran year round and were not dependent on water levels, wind or ice. But bulk transportation of prairie grain by barge down to the port of New Orleans is cheaper than rail if the federal government finances the roadway (the dam and lock system on the river). It was claimed sales of hydroelectricity from the dams and barge tolls would pay for the construction and maintenance of the lock and dam system. (They didn't.) In 1894 the hydroelectric station at Niagara Falls, New York, helped bring industry to Buffalo, N.Y. Through the middle of the twentieth century Buffalo milled prairie grain and made steel, thanks to its electricity and also its position on the Great Lakes and on a major railroad corridor. Hydroelectric power was also considered important because in the early twentieth century the United States was thought to be running out of coal.

Keokuk Dam at Alton, Illinois, was the first dam on the Mississippi's main stem. The site was suggested by Major Stephen Long during a survey of the river in 1817. (Long also called the Mississippi upstream from St. Croix marvelously clear, an echo of the traveling Mr. Jefferson's assessment of the Illinois.) Keokuk Dam was built for transportation and electric power. Electric power is generally the only use that will pay for modern dams; fees for irrigation and transportation will not pay for ongoing maintenance, much less capital costs, a measure of how cheap food and transport are. Public water supply and flood control provide what are considered essential services, and so a large part of their cost is usually borne by the public treasury. In general, flood control can be provided more cheaply and effectively by altering land use upstream than by building dams; or by keeping buildings out of reach of the flood; but this means altering land use on private lands, which means taking politically unpopular positions. Dams also create jobs, are a visible expression of the power of government, and are almost always the more politically palatable solution. Dams of course prevent fish migrations. Many North American fish and mussel species are co-adapted. Fish carry mussel glochidia (a type of larvae) in their gills. That is, the mussels use the fish, some of which eat adult mussels (and some of those release their glochidia into the fish's mouth as their shells are crushed) for distribution of their young. So dams cut off the movement of mussels and their recolonization of new territories. If overfishing hadn't doomed the mussels, the dams would have. Fish migrated around some dams on the Missouri in the 1993 floods and the levee breaking high waters of 1993 resulted in the best fish reproduction on the Missouri and lower Mississippi in several decades; but this is not a desirable situation on a controlled river.

The lands alongside large flatland rivers like the Illinois, the Wisconsin, and the Mississippi are overflow lands, with seasonal swamps, marshes and floodplain lakes, and require levees and drainage, or both, to function as cropland or buildable land. Such drainage schemes are expensive. Early private schemes usually failed and were completed, if at all, at public expense. Because of the expense of making them usable, riverside lands along the Mississippi (overflow land and swampland) had remained a forgotten part of the public domain. But it was government policy to convert as much public land to private land as possible and by the middle of the nineteenth century such lands began to be regarded as an opportunity. The Swampland Act turned them over to the states, which sold the land to speculators. The money from the sales was supposed to be used to finance drainage. The speculators resold the land to farmers, with or without providing drainage. The general notion was that once turned into farmland, the lands would enrich farmers and other landholders, increase tax revenues to localities, and give another turn to the upward spiral of wealth; but the money raised by land sales was usually not enough to provide drainage. The land available was not small; the Mississippi alluvial valley, all of which was once subject to flooding, varies from 20-80 miles wide from its junction with the Ohio River to the Gulf of Mexico, a straightline distance of 600 miles (about 1200 miles by river, before navigation improvements). The land rises about 6 inches per mile from sealevel at the gulf to 300 feet in southeastern Missouri, and so is essentially flat.

After the Second World War, the federal government took over management of the Mississippi River and slowly the wide flatland rivers of the Middle West were dammed and their floodplains were leveed off. Thanks to political pressure, the levees were built much closer to the main stem of the Mississippi than the Army Corps of Engineers wanted. The Corps wanted to set them back a mile on each side, which would have made the river much easier to control. Channeling eventually drained 17 million acres in the alluvial valley (that is, 70% of it) and a total of 120 million acres along the rivers of the Mississippi drainage.

The annually flooded lands are an important part of the riverine ecosystem. Most of the productivity of large river-floodplain ecosystems is in the floodplain. A floodplain greatly enlarges the area available to fish for feeding and spawning. A floodplain seasonally expands a river's littoral zone, the shallow water along its margins. The littoral zone is usually much more productive than the deeper waters of the river channel. Its water is well oxygenated and warmer. Sunlight reaches to the bottom. In a river-floodplain ecosystem the advancing margin of the floodwater creates a moving littoral zone over the width of the floodplain, the actual width depending on the topography of the floodplain and the height of the flood. (The Illinois had an exceptionally flat floodplain, with an exceptionally long flood, and was a very productive fishery.) The flood pulse controls the productivity of the floodplain; great floods re-arrange the plants and landscape and may result in major spawning and recruitment events for fish. The nutrients that are released from the flooded soils stimulate the growth of green plants, algae, zooplankton, aquatic insects, and various other invertebrates and the animals that feed on them. The firm terrestrial soils and the vegetation of the floodplain make better spawning sites for many fish than the soft sediments of the permanent lakes and backwaters. As the flood recedes, the fish, including the young fish of that year, return to the permanent backwater lakes and the river. In a river/floodplain ecosystem, the main channel of the river serves as a migratory pathway for fish and as a refuge and feeding area during periods of low flow, while floodplains with their lakes, backwaters and marshes provide new sources of food and spawning and nursery areas for fish and greatly increase the productivity of the main channel. Especially in turbid rivers, where silt reduces the penetration of light, the channel may be the least productive part of the ecosystem. Floods are thus essential to the system, providing large episodes of fish recruitment and replenishing floodplains with nutrients and water; and rearranging floodplain vegetation and the river channel. In the 1920s, before many riverside wetlands were drained, and the physical, chemical and biological integrity of the system began to break down, fish rescue operations were mounted to return stranded juvenile fish to the river from the drying pools of the floodplain (people competing with the raccoons, minks and herons). For successful fish reproduction, the flood must be high enough, last long enough, and occur when temperatures are favorable for spawning.

The permanent floodplain lakes and marshes along the upper Mississippi and its tributaries like the Illinois were spawning grounds for northern pike, large-mouth bass, and yellow perch, and were used by big-mouth buffalo and bluegills; they were feeding and resting places for migratory waterfowl; nesting sites for dabbling ducks, flycatchers, rails and herons; home for furbearers like muskrats and mink; habitat for turtles. The fisheries in such natural rivers were extremely productive. In 1900 the Illinois produced approximately 10% of the freshwater commercial fish catch in the United States, or 24 million pounds. It also produced about 500,000 pounds of snapping and softshelled turtles. This comes to about 170 pounds of fish per acre of permanent water (and an unknown fraction of the standing crop of fish). Such numbers are not unusual: the Tippah River in Mississippi had a standing crop of 241 pounds of fish per acre before channelization; and 5 pounds per acre after channelization to turn it into a waterway destroyed the riverine habitat. (Similarly, catches in the Illinois fell to 4 pounds per acre by the 1970s, as the effects of pollution and riverworks took firmer hold.) Whether a catch of 170 pounds per acre on the Illinois was sustainable isn't known but it isn't an impossible number for third and fourth trophic level fish (the wolves and tigers of the water) in a productive riverine environment.

The banks of these Middle Western rivers were wooded. In the 1820s streamside forests began to be cut for fuel for steamboats. Steamboats required a lot of wood and cutting soon deforested the banks. After deforestation, or a destructive flood (one that killed trees), cottonwoods and willow would resprout, sycamore and silver maple seed into the new mud; other trees would come later in the succession. The tallest trees in the floodplain forest topped out at about about 100 feet and were used as rookeries by herons and as nesting and roosting sites by hawks and eagles. In spring and fall migratory ducks would eat the mast of the nut trees. Trees on the banks would fall into the river, float away, jam against the bottom, collect other trees. Stable jams might become islands. Large logjams were common on Middle Western rivers. They extended the floodplain, raising the height of the flood.

By the 1920s pollution from industry, sewage plants and oil refineries; siltation from agricultural erosion, from bank erosion by barge traffic, from sewage solids; damage from dredging; and over-fishing would eliminate mussels (or fish) as an economic resource, or a resource of much biological importance to the rivers. During the booming development of the 1920s, industrial rivers often caught fire. (By the time the Cuyahoga River in Cleveland caught fire in 1969, helping to energize the modern environmental movement, that sort of development was coming to an end.)

The Illinois, a major tributary of the Mississippi, is an example of a middle western industrial river. The Illinois drains west central Illinois, rising at a height of land south and west of Chicago and entering the Mississippi above St. Louis. In 1848 a canal was dug to connect the Mississippi to Lake Michigan, *via* the lower Illinois. The canal was 36-48 feet wide and 6 feet deep and doubled the flow in the Illinois, permanently flooding some backwater swamps and killing some riverside vegetation. For a time, the additional water, together with the fertilizing effects of the nutrients from Chicago sewage and stockyard waste, probably increased fish production. In 1900 the Chicago Ship and Sanitary Canal, a larger waterway, opened and began the wholesale transfer of sewage and stockyard waste from the City of Chicago to the Illinois River. Algal blooms fertilized by this material grew, died and sank to the bottom of the river, where the algae decayed, using up the oxygen in the water. Navigation dams stopped the river from flowing in low water, turning it into a series of polluted pools, making the condition worse. (As long as the water was moving, the dams, by oxygenating the water that flowed over them, may have helped the situation, but when the flow slowed and stopped, the water in the pools stagnated.) By 1910 the river was anoxic for much of its length. The low levels of summer oxygen eliminated a normal bottom fauna (including the mussels) and killed most of the rooted vascular plants in the river. Levels of decayed material (algae, sewage sludge) accumulated on the bottom of the river. The sediments became too soft for aquatic vegetation to root or for mussels to anchor. Both were replaced by tubifed worms, which live in mud.

Agricultural runoff peaked in the 1920s and 1930s in northwestern Illinois and added heavy silt loads to the river. During the droughts of the 1930s, when erosion was enormous, stream beds in the Middle West were raised 10-30 feet by soil erosion from farmland, making streams wider and shallower and more prone to flooding. Dredging for navigation removed the accumulating material, both silt and sludge, but placed it on the floodplain, where floods returned it to the river, and where constant barge traffic kept it in suspension. Resuspending the sediments greatly increased the oxygen demand in the water, supporting the anoxia. Dredging also negatively affected the winged water insects (the naiads), whose larvae inhabit bottom muds. Naiads, such as mayflies, are insects of clean lakes and streams, whose larvae and adults are food for waterfowl and fish. They are sometimes extremely abundant. The return of mayflies to Lake Erie late in the twentieth century was a sign of the lake's recovery from massive nutrient pollution.

In 1948 the last factory making buttons from mussel shells on the Illinois closed. Diving ducks (ring-necked ducks, canvasbacks, ruddy ducks, lesser scaup: fish and invertebrate eaters) had begun to disappear from the river. By 1955 the fall population of scaup was zero, probably because organochlorine pollution had eliminated the fingernail clams on which they fed. By 1965, 1200 metric tons of chlorinated hydrocarbons were being spread on farmland in the Illinois basin every year. Much of this ended up in the river and the floodplain lakes. While many of the persistent organochlorines were phased out in the 1970s, abundant applications of insecticides, herbicides and fungicides continued. In the late 1980s annual applications to cropland in the Mississippi basin reached 100,000 tons of pesticides and 6 million tons of nitrogen fertilizer. These figures don't include fertilizers and pesticides applied to local lawns, which may increase that number by 30%-100%. Many modern herbicides are biodegradable, but break down more slowly in water than in soil, and so effect vegetation in the rivers and in the marshlands of the Gulf of Mexico, into which the Mississippi flows. They affect riverside trees during floods. Rooted underwater plants more or less completely disappeared from the lower 200 miles of the Illinois and its well-connected backwaters in the 1950s and with them the dabbling ducks (mallards, pintails, widgeon, teal: the plant and insect eaters).

The water quality problems caused by sewage sludge, stockyard waste and toxics slowly improved on the Illinois during the last quarter of the twentieth century, thanks partly to the Clean Water Act of 1972. Stockyards moved from Chicago to the High Plains. The federal government began subsidizing the construction of sewage treatment plants. Some toxics were banned. However the discharge of many toxic chemicals into the river (so-called permissible discharges) continues. Industrial use of Illinois water comes to approximately 1.5 billion gallons a day. Some writers claim the effect of this use is more affected by the quality of the effluent (what comes out with the water) than by the volume of use, but water used industrially is likely to be more or less sterilized of small fish, fish eggs and larvae, and the larvae of aquatic invertebrates, returning to the river as a nutritious, but dead, soup. Between 1974 and 1989 there were 350 spills of hazardous materials in the river, that is, about 1 every 2 weeks. The use of the river for shipping continues. (The valley of the lower Illinois ships more grain per mile than any other midwestern river.) Dissolved oxygen is still low, tubifed worms have replaced fingernail clams and naiads in the more polluted stretches of the river, and heavy metals are found in the mussels that survive; some mussels are still collected and their shells shipped to Japan, where they are ground into the grains that seed cultured pearls.

Agricultural erosion strongly affects the rivers of the Mississippi drainage. The Illinois basin is almost entirely agricultural. (The State of Illinois is 96% farmland.) From 1945 to 1986 land in row crops (corn and soybeans) increased by 67% in the Illinois basin. Hedgerows, once an important component of the farm landscape in Illinois, were removed to make fields larger and more easily cultivable. Most row crops require spring work to be done at the same time, so speed is essential. With 10 miles of hedgerow per square mile of farmland, hedgerows were important habitat for birds, insects and small mammals. They reduced wind erosion, slowed water running off fields, and assisted in aquifer recharge. Contour plowing, developed in 1700s Virginia by a nephew of Thomas Jefferson to slow erosion on hillsides, and rediscovered in the 1930s, was used less and less after the Second World War. (It also slows things up.) Fall plowing (good farmers plow in the fall so as to have the fields ready to plant in the spring) and growing the summer crops of soybeans and corn left the ground bare for two-thirds of the year. Improved farm machinery let farmers square off fields by channeling the streams that drained them. Channelization, by shortening a stream's length and increasing its slope, increases its velocity and erosive power. The bare banks of channelized streams, no longer protected by grasses and by bird-planted trees and shrubs, erode. Bank and bed erosion in channelized watercourses produces 50% of the annual sediment yield of Illinois streams. The sediment is carried down into the marshes and forested deltas where feeder streams enter the Illinois, raising their beds and eventually causing floods upstream. To control the flooding, the marshes are channelized. Channelization destroys their ability to trap sediments and nutrients, sending both into the Illinois.

The Illinois floodplain originally occupied 400,000 acres of the 18.5 million acres in the lower Illinois basin, or about 2% of its area. Half of the original floodplain was leveed off for agriculture, including areas below the level of the river, which have to be kept drained by pumps operating much of the year. This meant the increased yield of sediment was concentrated in the remaining oxbows and in the floodplain lakes and backwaters. Some floodplain lakes lost half their depth. From 1958-1961 the remaining clear and vegetated backwaters and lakes became turbid and barren. Their use by gamefish and ducks declined drastically. Lakes tend not to cleanse themselves but to collect and recycle nutrients and pollutants. Ecosystems can flip to different productivity states, some of which (usually those lower in the succession) have markedly less biotic regulation of energy flow and biogeochemical cycles. Turbid, shallow, eutrophic lakes, with heavy algal blooms (an early successional state), are a biologically stable alternative to shallow, clear lakes with rooted aquatic vegetation.

Turbidity is a major problem in most midwestern rivers. Increasing turbidity during the 1950s came partly from the increased amounts of silt in runoff, partly from algal blooms caused by nutrients in the water, partly from boat traffic and natural wave action, partly from rooting in the bottom by the introduced European carp. It caused the loss of rooted underwater plants. Rooted aquatic vegetation helps control turbidity, but excessive turbidity reduces the light available to vascular plants and thus reduces photosynthesis. This weakens and kills the plants in the deepest (and dimmest) parts of the rivers and their lakes and backwaters. No longer damped by plants, the waves from windstorms and from barge traffic become stronger, further increasing the turbidity, and killing more plants. Rooted vegetation goes into a downward spiral. Once killed, the vegetation is difficult to re-establish. Fine-grained sediments take 7-12 days to settle out after a windstorm, but in the Illinois basin strong to moderate winds occur every 7 days. Waves from barge traffic are undamped and rooting in the bottom muds by carp (part of its feeding behavior) also raises sediments. So the excess nitrogen and phosphorus from farm fields that might have been turned into useful biomass by functioning riverside swamps and submergent vegetation, by trees in the floodplain, and finally by the vegetation in the wetlands of the Mississippi Delta, turns into algae in the lakes and rivers, suppressing the growth of submerged vascular plants, and creates the dead zone in the Gulf of Mexico. (Even in this degraded system, bacteria in the remnants of the floodplain and the still pools of the reservoirs remove about 35% of the nitrogen that enters the Mississippi.)

The lack of aquatic plants and their associated animals means little habitat remains for waterfowl or furbearers. Up through the 1950s furbearers, especially muskrats, but also skunks, raccoons and mink, provided extra income for families on many small Middle Western farms. The fish population shifts from sight predators and nest builders (the traditional gamefish) to fish that locate their prey by scent and scatter their eggs over the bottom (such as the bottom feeding carp). Northern pike, predatory fish of clear lakes, inhabit the interface between rooted aquatic vegetation and open water. A pike population needs 25% plant cover to maintain sufficient biomass to control bottom feeding fish, like carp, in a lake, and so reduce their effect on turbidity. (Carp can also be controlled by drawing down a lake at spawning time, exposing their eggs and fry in shallow pools. This is a more drastic solution and a temporary one if the lake is connected to other bodies of water with carp.)

The Illinois is a satisfactory industrial river. It is a highway for barge traffic, a source of industrial water, and a dilution basin for industrial discharges. Its fish are not as unhealthy as the bullheads in the Anacostia River of Washington, D.C., many of whom have skin lesions and half of whom suffer from liver cancer. Its floodplain lakes, despite their blooms of blue-green algae, are used for water-skiing. The return of a healthy riverine environment would mean eliminating barge traffic and might in some ways be a nuisance: the mayflies that populate healthy midwestern rivers splatter on windshields during their hatches, collect on sidewalks and make roads slippery. Millions of migrating waterfowl shut down airports, if only for periods of a few hours. (Ninety million ducks headed south in the fall of 1997, when the continental population had recovered from the droughts of the 1980s and shut down some airports for a few hours at a time.) Shipping out grain by truck or rail to some point on the Mississippi or the Great Lakes that can better take barge or deep water ship transport means more truck and rail traffic. (The Mississippi at Vicksburg, a natural deepwater port, is 2000 feet wide and 60 feet deep.) As for the other environmental problems of the Illinois, the withdrawal of industrial water is relatively easy to manage satisfactorily: at a small increase in cost, most of it could be recycled or eliminated. The release of toxics could be much reduced or eliminated, by regulation or by modifying industrial chemistries.

Sometime in the 1990s, the Illinois was invaded by two species of Asian carp that had been introduced to southern catfish ponds and sewage treatment flows to control algae. Imported in the 1970s the carp soon escaped in floods to the Mississippi. Now one of them, silver carp, dominates fish biomass in the Illinois. Their numbers approach 8000 fish per mile. Silver carp grow to 60 pounds or more (their filter feeding relatives, the big head carp, get larger) and consume 20% of their weight in plankton a day. (Like the alien zebra mussels, they functionally replace native filter feeders like mussels.) The introduced carp do so well compared to native fish because the Illinois is a changed habitat: too many nutrients, too much turbidity, too few mussels and water plants, too much plankton. Silver carp are a common food fish in China. One way to deal with the Illinois is to accept the present situation, clean up the toxics in the river so that its fish are edible, and fish for the carp. (Since they are filter feeders, they won't take a bait and must be shot with arrows or netted.) Silver carp jump out of the water to escape predators, so a motorboat on the Illinois is surrounded by leaping 20-50 pound fish. Waterskiing in waters with silver carp is not safe. Similarly, we could control toxic runoff into the rivers of the Pacific Northwest, and accept the falling numbers of salmon and sturgeon, and the abundant eastern shad and other introduced fish, in the altered rivers there; and the rainbow trout below the dams, the striped and largemouth bass in the reservoirs, and the falling numbers of native fish in the Colorado. That is, we could embrace the new world we have created.

———

Modern rivers are maintained according to standards developed during the first half of the nineteenth century in Germany, the country that also developed the modern chemical industry, with the production of dyes for cloth. The nineteenth century channelization of the Rhine was the first great modern riverwork. The ideal industrial river is a controlled, single channelled stream that stays within its banks, and is useful as a waterway, a source of water and a drain. Gangs of men dug by hand new, narrow, straight channels for the Rhine (a notoriously meandering river), in order to speed its flow and make it dig itself a deeper bed. The natural drainage this created allowed former riverside marshlands to be cultivated and reduced floods upstream. It lowered water tables in nearby fields (not always an advantage and for this reason some communities resisted riverworks), made navigation easier, and eliminated malaria, salmon and the gold sifted from its gravelly banks (about 10 pounds per year in the early 1800s). The improvements, by sending more water more quickly down the river, increased flooding downstream in Holland. Channelization changed the Rhine from a slow, meandering, silty river with riverside swamps and forests into a fast, erosive stream, starved of silt, bordered by roads and railways, that constantly erodes its bed. Earth and stone have to be continually dumped in some reaches to prevent the river from eroding its bridge abutments. As for salmon, with some further improvements the Rhine could now support a run of 6-12,000 fish, about 2% of the run in the natural river. (Tributary reaches once used to absorb phenols from the coal processing industry are available for spawning.) Similar changes to the Mississippi for navigation above its junction with the Ohio have reduced the river's surface area by 33%, its island area by 25%, and its riverbed by nearly 25%. The river was narrowed to increase its depth and scour, to make it more fit for barge transportation.

Natural rivers flood, migrate across their flood plains, open two or more channels, dig cross over flows and cutoffs, and are full of snags (about 1 every 10 feet on large rivers). Riverwater spirals downwards through the watershed, using and acquiring nutrients (sometimes recycled within 100 yards), abruptly forming new habitats, as when pools give way to riffles and shallows. Rivers change as their slope changes, as they flow over different substrates (gaining or losing chemicals), as tributaries enter them, from exchanges of gases with the atmosphere, of nutrients with their connected groundwaters. The wooded, more easterly rivers of the Mississippi drainage, such as the upper Mississippi, the Illinois, the Wisconsin, and the Ohio, were clear in precontact times; and under development became more silty (and warmer in some reaches, deeper and colder in others). Many plains rivers, such as the Missouri and the Arkansas, were naturally warm and muddy, with populations of fish that were adapted morphologically and behaviorally to such conditions. Channeling changed the Missouri (at whose entrance the Mississippi changed from a clear to a silty river) from a warm water stream with high turbidity, wide seasonal variations in flow, and a braided channel that constantly changed course, to a narrow, cold, fast moving stream, with (below its dams) excellent trout fisheries. The gallery cottonwood forests that had lined the river for 1000 miles died from lack of water. Ninety-three percent of the emergent wetlands, backwaters and sloughs along the Missouri were converted to agriculture or dredged for channels.

Many plains rivers are used for irrigation. Largely because most of its water was taken to grow corn, the North Fork of the Platte River, a major tributary of the Missouri, changed from a meandering, braided channel 2500-4000 feet wide to one narrow, well-defined channel 200 feet wide. (The mean annual flood of the Platte fell from 13,000 cubic feet per second to 3,000 cubic feet per second and the mean annual discharge from 2300 cfs to 560 cfs after its development for irrigation: that is, about 75% of the flow was removed.) In dammed rivers, most of the silt and organic material end up being stored in the river, instead of feeding riverine wetlands downstream. Two-thirds of the Missouri's silt ends up behind dams (considerably shortening their lifetimes: there are 60 dams on the Missouri and its major tributaries). The Ohio, once a wide, clear, shallow stream that flowed through a watershed that was almost entirely forested, was changed by dams (for water supply, electricity and river transport, mostly of coal for power stations) into a deep, cold river, its mussel population a remnant, with new fisheries for walleyed pike below its dams. The silt the Ohio contributed to the Mississippi increased 10 times thanks to agriculture and industrial and urban development in its watershed, but the Ohio's increased silt loads did not make up for that lost from the plains rivers.

East of the Appalachians, in the Northeast, where rivers flow into the Atlantic Ocean, single channel rivers without extensive wetlands connected to the river, seemed normal to twentieth century observers, though some remarked that rivers in undeveloped areas formed multiple channels and had connected wetlands. It is now thought that eastern rivers had been transformed by agriculture and dams soon after settlement. By 1800 most eastern streams had mill dams at every suitable site (18,000 in Pennsylvania alone). These were low wooden dams, 6 to 8 feet high. They were set on long pine or hemlock timbers laid into the riverbed in the direction of the current. Braces mortised into the timbers held up two parallel wooden walls that crossed the stream. The space between the walls was filled with clay, with clay and stones in the center. On the upstream side planking angled down from the top of the dam to the streambed to let ice and trees slide over the dam. The weight of the earth fill held the dam in place. These dams flooded any adjacent wetlands or side channels. As the countryside about them developed, their ponds slowly filled with silt eroded off the surrounding watershed. The rivers became a series of long pools. When the dams, abandoned from the late 1800s thanks to other sources of power, eventually failed, the river cut itself a single channel through the accumulated sediment. It was now too far below its floodplain to have any connected wetlands. When in great floods the rivers try to re-arrange themselves over the floodplains (and construct new channels or excavate wetlands) they are put back in their former places since too much human development now occupies what has become dry land.

About 250 miles from the Gulf, the Mississippi starts distributing its water, that is, rivers start to flow out of it rather than into it. Much of the land along the river, and for many miles to each side, was once bottomland hardwood forest that accumulated sediment at 10-20 tons, a fraction of a millimeter, per acre per year. Some was marshland, which was inhabited by alligators. The holes the alligators dug were refugia in winter and also during summer droughts for amphibians and fish. Their nesting mounds (a mix of mud and vegetation 5-7 feet wide, and 3 feet high) were egg-laying sites for turtles. Dry spots in the swamp, they were colonized by plants and became nesting sites for birds. Eventually, mast bearing trees like oaks rooted in them. Thus alligators, like the accumulating silt, varied the habitat. Because of the build-up of silt, the river constantly changed its course through the delta, creating new river cutoffs or abandoning old ones. (In contrast, the Mississippi and its major tributaries upstream of Cairo, Illinois, which carry much less silt, have maintained more or less stable channels over the last 2500 years.)

Downstream the river met the sea in a maze of channels and islands, among which it dropped its final load of sediment. The main distributaries have shifted several times but the wetlands at the river's mouth have continued to grow at about 1.5 square miles a year for the last 5000 years. The delta shoreline is a balance among sedimentation, sealevel rise, and subsidence of the land underneath (both natural from the weight of the accumulating sediment and manmade from the pumping out of water and petroleum). The wetlands become more fresh as one moves inland, until trees (including cypresses, which live 400-1000 years) grow in the marshes, anchoring the islands and providing better protection for developed areas from storm surges and hurricane winds. The delta wetlands provide nursery areas for much of the marine life in the Gulf of Mexico (perhaps 80% of a $5-$7 billion fishery, which represents 200,000 jobs and 25% of the U.S. fish catch). The plants of the freshwater marshes are eaten by migratory waterfowl.

The Mississippi now carries 35-40% of the silt of 150 years ago (half of 50 years ago), despite more erosion in the watershed. The balance is stored behind dams, most built in the 1950s. The silt it now carries however, if allowed to spread over the delta wetlands, would cover 60 square miles 0.5 inch deep each year. Instead, river control structures send the silt, and the nutrients carried with it, out into the gulf, bypassing the marshes, which are now losing land at something like 50 square miles per year. (The rate increased in the 1990s; about one-sixth of the Louisiana marshes has been lost to the sea over the last 150 years and the delta is now subsiding at about 0.5 inch a year. Relative sea level has risen 3 feet in Louisiana in the last hundred years.) The fresh river water floats on the sea's surface, tending to stratify it and prevent overturning, especially in summer. The silt, biocides and fertilizers—partly through direct action (pesticides killing zooplankton or shrimp larvae), but mostly through fertilizing the algal plankton, which grow, die, sink, and decay, and so deoxygenate the bottom waters—create a so-called dead zone with lethally depleted levels of dissolved oxygen covering several thousand square miles of the gulf bottom. Currently the dead zone covers an area the size of New Jersey. Fish trapped here suffocate, as do the shellfish, soft corals, sponges and worms that live on the bottom. An artificial shipping channel (the Gulf Outlet Canal, not much used; most ships use the natural river) has been widened extensively by storms and tidal action and lets salt water intrude upriver into the marshlands toward New Orleans, making them less desirable to waterfowl and killing the cypresses. Ten thousand miles of abandoned oil exploration canals, also constantly enlarged by tidal action, help erode the wetlands. Letting the Mississippi's silt flow into the wetlands through siphons or gates would make it do useful work in rebuilding them. The fertilizer carried with it would grow useful biomass. The Outlet Canal and the oil exploration canals can be filled (if possible), or dammed with earth barriers (in the second case, the work paid for by oil companies). The value of the Louisiana wetlands in the Mississippi Delta have been calculated to be worth $846 per acre per year in fishing benefits, $401 in furs, $181 in recreation, and $7,500 in storm protection benefits (the last perhaps too high, the protective value of marshes in storm surges may be over-rated). These are coastal wetlands so their value in hurricane protection and fisheries is more than that of upstream floodplain lands, though other calculations put the value of any natural wetland at $6,000-$8000 per acre.

Most modern rivers are controlled by dams. Dams create discontinuities in riverine ecosystems. They prevent fish from using all of a river. Below a dam, the warm silty water of a plains stream emerges clear and cold, good habitat for the introduced trout, but not for native fish, which are adapted to the warm silty summer flow. The water is cold because it is released from the bottom of the reservoir. By storing much of the silt and organic material moving through river systems, dams impede the movement of material and energy through a drainage basin. Thus dams inevitably cause the erosion of riverbanks, riverbeds, deltas and the sea beaches at river mouths, and change the habitat of estuaries where marine fish spawn. They alter the thermal regime to which a river's fish and invertebrates are adapted. Temperature is often used as a cue for egg hatching and larval development; the wrong temperature makes eggs hatch or larvae develop at the wrong time. Cold flows from dams may be too cold for the larvae of native fish. So by changing a river's thermal regime dams change its faunal structure.

Dams alter the interface of land and water. Bennett Dam in British Columbia stopped the flooding of the Peace/Athabasca Delta in Lake Athabasca and thus eliminated the muskrat habitat (and with it the muskrats) on which the native peoples depended. Dams for the La Grande hydroelectric complex on James Bay more or less eliminated the char populations of the rivers and also reduced the numbers of lake trout. Whitefish, pike and chub increased. None of the fish will be edible for the next 20-100 years because of methyl mercury released by bacterial decomposition of rotting vegetation in the reservoirs. Because of the fluctuating water levels, the margins of the reservoirs, unlike those of the rivers that preceded them, are rarely used by wildlife like ptarmigan, beaver or moose. About 10,000 pairs of waterfowl lost nesting sites in the flooded riverine wetlands.

The reservoirs behind dams are often centers for transmission of fish diseases; and are inhabited by unstable assemblages of fish. Typically the fish population booms in a new reservoir with the release of nutrients from newly flooded soils, and then falls. But some fish introductions have been successful, such as the trout fisheries below the high dams on the Colorado; or (for a while) the striped bass fishery in Lake Mead. (That fishery was based on an introduced shad as a forage fish and on algae fertilized by the river water. Building Glen Canyon Dam on the Colorado upstream of Lake Mead shut off much of the supply of phosphorus the algae needed and the bass fishery precipitously declined.) Introduced fisheries are often successful at the expense of the native species in the river. Trout, for instance, eat young razorback suckers and pikeminnows in the Colorado River and make recovery of their populations, already hindered by the colder water, the loss of upstream habitat, and the loss of floodplain backwaters, more difficult.

Dams slow the flow of water to the sea. Five times the water volume in streams is now held behind dams worldwide, lowering sealevel by a little over an inch and slightly slowing the rotation of the globe. It takes several times as long for water to move through a dammed river's basin (up to a year, in some cases, rather than days to weeks). Reservoirs emit methane, a greenhouse gas, from the decomposition of organic material washed into them, or left in them when they are filled (for instance, uncut forests). Methane emissions from reservoirs are largely a tropical problem. Half the reservoirs in Brazil emit more greenhouse gases than would a coal fired power station of the same output. Flooding low gradients like those of the Mississippi Valley to generate electricity uses 200 times more land per kilowatt hour than solar collectors, but hydroelectric power stations with reservoirs produce 200 times the energy that goes into them (the energy used to build, maintain, and fuel the generating equipment), compared to 40 times for wind turbines, 20 times for heavy oil, 10 times for coal, probably less than 10 times for photo voltaic solar collectors. (So in this case, photo voltaic collectors are more efficient users of land but not of energy.) The reservoirs behind dams are kept full with spring runoff so as to generate electricity, supply irrigation water, and for water supply (all demands that rise in summer), so they are not much use in summer floods. During the 1993 floods on the Mississippi 12 times the storage capacity of the reservoirs moved through them. The reservoirs reduced flood peaks by a few percent (a few feet) of the total rise. There has been a statistically significant increase in warm season floods in the Mississippi Valley from 1940 to 1990, thanks largely to an increase in the frequency and magnitude of heavy rainstorms. More frequent and heavier precipitation is predicted for a warming world. Insurance companies may favor more realistic methods of flood control, such as letting rivers occupy more of their former floodplains. Controlling floods this way doesn't take up a large amount of land (perhaps 3% of urban and agricultural land in the lower 48 states). Large dams are engineered to last 250 years but have an economic life of 50 years. Then, as their concrete starts to deteriorate from contact with the water and their reservoirs fill with silt (1% per year on average, but a half life of 20 years on China's Yellow River), they reach the age of repair. In the United States dams are licensed for 50 years. Many licenses are coming up for renewal, which provides us with an opportunity to live with our rivers differently.

Some biological systems are difficult or impossible to manage, especially if they exist in alternative stable states. We see managed biological systems as stable, but natural systems experience (and may require) fluctuations, some seasonal or episodic, some chaotic. Stabilizing water levels in wetlands reduces the diversity of plant species and results in a long-term decline in productivity (as with reservoir fish); stabilizing water levels in prairie potholes reduces the wet meadow and marsh zone, important in spring for some species of ducks, reduces the overall diversity of plants and lets cattails take over the wetland. Stabilizing water levels in the Everglades also lets cattails take over. Important seasonal habitats may not appear attractive: dabbling ducks use flooded sections of plowed fields (vernal pools: full of newly hatched invertebrates) in early spring and move to more permanent water in summer. In many ecosystems drought, flood and fire are necessary agents of biological renewal. Such catastrophic events interfere with human economic lives and are not consistent with our ideas of "nature."

For the most part we have not tried to see rivers as biological systems connecting the hills to the sea, with all their interconnected productivities, but as untamed beings to be turned into stable and benign industrial entities: straighter, with a constant water level, useful for navigation, water supply, as a source of power. River control schemes were part of the Progressive Movement (1890-1920), that sought to transform the natural world into an efficient, productive and controlled system for maximum human benefit. Whether in the end controlled rivers are worth more than natural ones is a matter of debate, and depends on what the ecosystem services of a natural river are worth, compared with the river's value for navigation and electric power, the use of its floodplain for farms, factories and houses, and the cost of maintaining a controlled river, including the cost of rebuilding after floods (currently something like $2-$4 billion a year in the United States). The ecosystem services of a natural river include lower downstream floods; the turning of riverine silts and nutrients into trees, waterfowl and fish; cleaner river water; abundant riverine and ocean fisheries; expanding deltas and offshore islands; healthy coral reefs; growing beaches; protected coasts. Real estate values rise near swimmable rivers with fish and eagles. Luckily, one can keep many of the services of a wild river and still use some of its water and some of its power and farm some of its floodplain fields.

The United States has 620,000 miles of controlled rivers. Some of the dams that control them (those that are old, abandoned, or of less economic value than the fisheries they displace) may be removed as their licenses expire (as was Edwards Dam on the Kennebec in Maine), but most will not be because of fear of a different environment, pressure from affected industries, and lack of understanding of the natural world. The Napa River in California has flooded periodically for decades. After a flood in 1986 caused $100 million in property damage and the evacuation of 5000 people in Napa City, along with 3 deaths, the residents of the City of Napa decided, rather than further reinforce the levees, to let the river reclaim part of its historic flood plain. The Corps of Engineers removed dams and levees along 7 miles of river. Approximately 650 acres of wetlands were restored. The expenses not covered by the federal government, which included buying up some of the floodplain and moving a set of railroad tracks and some buildings, were paid for by a 0.5% rise in the sales tax in the City of Napa. In this case, the rise in real estate values in the city from ending the risk of flooding, turned out to be considerably greater than the cost of the additional taxes. Napa residents also avoided the costs of flood insurance, the inconvenience and danger of flooding, and the occasional death.

Once a river is dammed, engineers favor damming the whole drainage so as to control flow all along the river. The best way to restore the rivers of the upper Middle West would be to remove all the dams (and thus barge transportation) on the Missouri from its junction with the Mississippi; and all those on the Mississippi and its tributaries above St. Louis. Freight such as grain would be shipped by rail on new lines, their construction subsidized, to natural deep water ports further down the river or on the Great Lakes. Coal would move by rail up the valley of the Ohio. Levees would be moved back and a sufficient amount of the original flood plain bought up to absorb floods. Watersheds would be managed to reduce the drainage rate into streams. Less land would be irrigated from the rivers that cross the plains. (As a writer has pointed out, if crop production is the goal, irrigation of humid eastern fields with water from spring runoff stored in riverside pools, would take much less water and be much more efficient; and then the Platte Valley could raise buffalo.) This isn't going to happen, but controlled rivers can be managed so as to take better advantage of their potential services. Riverside fields that require pumps to keep them dry can be returned to the floodplain. Lowering levees to let 10-year floods take over farm fields (which could be farmed 9 years out of 10 and the lost year compensated for) would let the river produce more fish and the fields and floodplain remove more nutrients. Run-of-the-river flow schemes time water releases from dams to mimic natural cycles (spring floods, summer low flow), and try to maintain a suitable level of flow, against the demand for industrial water (for power generation, moving barges through locks, irrigation), and so restore aquatic habitat. Under them, populations of insects, mollusks, invertebrates and fish all rise and their growth rates increase. Letting rivers like the Illinois flood during their normal flood period rather than storing the water behind dams, and letting them reclaim much of their floodplains, would improve the functioning of the riverine habitat. The floodplains would absorb the normal fall and winter flood and the warm season floods from thunderstorms. Scheduling lock and dam repair for low water, in low water years (being flexible), would make such work interfere less with the natural flow of the river. Sequestering dredge spoil (full of toxics and nutrients) would keep it out of the river. Ideally, dredge spoil would be processed to remove the contaminants and the remaining soil spread on farmland, where improved cultural practices would keep it. Ending industrial releases of toxics and increasing fines for spills would make future dredge spoil less toxic. Changing agricultural practices, such as rotating row crops with legume hays, strip cropping, the use of winter cover crops, and vegetating the banks of channelized streams would reduce the flow of silt and nutrients into the river. The ecological services of a healthy river benefit everyone but are owned by no one. But many uses of private property connected with the river will reduce them; and thus reduce the general good.

Restoring shallow floodplain lakes, such as those along the Illinois, is more difficult. Lakes tend to store nutrients, silt and pollutants. The muds of the lakes are now saturated with phosphorus. Release of stored phosphorus from these sediments is likely to maintain algal blooms in the lakes for a long time after agricultural input of the nutrient falls. In summertime, the lakes rapidly stratify, with warm, oxygenated water on top and cool, oxygen-poor water below. Stratification is accompanied by the rapid development of anoxia in the sediments and the release of more phosphorus into the water column. Every 7-10 days winds break up the anoxia and mix the phosphorus throughout the water column, setting the stage for another algal bloom (which returns much of the phosphorus to the sediment in the bodies of the algae). Bottom-feeding carp and brown bullhead in the lakes transform the particulate matter of the sediments into soluble nutrients through their digestive systems, also at a rate sufficient to maintain the algal blooms. The blooms help prevent the re-establishment of rooted underwater plants, and thus help prevent predation on the carp, whose population would be partly controlled by a healthy population of pike. Floodplain lakes can be dredged and their nutrient rich sediment (if not too toxic) pumped onto nearby farmland. Together with agricultural practices that reduced nutrient losses and erosion this would solve much of the nutrient problem. Permanent improvement means re-establishing aquatic vegetation. Since the turbid lakes are now in a relatively steady biological state, and are kept turbid by periodic prairie winds, this is difficult, but probably not impossible.

Control of a river allows its overflow land to be farmed or otherwise developed (perhaps 120 million acres in the Mississippi drainage). The economic return from agriculture on drained land in the greater Mississippi Valley (about $36 billion at $300 per acre gross return) is balanced by the costs of maintaining the controlled river (dredging, dam and levee construction and maintenance; a few billion a year) as well as the other costs associated with a controlled river, such as flood insurance and the yearly $2-$4 billion in flood clean-up costs, an 80-90% loss in riverine fisheries and an unknown one in ocean fisheries (another few billion), more expensive water purification systems (this cost is unknown, but in general, $1 spent on protecting water resources saves $7-$200 in the cost of water treatment and filtration facilities). The social profit involved in developing the most vulnerable overflow land (that nearest the river or with the highest water tables) is less than zero. Commodity agriculture is not very profitable in the first place (returns on capital are usually under 3%). Government-supported loans, grants for farm improvements and price supports for crops make it work. Subsidies to farmers in the industrialized world amount to a billion dollars a day. Farming is so little profitable because so much of its income goes to the larger economy: to steel makers, oil drillers, chemical manufacturers, farm equipment makers, car dealers, grain brokers, railroad and barge transportation systems, electric power generators, insurance salesmen, banks, seed companies, bio-tech laboratories, and so on. So agriculture contributes to the growth of the larger economy. Its total contribution to the U.S. Gross Domestic Product is estimated at 12.5%. Some of this is lost if overflow land is not drained.

One could also argue that in a society in which energy is essentially unlimited, the free economic goods provided by a less disturbed landscape, such as flood control, clean water and abundant wild fish, are less economically desirable than the active management required by the current riverine landscape (levees, locks, dams, yearly clean-up costs, flood insurance, more expensive water purification systems, fish farms). The more manipulated a landscape is, the more work it provides. (Of course, such complete control may have unpleasant consequences, such as uncontrollable floods.) Farmers and riverside communities cannot afford the cost of maintaining the river, so money must come from taxes on the economy as a whole, but the economy as a whole (those steel-makers, car manufacturers, and so on) benefits, and the tax monies are redistributed to the economy through the work required to maintain the man made river. A similar logic would lead us to finish logging off the remaining old-growth forests in the United States; to log off all tropical forests; or fish out all wild ocean fisheries (an immensely lucrative pastime, heavily state supported). After all, global warming may destroy these ecosystems anyway. While economic arguments may not be sufficient to justify maintaining working ecosystems, it would be hopeless to try to replace them with systems under human control. We couldn't do it and we couldn't afford it.

Economic arguments favoring development work because ecological goods such as wild fish and clean water are not given much economic value. Ecological services acquire economic value when their loss causes economic harm, say, when rivers flood developments. While the economic value of a landscape varies enormously over time, its ecological value doesn't change (or changes over very long periods of time). A newly cutover piece of timberland is worth little as timberland (who wants to pay taxes on it while the trees grow?) while 200 years later its timber may be worth a fortune; but all along the streams that flowed through it depended on the state of that landscape for their health; and animals depended on the forest for their livelihood. The economic costs of ecological conservation measures are in most cases trivial compared with their economic benefits, but interfere with our attitudes toward nature—why shouldn't we levee off a river's flood plain in order to farm it or build on it?

In the end the economic argument misses the point. We could not live without clean air, clean water and fertile farmland. Many ecological benefits (swimmable rivers, birds that return in the spring, the roar of the spring flood) are difficult to value: perhaps they are incalculable. Others are too remote from human notions of value to be easily assessed in monetary terms. To replace the cycling of nitrogen by bacteria, for instance, we would have to manufacture the nitrate, spread it over the surface of the earth and the sea, and then control its leaching over the whole terrestrial planet. To replace the work done by cellulose-decomposing bacteria we would have to deal with the 100 billion tons of plant material that die every year, 10 times the mass of the fossil fuels extracted yearly. Luckily (or as far as we know) we have not yet much interfered much with the control of the planet by microbes. The rising seas associated with global warming will wipe out much of the real estate value of the U.S. East Coast ($250 billion in Miami alone, much of it no longer insurable), as well as the remaining ecosystems the land once held (the estuaries, sand

dunes, mud flats, tidal marshes). A 2 foot rise in sealevel will change half
the freshwater marshes of the Everglades to salt marshes. But waves and
rivers will create new and similar ecosystems inland, over the foundations
of our former settlements. Did we do the right thing in extinguishing
the value of coastal ecosystems in order to create all that human value
since it's now going to be lost anyway? (The 'right' thing, an economists
frowns?) A prudent person would never have placed permanent structures
on many coasts, barrier islands, or river floodplains. Hurricane Katrina
sent a storm surge 35 feet high over the barrier islands and 18 feet high
into the coast of Mississippi, wiping out 90% of the structures within
a half mile of the coastline. The same buildings had been wiped out
37 years before by Hurricane Camille. Such coasts should never have
been settled, but left as habitat for fish, shellfish, deer, eagles, cougars, as
wooded refuges for exhausted, northward migrating songbirds.

Since the quest for economic benefit is the only thing that causes significant, efficient change in a capitalist society, many people look for a way
to give ecological goods (like empty land) a reasonable dollar value, in order
to prevent a market economy from further degrading the natural world.
In this way one tries to guide the market economy toward improving the
natural world. Whether this is possible seems problematic. The average
value of terrestrial ecosystems is put at $466 per acre per year, probably a
trivial number to a suburban developer. At 5% interest, it would require a
payment of $9320. per acre into a fund to ameliorate the loss of ecosystems
services by the development (the loss of public good). But such numbers are
the beginning of a negotiation.

Numbers put on the value of ecological services are still somewhat
arbitrary. (I am thinking of the value of flood control, of clean water, of
carbon capture by grasslands or forests, of the protection of coasts by mangroves and coral reefs.) Some calculations put the value of global ecosystem
services equal to that of global annual human production, some put it at
10% of that. Land in a modern economy is still cheap, compared to the
costs of human services, like pensions or medical care. About 12% of the
earth's surface is currently protected, much of it uninhabitable mountains
and deserts. A writer estimates that raising the area of protected landscapes
to 10% of every major region would cost $23 billion annually for protection
and policing; buying the land, another 2 million square kilometers, would

cost $4 billion. Such costs seem impossibly low; a plan to put 12-25% of the United States into permanent conservation status is estimated at $360-$980 billion for purchasing the land or easements on it. The cost of such schemes, especially in developed countries, and their necessary size (20-40% of each landscape is better than 10%), makes one realize that—even if such schemes happen—a gentler way of using the human landscape is probably the only way a functioning natural landscape, overall, will survive.

An advantage of natural systems is that their return, while small compared to man-made systems (probably 2-3%), is very long-term. If lands are selected carefully, their ecological value in human terms is endless, and if they are sufficiently connected, their value continues through periods of changing climate. While investment in human systems must be renewed at intervals of 30 years (roofs) or 250 years (large dams), investments in natural systems renew themselves. Overflow lands constitute about 7% of the lower 48 states, and probably half of that is necessary for healthy riverine ecosystems. As the world develops, the value of functioning rivers, woodlands and coasts rises; this is scarcity value, expressed in real estate value. In an edible world, such landscapes provided food, clothes, drinking water, building material. These values are lost in a modern marketable landscape, but the effect of being near living biological landscapes on real estate values still grows. Since the degradation of ecological function always affects someone somewhere (the value of his lands, the condition of his water and air, the state of his health), more and more schemes to protect the natural functioning of landscapes are probably in our future.

Ecosystem services are part of the general good of nature, and of the human world, not owned by anyone, but heavily influenced by human economic behavior. So-called stakeholder capitalism attempts to take account of the needs of the whole capitalist community: a company's stockholders, its workers and the communities in which it operates. We should add to this the welfare of the larger natural environment on which any civilization floats, the welfare of our agricultural soils and of our farm animals. Expanding the capitalist view is not always a simple matter: Henry Ford was sued by his stockholders when he raised his worker's wages by $1 a day, even as the time taken to manufacture a Model T fell from 12 hours to 2.5 hours. But views evolve. Protecting ecological function is a matter of cultural survival. Biologically functioning landscapes also give one hope, as well as a connection with the non-human world in which we lived up until several millennia ago, and to which, if our civilization collapses, we will return.

Chapter 14
The Problem of Economics

It is hard to overestimate the value of natural resources (timber, minerals, farmland) in the development of high civilizations. Abundant resources, expanded by human ingenuity through technical innovations like pottery making, metal smelting or irrigation, supported the high human populations of these civilizations. A writer has pointed out that agriculturally based civilizations (which constituted most of the world until 1900) depend on their landscapes for their maintenance and growth, while modern economies depend on the application of capital to new products or new markets. (One could argue that the growth of modern trading cities like Singapore and Hong Kong show that a land base is irrelevant to development. But behind these trading cities, or any modern economy, is a huge footprint in the larger landscape.) Converting natural resources to human use, that is, to capital (commodifying them) reduces the ecosystem services provided by the landscape. Very generally, such services include the following: regulating climate; maintaining the hydrological cycle, including the rate at which rain falls and its distribution throughout the year (thus the level of streamflow); filtering out dusts and gases from air and silts from water and turning the nutrients in air and water into biomass; maintaining the gaseous composition of the atmosphere; regulating daily temperatures and windspeeds; forming and maintaining soils; storing and cycling essential nutrients such as calcium, nitrogen, sulfur and carbon; immobilizing or detoxifying pollutants; pollinating crop plants; maintaining landscapes as functioning, nutrient conserving wholes. Ecosystem services come down to things like the work of forests in preventing floods; and the work of bacteria in breaking up the bits of lettuce that go down the drain. The conversion of natural landscapes with their free ecosystem services to man made landscapes in which those services are eliminated or reduced is a condition of development.

During the process of natural succession the physical environment is modified by the biotic community until a more or less stable system is reached. (Stable, but vulnerable to perturbation and rapid change.) Such a system (a redwood forest, an oakwood, a prairie) reaches its maximum possible biomass partly by exerting substantial biotic control over its environment. So called primary ecosystems have numerous symbiotic connections among their inhabitants, with many predator prey relationships which help maintain population stability, and recycle each molecule of nutrients and water many times. By absorbing and transforming much of what flows through them, such systems provide ecosystem services. Such systems conflict with the human goal of maximizing yield from the soil, a function of younger, more leaky ecosystems, such as frequently logged forests and grain fields.

Worldwide, about 15% of the land surface not covered by ice has been entirely remade by humans into fields, housing developments, industrial landscapes and transportation corridors; about 55% of the ice free landscape has been changed by direct human action, including such uses as grazing and logging. Humans appropriate 30-40% of net annual primary productivity on land (net growth of land plants); compared to 5% in 1860. People and their animals directly consume about 4% of this; the rest is recycled or wasted (let rot; landfilled; dumped in rivers). We use maybe 35% of the productivity of the oceans; and over half of all rainfall. (Approximately 90% of this is for irrigation.) Fallout of industrially produced metals, hydrocarbons, and oxides of sulfur and nitrogen affect the whole planet; as do the anthropogenic gases that affect the ozone layer and global climate. The conversion of natural resources to commodities and the resulting destruction of the natural world has made the modern world possible. Development makes landscapes and their resources saleable; and thus makes landscapes produce income. In the case of the United States, the debt incurred from fighting the Revolution, and for purchasing the Middle West from France (the Louisiana Purchase), was paid off through sales of public land. American railroad companies were granted 10% of the land area of the continental United States in return for the construction of the transcontinental railroads. (Under better management the flow of income from the land would still pay for maintaining the roadbed.)

Human development of land degrades it by breaking down its nutrient recycling abilities. An ecosystem is defined by its boundaries; a great deal more nutrient exchange occurs within these boundaries than occurs through them. Leakage of nutrients to the great enveloping fluids of air and water is minimized. Climax or primary ecosystems may support great amounts of biomass on rather small inputs, largely that needed for respiration and maintenance. If these systems developed over long periods of time in infertile landscapes, such as the eucalyptus forests in parts of Australia, or the conifer forests on the dry eastern slopes of the Rockies, they may not be easily replaced once removed. Development, by setting back succession's clock, makes the landscape shed more water, soil and nutrients. Some changes in a developed landscape are obvious, such as the replacement of forests by fields; changes may be less obvious in a landscape's watercourses. We are used to modern single channel rivers, not rivers that split into several channels and meander through wetlands, which are largely man made. Long Island Sound looks as beautiful in the moonlight today as five centuries ago (or one century ago, when painted by Albert Pinkham Ryder), despite the fact it is slowly dying from too much man made nitrogen. This became obvious in July 1987 when lobsters began to crawl up on shore to avoid suffocation in water with essentially no oxygen. On summer nights the sound may give off a whiff of hydrogen sulfide, from decaying fish trapped by low oxygen levels in the recurrent summer anoxia. Soil and nutrient runoff from agriculture, nutrients from sewage treatment plants, from pet droppings, from suburban development, from lawns, all flow into the sound and encourage the algal growth that lead to the anoxia. The sound's dammed rivers no longer support healthy spawning populations of forage fish. Alewife populations in the sound have fallen to 3% of historical numbers. So populations of birds and fish that depend on forage fish have fallen.

In general, under human settlement, water quality deteriorates. Polluted by street runoff and atmospheric fallout, surface waters are no longer safe to drink. Ground water levels fall as landscapes are cleared and as aquifers are pumped for drinking water and irrigation. On farmland, productivity declines over time, as the nutrients stored in the soils by the former vegetation are used up by crops, by the increased microbial activity on ground warmed by the sun, or are lost with eroding soil and leached out by rain. Modern agriculture, even organic or regenerative agriculture, is hard on soils and some soils support it better than others. Rates of erosion of less than a ton of soil per acre per year are considered good (they are good, as they approach replacement levels that are, on average, half that, or about an inch every 250 years); 10 tons per acre is more usual, or a ton of soil for a ton of grain. Some of this soil ends up in roadside ditches, some in the atmosphere, some in waterways. Logging, mining and grading for construction result in large surges of soil, nutrients, water, toxic metals and sulfates into streams. (Mining waste amounted to 10 tons per person— almost 40 tons per family—in the United States in the 1980s.) Grading destroys the soil's profile and its nutrient structure, and thus changes its relationship with plants and with surface waters; under natural conditions the profile takes 1000 years to redevelop. (The process can be speeded up by planting deep-rooted grasses and shrubs.) Land development also affects the atmosphere; but the atmosphere lacks the discreetness of surface waters; mixing is more rapid and complete, so changes are harder to see; and effects may be remote from causes. In general, terrestrial changes must be large to have a measurable effect. Convective rainfall is said to be reduced by deforestation when it reaches 100,000 square miles (an area a little more than 300 miles on a side). Plumes of dust from the Sahara, or from spring plowing in China, measurable in Hawaii several days after it begins (the aluminum in the dust was the clue), affect both terrestrial and oceanic environments (the dust has nutrients) and the earth's heat balance. Dust from the Sahara has increased 5 times since the 1970s because of drought and population growth in the Sahel; and from increased traffic over the desert, which breaks its surface crust of lichens or wind-swept gravel and exposes the sand below. The dust carries fungal spores and bacteria and may be a factor in the decline of some coral reefs. Rich in iron, the dust raises ocean productivity and the productivity of tropical forests. Enough dust in summer will cool the surface of the tropical Atlantic sufficiently to reduce the frequency of Atlantic hurricanes.

Some scientists claim that changes in land use in the Northern Hemisphere over the last 300 years, such as replacing forests with farmland, and building urban areas that act as heat traps, have raised the midwinter temperature of Europe by 3° C. and speeded up jet stream flow in the Northern Hemisphere by 6 meters per second. If true, such changes would rival those claimed for global warming. Forests in temperate regions tend to cool the atmosphere in summer, as water evaporating from their leaves absorbs heat. Turning forests into urban areas or farmland thus warms the atmosphere in summer. In winter, in snowy areas, forests tend to warm the atmosphere, since they absorb more sunlight than snow-covered ground. Boreal forests are thought to warm the atmosphere by about 1° C. in winter and summer. Locally, this warming is thought to be greater than the cooling caused by their absorption of carbon. Irrigation over a large area, as in the Central Valley of California, makes the atmosphere more humid, and thus changes the microclimate of both winter and summer, making summer more oppressive, and hiding the view of the Sierra, on which John Muir remarked. Most landscapes that we see as natural are in fact severely altered.

The effects of development on the ecological functioning of a landscape can be reduced. Belts of trees along streams, if wide enough and located properly, will convert much of the overland runoff of water into subsurface flow, catch topsoil, and reduce the nutrient levels in the runoff water. The suggested widths of such belts vary from 50 feet to 300 feet on each bank. Small trout streams, whose ideal water temperature is 55° F., need 200 foot buffers. (At 55° F. brook trout eat half their weight weekly, mostly in the larvae of aquatic insects; they eat less at higher or lower temperatures.) Such areas are most effective if they include any adjacent wet habitat, such as ponds, wet grassland, swamps, or wet woodland, that is connected with the river. The larger numbers usually involve larger streams (the natural floodplain of the Mississippi ranged to more than 100 miles in places); and the creation of a corridor through which larger animals (mountain lions, elk, wolves) can move, and in which shyer birds and animals (fishers, eagles, owls, some neotropical migrant songbirds, red-shouldered hawks) can breed. California coho salmon are at 1% of historic levels, and recent California regulations to protect coho salmon and steelhead trout call for a 150 foot buffer on each side of streams with fish. Loggers must leave 85% of the canopy within 75 feet of the stream, 65% of the canopy in the remaining 75 feet. This lets them cut the largest trees, which may be useful genetically as seed sources because they are fast growing; and physically useful as nesting sites for raptors. All such regulations are compromises; and no protection was given for streams now without fish, or for dry gullies that carry water during the rainy season, and feed silt into perennial streams. There was no provision for leaving some downed logs on the forest floor, where they act as dams, collecting soil behind them. (Generally, neat park-like forests lack the ecosystem services and microhabitats of natural forests.) Belts of undisturbed grasses also reduce the flow of nutrients and soil into streams, and may be a better choice than forest in some areas. If large enough, they provide habitat for grassland birds. Studies in Tennessee have shown that 6% of a watershed under contoured forest strips will cut the runoff from agricultural land in half; 30% to 40% of the land area in forest will transfer all the surface runoff to the subsoil and stop erosion from the area as a whole. One could conclude (in a compromise) that 20% of formerly forested agricultural landscapes should be in forest. Some of that forest must be downslope of the fields, which usually means giving up farmland. In developed areas, if drainage water is left in unmowed ditches rather than confined to pipes, the cattails and sedges that grow in the ditches will slow the flow of the water and capture the nutrients and silt; the microbes associated with the stems and roots of the plants will remove much of the metals, nutrients and hydrocarbons in the water. Some of the water will sink into the ground, recharging local aquifers. Flashy runoff into the receiving streams, which excavates them and reduces their populations of invertebrates and fish, will be reduced, slowed and cleaned. If necessary, the ditches can be periodically dug out and the contaminants removed with the soil. Such ditches function as settling basins, which otherwise (to be effective) must take up 1-5% of a developed watershed.

Strips of forest or grassland that intercept runoff water take up land of potential economic value. Wide strips change land use on large areas of lowland soil. Much of the de-nitrifying activity in soil water flow, which returns soluble nitrogen in the soil water to the atmosphere as nitrous oxide or nitrogen gas, occurs in wet, oxygen-poor environments near wetlands or streams. These anoxic areas can be extensive or scattered. If not too much nitrogen is running off the landscape (from fertilizer, septic tanks, pet manure, car exhaust) sufficient streamside habitat can reduce it to reasonable levels before the runoff reaches the river. Phosphorus, which is usually carried by soil particles, is greatly reduced by a band of streamside vegetation, though less so in winter when the ground is frozen and the roughness of the silt trapping surface litter reduced. (Heavy rains, which let the runoff form channels through the woods, also tend to overwhelm the ability of filter strips to capture phosphorus.) Leaving wetlands alongside rivers keeps nitrogen out of waterways in more than one way; the water in the channel is part of a larger underground pool to the side and below; water flowing downstream spirals in and out of riverside wetlands, where it is also cleaned of nitrogen.

Most rivers in developed countries are supersaturated with nitrogen. The Platte River in Nebraska, whose watershed is lightly settled but heavily agricultural, leaks nitrous oxide to the atmosphere along most of its lower course. The nitrous oxide is produced by microorganisms that live in the water from soluble nitrogen coming into the stream from cattle feedlots, fertilizer and human sewage. Nitrous oxide is a greenhouse gas that currently produces 5-6% of the man-made greenhouse gas effect (40% that of global transport), and which also contributes to ozone depletion in the stratosphere. But the bacteria that chemically transform nitrogen can't eliminate it; they must return it to the atmosphere as nitrous oxide or nitrogen gas. If it returns as nitrous oxide, it adds to global warming. With a growing human population most of whom depend on crops raised with excessive amounts of nitrogen fertilizer, ways of reducing nitrogen use become more and more important. (Agriculture in 2000 contributed 14.9% of anthropogenic greenhouse gas emissions, transportation 13.5%.) A more benign agriculture is possible, but may imply a different diet.

The obvious reason why ecologically appropriate land development schemes aren't more popular is that they are seen as limiting income to the landowner. This may remain true until ecological services are given a value. Unless the clarity and nutrient status of the water that runs off his land have a value, no benefit accrues to the landowner from reducing the downstream impact of his land use. He may however benefit from changes in land use upstream. Australian farmers help subsidize the planting of trees to lower the level of salty groundwater on lands upstream of their fields; their payments are based on the transpiration rates of the trees; lowering the level of salty ground water keeps the river water, used for irrigation, less salty. And using vegetated drainage ditches rather than pipes for runoff water saves a developer money (perhaps $800 a house) as well as improving the quality of the water running off the development (an environmental benefit). Since a drainage ditch lets more water sinks into the ground, less irrigation is necessary for the householder's trees. Narrower side roads, less expensive to build, slow traffic and allow the tree canopy to close over the road, cooling the street and its houses, and saving the homeowners money on air conditioning. The additional cost of grading so roof and driveway runoff is captured in depressions, where it sinks into the ground, is insignificant. Building better insulated houses, with more efficient appliances, more efficient lighting, and properly sized, efficient heating and cooling systems, together with installing solar panels for generating electricity, increases the initial cost of the building, but lowers the buyer's monthly payments for electricity, heat, hot water, air conditioning; thus the buyer can afford the more expensive house, which has a lower impact on the landscape as a whole. The builder's profit and that of the bank, which depend on total cost, are greater. (In general, energy efficient houses, warehouses, buildings with low energy intensity manufacturing, and one story offices and institutional buildings receive enough sunlight on their roofs to power them if the energy were captured and converted to electricity.)

Undeveloped areas along streams take up buildable land, especially if they are wide enough so a variety of birds and animals can live in them, and if parts of them are left without paths, so their purpose is to remain a bit of wilderness as well as to protect the stream. Such lands also flood and are often wet, expensive to develop and unsafe or unpleasant to inhabit. Developers can be compensated by letting them increase the density of buildings away from streams. Then the presence of the natural area is likely to raise property values. For farmers, land taken out of production probably constitutes a loss. The unfarmed land may allow them to farm more successfully, by providing habitat for native insect predators and pollinators and for birds and mammals that control rodents. The trees or grasses the land produces may be useful on the farm. (In the South, pine straw collected under mature pine forests, is marketable at garden centers.) Undisturbed prairie or forest is worth $20 to $100 an acre a year in sequestering carbon. Natural land also has a public value in cleaning and storing water. (In a biological economy farmers could bid for contracts to provide such services.)

The appearance of a landscape has a moral aspect. Ecologically successful landscapes, often not very picked up, tend to conflict with ideas of what is right: the messy and dangerous wilderness versus the calm and settled landscape, with cows. Drainage ditches with cattails and frogs are considered unsightly and may also have mosquitoes. Neighbors sue each other over lawns of unmowed prairie grasses and stuff fliers from chemical companies into the mailboxes of homeowners whose lawns are bright with spring dandelions. Hanging out laundry to dry, or raising chickens, is illegal in many neighborhoods. Subdivisions in the southwestern United States have by-laws that forbid the use of solar panels on roofs; like laundry, the collectors are thought to depress property values. In our world (perhaps in any world) the appearance of the human landscape is as much a moral as an esthetic matter, perhaps more so: one's landscape, like the interior of one's house, is a reflection of one's self. A certain evolution of the landscape took place in the United States, which was considered morally correct: the primary forest was cut to make farmland and pasture. In nineteenth century prints the dark forest surrounding the stumpy field steadily recedes to sunlit pasture and meadow, and the rough log cabin becomes a two story Greek revival farmhouse inside a fenced lawn, sometimes with a seedling elm sprouting at the corner. Empty prairies become grainfields. The emptier plains, the North American steppe, now grows irrigated grain if water is available; or is used for dryland farming and cattle pasture if it is not. Cattle, wetland animals, hang out near water and exert their own ecological pressure on the landscape, eating streamside vegetation, eroding streambanks, polluting streams and altering the plant cover of the uplands.

The regrowing forest in the print's background is cut when profitable, generally without interim stand improvement. Such management may more or less work ecologically and economically in some forests, such as the sprout hardwood forests of northern Pennsylvania, which were never farmed, and are now cut at relatively short intervals for hardwood flooring and furniture. Hardwoods that sprout from stumps, such as oak, cherry and maple, have replaced the primary mixed forest of northern hardwood, hemlock and white pine that Audubon visited. The forest is not allowed to mature biologically but is cut at economic maturity, when the diameter growth of the stems slows down. Since old trees may rot or be knocked down by windstorms, letting trees mature beyond the point of maximum economic return is considered risky and wasteful. The lack of mature forest changes nutrient relations among parts of the ecosystem, but the landscape is still more or less covered with trees (invasion with grasses and ferns, probably because of the heavy cutting and of browsing of tree seedlings by deer, is a problem), and depending on how heavily and often it is cut, the forest may still have a reasonable hydrological relation with its watercourses and a reasonably tight internal nutrient dynamics. The economic value of the forest remains high, though not as high as it would be if let mature further. (Over half the commercial forests in the United States are under 55 years old; 6% are over 175 years old: that is, they approach the age of old growth for eastern trees.)

In the young, cutover forests of the Northeast, the South, the upper Midwest, and the Appalachian highlands, the development of new logging machinery, along with products like chipboard and finger jointed lumber, has made possible the exploitation of wood of lower and lower value. The new harvesting machinery and end products compensate for the lower value of the logs and their lower density in the forest. That is to say, the return to the logger has risen, and manufacturers continue to profit, but the profit for the landowner, measured per acre of land per year, has fallen; and the landscape is much more heavily used. (A man with a feller buncher might cut 1000 trees a day, a man with a chainsaw 40.) In woods clear cut at short intervals, such as Midwestern aspen, Southern pine, and Maine spruce-fir forests, no mature forests, nor the plants, birds, lichens, mosses or amphibians associated with mature forests, survive, and the streams in small, frequently logged drainages are ruined by siltation. Nutrients that would be returned to the soil in fallen logs or leaves are removed as whole trees. How sustainable such practices are in the long term is hard to say. Long enough, if studies of a light sandy loam in Montana are correct: the soil contains sufficient mineral nutrients to sustain current logging practices for 100,000 years: that is, until long after the climate has changed and the trees disappeared. Soil nutrients are not the whole story, and how fast the trees grow after a few rotations, the cumulative effect of successive clear cuts on soil fungi, insectivorous birds, insect grazers, and on the waterways that flow out of the forest, are less clear. The effects on streams are generally disastrous. Clearcut, west facing slopes in sunny, dry western forests can be almost impossible to reforest, once the mycorrhizal fungi that support the trees' growth are gone (the fungi disappear a couple of years after their host trees).

After the Second World War, settlement along the Atlantic beaches of the United States grew. Houses were also built along the beaches and atop the bluffs of the Pacific coast. Neither of these areas is stable in the long term, the long term here being several decades to several centuries. Atlantic beaches move inland at several feet a year, 6-8 feet on the Outer Banks of North Carolina, that is 60-80 feet in 10 years, something lighthouse builders knew; the average for east coast beaches is 2-3 feet a year. While moving inland, the beaches retain their approximate slope and width. Bluffs on the Atlantic side of Cape Cod retreat about 2 feet a year; again the shape of the bluff and of the beach below remains similar. So the path Thoreau took on his walk along the cape in the mid-1800s is now some hundreds of feet offshore.

The cause of this movement is wave action. Waves are created by winds. Winds are powered by the sun and the earth's rotation. Breaking waves transfer their energy to the beach, moving it around. Since waves rarely hit the beach at an angle of exactly $90°$, their impact creates an alongshore current that carries sand with it, generally south along the East Coast. This alongshore movement of sand drives the beaches inland. On any given beach, sand also tends to move back and forth between offshore sandbars, formed in storms, and the beach, replenished in calmer weather. Such movement, together with the particular sea-floor characteristics, geology, tidal action, and weather of a given site, help the beach keep its characteristic shape. Despite the alongshore movement that brings in more sand, there is a continual net loss of sand to the deep sea. Some is lost in inlets, where the sand is caught up in tidal flows and deposited inside or outside the beaches; some simply flows down by gravity into submarine canyons. In the modern world, inlets are dredged and the sand is dumped offshore. New sand comes from rivers and eroding headlands. On the East Coast, most of the riverine sand is deposited in estuaries behind the barrier islands and only slowly, if at all, makes it to the beach. The estuarine marshlands that receive the sand help hold the barrier islands in place. Most east coast rivers are now dammed so their load of sand is greatly reduced. (Sand being heavier than silt, it settles out preferentially behind dams.) Along the East Coast most new sand is supplied by eroding bluffs and cliffs. Armoring cliffs to protect clifftop homes prevents the generation of new sand. Armoring beaches with jetties and groins prevents the alongshore movement of sand, starving beaches downstream, which then shrink and recede. Armoring beaches with seawalls to prevent the loss of buildings that were built too near the sea, results in the total loss of the beach outside the wall. In California, the rivers that flow through the steep, erodable hills of the California coast ranges match headlands as sources of sand. Damming those rivers traps much of the sand behind the dams. Dams and channelization along the Santa Clara River reduced its estimated input of sand to the beaches of Ventura County from 600,000 cubic yards of sandy sediment a year to 150,000 cubic yards. It is thought that dams on California rivers now hold back something like 100 million cubic yards of sand annually. Beaches are also changed by catastrophic events; the 1938 floods along the Santa Clara brought down an estimated 8 million cubic yards of sediment, building up beaches downstream. Armoring California's sea cliffs to protect houses or roads also prevents the creation of new sand. So beaches shrink; and eventually change their profile. (In general, they become more steep.)

The notion of sand rights, thought up by a lawyer named Katherine Stone, is based on a provision of the Institutes of Justinian, a summary of Roman law compiled in the sixth century. The Institutes stated that any Roman citizen had a right to use shorelands or riverbanks to fish, tie up a boat, or unload cargo. This provision of Roman law was taken into English common law, which was brought by English colonists to America and became state law when the colonies became states. The notion is now known as the Public Trust Doctrine. Through the doctrine, a state has an interest in protecting shorelands, bottomlands, tidelands, navigable freshwaters, and their plant and animal life, for the use and enjoyment of all the people. Individuals may own such lands, but the interest of the state in them is inalienable, and when the state takes steps to protect or manage these resources it does so with the rights of an owner, not a regulator. Thus in theory no compensation is owed to the landowner. The public interest in such resources can be terminated, but only narrowly, and only in pursuit of a public interest that is judged to be greater. Thus construction for navigation, or for unloading of cargoes has been allowed. (In general, more development has been allowed than is consonant with modern interpretations of the Doctrine.) The Public Trust Doctrine has been expanded through lawsuits to include rights such as strolling, swimming, the esthetics of the shoreline, and environmental health. Out of these rights come the public right of access to the "wet" beach, the intertidal beach, in all coastal states except Massachusetts and Maine, where such rights were extinguished under the charter of the Massachusetts Bay Company. In some states, such as California, the public also has a right to a portion of the dry beach, over which passage is necessary to reach the wet beach.

The Public Trust Doctrine was the basis of the ruling that made the city of Los Angeles reduce the amount of water it was taking from Mono Lake. Mono Lake, in the desert east of the Sierra Nevada, receives runoff from the eastern side of the Sierras. A California state court found that the lake's wildlife constituted a public trust whose needs must be balanced with the need for water of the citizens of Los Angeles, who had purchased the water rights to Mono Lake. By the same token, if there are to be fish, there must be limits on water use and on water pollution; and if there are to be beaches, there must be sand. If the right to the shore is a common right, then so is the right to the sand that feeds the beach. States have not moved to regulate development so as to protect the rights of beaches to sand, but have left redress to property owners and municipalities, who file lawsuits to address the matter. Such suits have generally been upheld in state courts, where the expansion of the Public Trust Doctrine has occurred, but have never come before the Supreme Court of the United States. If enforced, sand rights would force people to make a more accurate assessment of the costs of beachfront or clifftop development; of navigation works; and of dams. Dams provide water for irrigation; for commercial and residential use; for hydropower; they provide for river navigation and flood control. They also destroy fisheries, increase the river's production of methane, change the ecological functioning of rivers and riverside wetlands, and intercept the flow of sand to beaches. Strictly speaking, the beneficiaries of dams, that is, cities, industry, agriculture and river navigation companies, should pay to rectify the damage to beaches, marine lands, and riverine and offshore fisheries. Dams can also be operated so as to reduce such damages; this would benefit everyone.

Sand rights are a powerful idea because they connect uplands with ocean beaches in a working physical system. Such connectivities are common in nature but little recognized in biological theory or human economics. Before modern times and the saturation of once nitrogen limited forests and grasslands with airborne nitrogen (a product, like carbon dioxide, of combustion and of agriculture), ammonia volatilized from seabird droppings is thought to have constituted a major atmospheric input of nitrogen to terrestrial environments; the input was substantial where seabirds nested along coasts, such as in parts of Alaska, the Pacific Northwest, and New Zealand, where it contributed to the growth of grasses and trees. Lightning was another atmospheric source of nitrogen, and helped fertilize the American plains. (The main source of terrestrial nitrogen is nitrogen fixing bacteria in the soil.) Including the natural environment as a whole in the Public Trust Doctrine would give one the regulatory tools to create, or re-create, a world in which people fitted into the working natural environment. To once again paraphrase Locke, the ecosystem services a landscape performs constitute its greatest intrinsic value, and thus justify regulating the economic world that affects them; in other words, this is another example of markets requiring supervision.

Many degradations of the environment are difficult to sue over, since specific causes are difficult and expensive to establish (the specific source of that chemical, or that fertilizer runoff); and many have been aggravated by government action in the economic interests of owners of riverfront or shoreline property. Should the fisherman of Texas and Louisiana sue Iowa farmers for the condition of the Gulf of Mexico? Should Louisiana trappers and property owners sue oil companies for the disappearing Louisiana wetlands? Should fisherman along the Mississippi sue riverside farmers, or the Army Corps of Engineers, the body responsible for the river, for the loss of overflow wetlands? Is navigation up the Mississippi to Minneapolis, up the Missouri to Sioux Falls and up the Ohio to Pittsburgh, a necessary use of those waterways? The annual barge traffic between St. Louis and Sioux Falls is worth $7 million; 93% of the emergent wetlands, backwaters, and sloughs along the Missouri were converted to agriculture or dredged for channels in constructing the waterway; the shape, chemistry, and temperature regime of the river have been changed; non-point, largely agricultural, contaminants such as chlordane, dieldrin, and PCBs are a hazard to fish and wildlife in the remaining floodplain and the channel; use of the river by ducks has declined dramatically and the fish catch along parts of the river has dropped 80%. If people have a right to clean air, should the inhabitants of New York and the New England states sue power plants in Ohio and West Virginia for the condition of their air? (This has happened.) Or the Inuit of Nunavut sue chemical plants in Alabama for the state of theirs? (The source of some of the chemicals in their body fat has been traced to chemical plants there.) Ground water moves, air moves, and if one looks far enough most things on the planet are connected. Including the natural environment with its ecosystem services in the Public Trust Doctrine provides a way, in our litigious society, to reach a working natural environment; a sort of public property right that includes the environment as a whole. The idea of rights to natural goods are not new: Egyptian cities had laws regulating the heights and placement of buildings, to preserve other people's rights to sunlight and air. Under English common law, property owners have a legal obligation to not use their property so as to inflict legally recognized injury on others; does this obligation include such things as lawn chemicals, that flow off with the rain and drift on the air?

Human use of the landscape runs along a continuum. Industrial civilization has to convert some of the landscape to human use, but not all of it, while economics of property ownership in a capitalist society will inevitably make us use all of it. Sand rights tend to pit one form of development against another. Sand rights have the potential to transform some land uses because loss of beaches brings into play financial interests that dwarf those of fishermen or conservationists. In fact, we would be better off if there were no development along beaches, in the floodplains of rivers, or on top of seaside bluffs. A reasonable, long-term, national, land-use plan would phase such developments out through buyouts and the termination of federal guarantees for flood insurance.

Of course (unfortunately!), any land development affects watercourses. Once impervious surfaces reach more than 10% of a watershed, streams suffer. (Impervious surfaces are usually thought of as roofs, roads and parking lots, but plowland is several times as impervious as grassland or forest, and mowed lawns are also relatively impervious surfaces.) The problem is the rush of water that comes off such surfaces in a rainstorm. The heavy flow carries chemicals, nutrients and sediment and causes channelization and sedimentation in the streams that receive it. When impervious surfaces reach 20% of surface area, only the hardiest species in the streams receiving the drainage water survive. In parts of New Jersey, impervious surfaces now approach 60% of the landscape, and flooding is a constant problem. Such problems can be ameliorated by modifying the pattern of human settlements. Settling ponds and artificial wetlands reduce the pulse of water reaching streams and trap silt, debris and pollutants. If the drainage water flows to these ponds and wetlands in vegetated ditches rather than pipes, the structures work better, the water that reaches them is cleaner and some of the flow returns to ground water. Ground water levels are usually considerably lowered in developed areas. Catch basins for roof or street runoff also return surface flow to ground water. Catch basins require 5-10% of the area from which the runoff comes; so a roof of 2000 square feet would require a catch basin of 100-200 square feet, say a shallow hollow 6-12 inches deep and 12 feet on a side. Catch basins work best in sandy loams with a high organic content. Such basins are planted with vegetation that can stand periodic flooding. Larger areas require approximately one basin per acre for runoff water. An acre requires a basin of 235-475 square feet, or a hollow about 20 feet square. (I saw one once off the parking lot of a motel in downtown Bakersfield, California, that was full of satellite antennas and singing frogs.) Connected drainage areas in the backyards of older, steeper urban developments would take care of roof and lawn runoff; road and sidewalk runoff would require other solutions. Permeable pavements on parking lots and roads let much runoff filter in; parking lots and roads can also devote 10% of their area, such as shoulders and medians, to catch basins, vegetated with cattails, giant reed, or shrubs and trees. Much of what comes off roads and parking lots is relatively toxic and includes motor oil, antifreeze, copper from brake linings, cadmium from automobile greases, and various toxic organic chemicals and particles from automobile exhaust and tire dust. Catch basins and artificial wetlands capture and process some of this before the water reaches streams. Such structures can be added when roads are rebuilt. A common location for recharge structures in new developments is the strip between the sidewalk and the road. So how to encourage the proper handling of water? What is the water that runs off suburban yards worth? A hundred dollars an acre? Four hundred? Homeowners whose runoff was satisfactory (no nutrients above the natural background, no toxic chemicals, most of the flow returned to ground water) would have the relevant amount taken off their property taxes. Alternatively, homeowners could be charged for their runoff, and the charge reduced as the condition of the water improved and its amount fell.

An ecological perspective puts us all in the same boat; we are all absolutely dependent on each other no matter who owns what piece of land. And the connections in the modern world are global; through the currents of the seas, the circulation of the atmosphere, the ships and planes of trade (some 25,000 flights a day within European airspace, 600 across the Atlantic). Some landscape changing organisms have been introduced to new environments deliberately, such as pigs, African grasses, rats, bamboo, songbirds, cows and mongooses to Hawaii. (Mosquitoes arrived as larvae in the water casks of sailing ships, avian malaria in imported birds.) Some arrive on their own, in aircraft (snakes that hide out in wheel wells, mosquitoes in cabins), in packing material (wood boring insects), in boxes of fruit (snakes, fungi, insects), in ship ballast water (larvae of many marine fish and invertebrates). Some of this can be controlled. But trade and imaginative manipulation of the landscape is part of our world: we are constantly trying to improve on nature.

Use of private land can be regulated, or shaped by economic incentives, such as tax reductions or cash payments. Land in developed countries is still relatively cheap compared to national incomes or social costs. In the United States a program to purchase lands essential to ecosystem services, such as floodplains, overflow lands, wetlands, streambanks, large tracts of native wildland, coasts subject to erosion or storm surges, would cost comparatively little. As with shifting to a low carbon, non-toxic economy, a long-term program is needed. A small annual appropriation, say $10 billion a year, for a century, would give the 3000 counties in the United States $3.3 million dollars a year to purchase lands. If land costs $3,300 an acre, a county could purchase 1000 acres a year, if $500 an acre, 6,500 acres. (In New York State, with 60 counties, this would result in perhaps 20 million acres of land being protected, or more than doubling protected landholdings. More importantly, the lands would be spread in low-lying ecosystems throughout the inhabited state, in places where much of the population lives.) Precedents exist for programs with willing sellers. County committees of ecologists, soil scientists and geologists would come up with lists of suitable lands; local environmentalists, land planners, land developers, farmers and other stakeholders would decide what lands to buy; an existing federal agency like the Soil Conservation Service would administer the program. One problem will be the loss of property taxes to localities. But any serious look at land use in the United States has to address the question of property taxes, which now are asked to support many more services, including medical care for the poor, than they should. Three quarters or more of county budgets in New York State, supported with monies from property taxes and sales taxes, go for social services. Such costs must be shifted to a larger base, if only because basing social services on such taxes results in tremendous inequality among counties (the same problem as with school taxes). Conservation Reserve payments to farmers are an expensive way to pay for the maintenance of ecosystem services (after 10 years, payments often amount to the purchase price of the land), but help keep small farmers on the land, and since the landowner continues to pay property taxes, have the advantage of shifting payment for local services to the federal government.

The adoption of agriculture made human use of natural landscapes more exploitative, though the effect on natural environments varied. I would argue the 'end of nature' arrived with extensive settled agricultural landscapes, and a view of the natural world as commodifiable, more useful as fields and grazing land, or as a source of timber and minerals, than as a landscape that (in its entirety) provided the means of living. Some systems such as Tiahuanacan raised beds, Mayan ditched swamps, the shifting slash and burn agriculture of the tropics or the temperate zones, were sustainable indefinitely, as long as they weren't overused. While these systems changed their surroundings, they didn't degrade them. A large part of the sustainability of these systems depended on extensive areas of nearby land remaining undisturbed. The natural vegetation was necessary to sustain water flows, to renew soils for slash and burn farmers, to conserve fisheries, to reduce erosion, and to provide other things of economic value: medicines, foods, building and basketry material, fodder. Upland agriculture that used permanent cleared fields, and irrigated agriculture, tended to have more serious ecological effects, though as long as populations were small and technologies low, economic destruction of whole landscapes took much longer than at present. "High" civilizations that grew in population were another matter. Their constantly expanding demands for raw materials, food and fuel stressed their agricultural systems and the surrounding uncultivated lands, which were cut, pastured, or brought under agricultural development, often with disastrous consequences (soil erosion, soil nutrient depletion, waterlogging, soil salinization, drying up of streams, flooding, growth of deserts; grazing animals turning upland forest into rocky scrub). Upland agriculture can work within a larger ecosystem, and not degrade its surface waters or its soils, if its sites are limited and its methods take account of the larger environment. The Amish of Pennsylvania and Ohio, who farm with horses and use no manufactured fertilizer or pesticides, have left their soils more fertile than they found them, though their effect on streams may not have been ideal. Modern no-till agriculture, which keeps the soil surface constantly covered, reduces soil erosion by 75-90%, and nutrient losses by similar levels. Soils under no-till cultivation store atmospheric carbon and slowly rise in fertility.

In general, the demands of larger and larger populations have forced certain patterns of settlement on landscapes; the energy supplied by fossil fuels and the accumulation by societies of enormous reserves of capital has made this process faster, more intensive, and more extensive. Landscapes have not been settled with their own (the landscapes') interests in mind since the development of large settlements 8000 years ago. But human settlement for the last 6-8 millennia has always existed along a continuum of use. Until recently (1860? 1900? 1950?) many parts of the world remained relatively untouched by agriculture or industry. Eugene Odum thought we should leave 40% of any landscape undeveloped, to allow room for nature to work. In many North American landscapes this is no longer possible, though such matters can be (somewhat) reversed. Without intervention, it is the ineluctable fate of every piece of private land in a market economy to be developed. The nature of the economy, together with our current property tax system, force this upon us. Development is the only rational use for land in the modern world.

What made our world possible was the exploitation of fossil fuels. Fossil fuels freed us from the cycles and limits of natural production, and so the limits on what was possible vanished. The use of fossil fuels enabled a tremendous expansion in the human habitat and population. But the use of fossil fuels did not change our dependence on nature; it simply changed its scope. Fossil fuels are thought to have been laid down several hundred million years ago, on a warmer planet, with an atmospheric carbon dioxide level 2-5 times ours, and an oxygen level of 1.5 times ours. The rate of photosynthesis may have been several times ours. The carbon dioxide in the atmosphere was turned into plant tissue, and that tissue preserved in great wetlands, as coal. (Hard coal: bituminous coal and anthracite; brown coal developed more recently from peat.) Oil arose as massive algal blooms in shallow seas. While some scientists think oil is not derived from plant tissue, but from organic chemical reactions deep in the earth's mantle, the more general view is that oil is derived from phytoplankton that settled to the bottom of shallow seas. The carbon dioxide that was removed from the atmosphere by the formation of oil and coal resulted in a substantially lower level of atmospheric carbon dioxide; partly, this has created our cooler modern planet. Partly, because carbon has also been stored in forests and grasslands, in soils, in ocean sediments, as methane in frozen Arctic tundra, as methane clathrates under shallow polar seas, in the immense amounts of carbon hungry rocky debris created by the Himalayas' pushing up the Tibetan Plateau (among other geological cataclysms); and partly because solar radiance, the aspect of the earth to the sun, and the positions of the continents on the globe also affect climate and have all changed in the last 200 million years. Fossil fuels, limestones, phosphate deposits, coral reefs, soils, the composition of the atmosphere, some iron and copper deposits, are all manifestations of life. The deposits of iron ore that made the Industrial Revolution possible began to be laid down about 2.5 billion years ago, when the first photosynthesizers (green algae and blue-green cyanobacteria) started to raise the oxygen content of the atmosphere. The soluble iron in the oceans precipitated out as deposits of insoluble iron oxides: rust, or iron ore, which we heat in the presence of carbon to eliminate the oxygen and so produce iron metal.

In our use of fossil fuels we are completing a circle. We are returning the stored carbon in fossil fuels to the atmosphere as carbon dioxide. Sooner or later this would have happened anyway as the planet's surface was slowly recycled through the mantle. While this process takes tens to hundreds of millions of years, industrial combustion will take less than a thousand. Rapid releases of greenhouse gases have happened before. About 55 million years ago the intrusion of molten magma into carbon rich mudstones under the Norwegian Sea, and perhaps also into coal deposits in South Africa and Antarctica, decomposed the solid carbon into methane, which entered the atmosphere. This was accompanied by an increased emission of carbon dioxide from volcanoes. As the ocean warmed a few degrees in response to these gases, methane burst up from underwater clathrates. Perhaps 1200 billion tons of methane, of the 10,000 billion tons thought to be stored in clathrates in the permafrost and under the seabed (twice fossil fuel reserves), entered the atmosphere. Global temperature warmed about 13° F. over 30,000 years—some say in a much shorter time—then fell as massive algal blooms in the ocean, fed by the rising temperatures and the increased nutrients running off the land from the increased rainfall, absorbed the carbon and returned it to ocean sediments. The complete cycle took about 60,000 years, or 6 times longer than agricultural civilization has lasted.

Warming the planet returns the earth to an earlier state, in which, for instance, algae and jellyfish dominate estuarine production instead of sea grasses and turtles. Similarly, metal smelting and the burning of fossil fuels are distributing metals and oxides of sulfur and nitrogen over much of the terrestrial landscape, slowly making it less hospitable to the current vegetation. Other industrial chemicals, such as the chlorofluorocarbons and compounds of bromine, released in relatively small quantities into the atmosphere, have reduced the effectiveness of the stratospheric ozone layer that shields the planet's surface from ultraviolet radiation. The development of an ozone shield a very long time ago was one of the things that made life on the surface of the earth possible. The chemical reactions that deplete ozone take place at very low temperatures at the end of the polar winter, with the returning sun. The carbon dioxide mediated greenhouse warming of the lower atmosphere necessarily cools the upper atmosphere (the stratosphere). Only so much heat is radiated out from the earth. If more is captured by the lower atmosphere, less is available to warm the upper atmosphere. The cooling of the upper atmosphere, by favoring the reactions that deplete ozone, reinforces the effect of the ozone depleting chemicals. (It also makes the stratosphere more dense, allowing satellites to stay up longer.) By poisoning soils and waterways with nutrients, heavy metals and chlorinated hydrocarbons; exchanging calcium in forest soils for metals like aluminum; by increasing ultraviolet radiation at ground level (UV-B is up 15-20% at 40° North Latitude, the latitude of the Pacific Northwest, southern Canada, New England and upstate New York; worldwide the ozone shield is probably reduced by 6%); and warming the planet at a rate too rapid for many organisms to move or adapt, we are changing the conditions under which current life exists. We are making the world more hostile to organisms like us, but perfectly suitable for bacteria and other microorganisms that evolved in more extreme worlds and whose rates of reproduction (minutes to hours), and ability to exchange genetic material among species, let them rapidly evolve to take advantage of new conditions. If we left for Mars, few organisms would miss us (cows, cockroaches and ragweed, perhaps); but the departure of microorganisms would mean the end of the biosphere.

Carbon dioxide is not the only greenhouse gas that raises the temperature of the lower atmosphere. The other greenhouse gases equal it in effect. Nitrogen fertilizers are decomposed by soil bacteria to release nitrous oxide to the atmosphere. Nitrous oxide is another greenhouse gas, though one with a shorter residence time in the atmosphere than carbon dioxide. Nitrous oxide also attacks the ozone shield. It would be hard, but not impossible, to feed the current world population without the use of nitrogen fertilizer, but its use could be cut substantially. About 40% of the nitrogen fertilizer used in the United States is used on lawns. There, use could be cut to zero. Composts, applied to mixes of grasses and low clovers, produce a healthier lawn. Or vegetable gardens and native vegetation can replace lawns. Methane from coal mines, from leaky natural gas pipelines, from irrigated or flooded lands, from reservoirs, from human sewage, from the guts of ruminant animals (cows, goats, moose, sheep), also helps warm the planet. People have probably doubled its concentration in the atmosphere. It would be relatively easy and inexpensive to cut methane emissions substantially, and cutting them would have a substantial effect on global warming; but we have no incentive to do it. The chlorofluorocarbons that cause ozone depletion are extremely potent greenhouse gases and so have an effect on climate even at their current miniscule concentrations; the chlorofluorocarbon refrigerant that most affects the ozone layer has been banned in industrialized countries, but its use in the underdeveloped world continues. New, similar compounds that cause warming (fluoroform, nitrogen trifluoride) are let accumulate in the atmosphere until sufficient pressure arises from environmentalists to control them. That such chemicals are likely to be harmful is no secret.

For the rest, soot (from burning coal, from burning wood and dung in cooking fires) absorbs sunlight and warms the air around it, and when it settles out on glaciers or sea ice, helps the ice melt. Its contribution to

global warming may be substantial. The many cancer-causing and mutagenic compounds of combustion, the neurotoxic metals, the chlorinated
hydrocarbons that mimic human hormones, cause cancers and interfere
with fetal development, enter the food chain through air or water and are
concentrated as they move up it. Fat soluble, they are stored in the fatty
tissue of plants and animals. Such chemicals are lost during nursing and
excreted in feces. We can hope these chemicals will provide more food
for bacteria, which will decompose them and thus render them (probably)
harmless, that is, harmless to us, no longer useful to the bacteria. Earth-
moving by people, now estimated at 40 billion tons per year, surpasses estimates of material released from seafloor volcanoes, and is comparable to the
earth moved by rivers. This material is a source of more nutrients, metals,
dust and acids that end up in the atmosphere or water. (On a more cheer-
ful note, the mass of material also soaks up carbon dioxide.) Our effects on
nature are no longer limited to matters like the misuse of agricultural soils,
or the unsustainable harvesting of renewable products like fish or timber,
though these remain, but include global temperature, global sea levels, the
acidity of the sea, levels of calcium in soils, and levels of toxic materials in
animals such as the beluga whales in the St. Lawrence estuary, dolphins in
the North Sea, common loons in Maine and humans. Many of the current
effects of human development are subtle, and invisible to the naked eye.

Economics has given us this world and, under the right direction,
economics can change it. The fundamental mistake of economics is the assumption that what is made doesn't matter. This simplifying assumption
disconnects economics from the real world. Whether people are building
coal fired power plants or installing solar panels matters; whether the modern chemical industry is based on the manufacture of bio-accumulating
chlorinated hydrocarbons or on more benign chemicals matters; whether
bureaucrats push pieces of paper around, take bribes or accomplish something of value matters. It is the duty of a benign government to realign
economics with the needs of the world. So I would like to consider settling
a landscape with the landscape's interests in mind. That is to say, suppose
a landscape were settled so that its native ecosystems, or some analogue of
them, continued to work, so that settlements didn't pollute the streams
that ran through them, or unduly disturb their temperatures or patterns of
flow. Biologically speaking, this means keeping much of the native nutrient

recycling systems in place. (We are talking of bacteria, soil invertebrates, soil shading and root holding systems here, not necessarily old growth and wolves.) Of course this is only partly possible in agricultural or logged ecosystems, or in suburbs, and may not be possible at all in more heavily settled locales. Places like Manhattan must control their nutrient output, and their interference with natural patterns of water infiltration and flow, with industrial technologies like sewage treatment plants, storm water treatment plants, the use of man-made marshlands to strip the waste water of its remaining metals, nutrients, hydrocarbons, hormones, bacteria and viruses. Urban storm water may contain nutrients on the level of sewage effluent. Use of low flush toilets and low flow showerheads reduces waste-water flows and lets sewage remain longer in treatments plants, greatly reducing the nutrient load of the outlet water. The sewage treatment plants of New York City constitute Long Island Sound's fourth largest freshwater tributary (others include the Hudson and Connecticut rivers). Constructed salt marshes for cleansing effluent are cheap at $20,000 per acre.

Reducing car use in urban areas helps. In general, reducing combustion helps. So, too, does composting food waste and installing solar electric panels. Ultimately, discharges from urban systems must be converted to inputs that sustain local or distant ecosystems. The city itself will not work well as an ecosystem, but relationships are possible: between peregrine falcons and pigeons; Canada geese, coyotes and parklands; between parklands and migrating birds; between sewage sludges and Southwestern farmlands, or strip mined lands; the use of processed human urine as nitrogen fertilizer anywhere. Ecosystems and their animals and plants are resilient and adaptable. Striped bass now spawn under the piers in New York City, whose removal (for this reason) was halted. Oxygen levels in New York Harbor have recovered to the point that wooden pilings are attacked by marine boring worms and blue crabs are caught in Newtown Creek. A system of animal overpasses and underpasses, together with more protected habitat, might let foxes, mink, weasels, barred owls, salamanders and frogs occupy more of their natural range in the counties surrounding the city, perhaps reducing the white-footed mouse populations that carry Lyme disease. (For unknown reasons, deer ticks in parts of California with good fence lizard habitat—and thus many fence lizards—have almost none of the bacteria that carry Lyme disease, leading biologists to speculate that the more animals the ticks can bite that don't carry the disease—are not competent hosts—the better.) Energy efficient, non-toxic green buildings cost slightly more to build but have operating costs 8-9% lower and produce a 6-7% greater return on investment. A city can reduce its impact on the surrounding ecosystem to tolerable levels, even if the effects of dense human settlement are capable of only so much amelioration.

Part II
Two Capitalist Landscapes

Chapter 15
The Buffalo Plains

As one goes west from the Mississippi River, which up to the toe of Illinois, 600 miles from the Gulf, is little above sea level, the land rises. It becomes higher and drier as one approaches the rain shadow of the Rockies. The plains west of the 100th meridian (the line where the West starts) range from 2000 to 4000 feet in elevation and are quite flat. The soils, subhumid and little leached by rainfall, are fertile. Winters in the northern plains can be extremely cold, with severe blizzards. The native grasses are short and the grass cover is discontinuous. Sunshine and wind are abundant. Rainfall is low and variable. Light rain showers are common but a third of the annual rainfall can fall in an hour. The plains experience droughts of up to a decade, during which the grass cover may retreat by 70%, exposing much bare soil. For farmers the long-term probability of rainfall in a given place is more important than the rainfall average over a larger area: that is, in some specific places, because of topography and airflow, rain is more likely. Without irrigation, farming on much of the plains is a gamble (some say impossible); but the Great Plains are huge, and until recently produced perhaps half the wheat traded on world markets (much of it irrigated, some dryfarmed: the original plowup of the plains to take advantage of the high price of wheat during the First World War—the result of the blockade of Russia by Turkish ships in the Dardanelles—resulted in the dust storms of the 1930s when dry weather returned; 100 million acres were plowed as farmers chased the boom and then falling prices and deteriorating weather; most of it eroded badly). Escarpments to the east and south lead down to the tall grass prairies and perennial streams of the lands nearer the Mississippi Valley.

Earnest Thompson Seton, a Canadian naturalist, estimated the extent of the plains biome in North America at 750,000 square miles. The so-called buffalo commons, a buffalo ranching scheme proposed a decade ago for the depopulating, poorer quality farmlands of the Dakotas, Texas, Wyoming and Nebraska was much smaller, 147,000 square miles. From

Nebraska, including a bit of southern South Dakota, south through Texas and New Mexico, approximately 220,000 square miles of the plains are underlain by an aquifer of fossil water, the High Plains Regional Aquifer. This is water stored from a previously rainy time, that is, fossil water, some of it glacial meltwater. Much this aquifer consists of the better-known Ogallala Aquifer. The depth to the water varies. Before pumping began, it lay within reach of hand-dug wells in parts of Nebraska: home-made windmills, of many different designs, were characteristic of that part of the plains. Some of the water irrigated truck crops. (A decade after Custer's defeat on the Little Bighorn, one truck farmer commented that if his egg-plants sold as well in the Lincoln market as they grew, he'd be rich.) Even now, with much of it pumped down (water pumped for agriculture is about half of what it was in the 1970s), groundwater flow from the aquifer pro-vides substantial parts of the dry season flow of the rivers that cut across the plains. Groundwater pumping has eliminated much of the habitat for fish, amphibians and mammals on the plains. Sections of some Kansas rivers now dry up in summer. A hundred and fifty years ago when small streams and seasonal wetlands were more common, the land was more eas-ily habitable. Vernal and autumnal pools provided habitat for shorebirds and amphibians; buffalo dug out tiny seeps for wallows; and prairie dog towns, with their avian, mammalian, insect and reptilian inhabitants, oc-cupied one fifth of the landscape.

The buffalo on the North American Plains constituted the greatest concentration of large herbivores since the Pleistocene; greater than that of any single species on the African plains. Seton estimated the North Ameri-can buffalo population at European contact at approximately 75 million animals. Before the Europeans arrived, buffalo ranged over the plains and the prairies and into the eastern woodlands, and were found in small numbers in the Great Basin, whose grasses are not adapted to continuous grazing, and on the southern parts of the Atlantic Coastal Plain. Seton's estimate was based on the carrying capacity of the different regions for cattle and horses. The prairie, which Seton called the best habitat, is now all farmland. (Whether it was in fact the best habitat for the buffalo is uncertain.) Prairie was said to support one animal per 7-8 acres. One acre of prairie was said to equal 4 acres of plains or woodland; and the esti-mate followed from these calculations. Seton's estimate was undoubtedly too high, but the modern rule of thumb is that grassland that will support 2 cattle will support 3 buffalo, despite their similarity in size. Buffalo are browsers as well as grazers and stand cold much better than cattle. During storms, they browse into the wind, rather than downwind like cattle, and by moving this way, historically for long distances, shorten the time the storm lasts. Unlike cattle, buffalo go into a state of reduced metabolism in winter, which reduces their need for food. Most writers now think the short, protein rich grasses of the High Plains supported a population of 25-30 million animals; and was the center of the buffalo's distribution. The short grasses of the plains are 15-20% protein when dry, higher in protein than the taller grasses of the prairie. They reproduce vegetatively and withstand grazing better. Buffalo need a minimum of 1 part protein to 6 parts carbohydrate, so the plains in winter, with their cured high protein grasses exposed on windy slopes, might not have been such a bad wintering ground. Most northern mammals are adapted to cold; food is what limits their distribution.

Buffalo are also migratory. Their hump acts as a suspension system for their legs and lets them maintain a steady, rapid gait (a canter-gallop) for hours at a time. Their capacity for rapid long distance movement is an adaptation for a steppe environment, where grasses grow thinly and the forage in any one place is limited in amount, where the pattern of high quality grasses forms a mosaic, and peak forage quality occurs at different times in different places. So buffalo, like other migratory grazers, were here today, gone tomorrow. Animals that eat fibrous forage also need water daily; buffalo will eat snow and drink from springs, but water is sometimes distant on the plains. Buffalo hold their heads low, an adaptation to a short sward. They probably moved east to the better watered prairies in times of drought. (During long droughts the short grasses also moved east, taking over ground among the taller grasses.) Buffalo dig out seeps in damp depressions to drink from (several per acre in good environments; this waters the landscape for other creatures) and can smell surface water from several miles away. After drinking from a stream, they move away from it to feed and rest, being much less an animal of stream margins than cattle, who need more water and tend to hang around on riverbanks. (Cattle are native to forest edges and streams.) Buffalo probably stopped breeding during droughts, and death rates would have gone up, so the population would have fallen. But like the short grasses, buffalo are capable of responding quickly to improved conditions. Various lineages of buffalo have probably grazed the North American Plains (which as a habitat has also moved around) for the last 200,000 years.

Perhaps 1.5 million wolves lived with the buffalo and killed up to a third of the calves yearly. (This is not unusual: in parts of Alaska grizzly bears kill 50-90% of newborn moose.) So buffalo were limited by predation by wolves on calves; by predation by people, with whom the modern buffalo evolved; and by weather. Drowning, falls, grizzly bears and fires also killed some buffalo. Before the Plains Indians had horses, their take of buffalo is thought to have been 300,000 animals a year. These animals were mostly killed in the summer with drives or fire surrounds. A healthy population of buffalo can withstand predation of 7% of its breeding females annually without being reduced. Out of 30 million buffalo, this is about a million animals, if half the population was female. Probably the majority of buffalo killed were young females, whose meat was preferred and whose more pliable hides were easier to process. Early observers on the plains thought the buffalo calved only every other year, since only half the cows had calves; but healthy populations of ungulates calve every year, and this observation was probably the result of a natural calving rate near 80%, with a third of those calves being eaten by wolves.

Where buffalo numbers were limited, as in the eastern woodlands, they may have been overhunted, as they were near the agriculturalists of the Mississippi Valley. Areas inaccessible to hunters because of tribal warfare served as refuges, whose surplus animals drifted into the more heavily hunted areas. Buffalo were not alone on the plains; they shared the landscape with several other herbivores: white-tailed deer in the river valleys; mule deer in the more broken landscapes; the plains elk, a summer grazer like the bison (elk are probably a plains animal that now live in the mountains); bighorn sheep, now also an inhabitant of the hills; pronghorn antelope in the drier sagebrush habitats; various species of grouse, including sage grouse and sharptails; jackrabbits; and in the lighter soils, prairie dogs. Prairie dog towns occupied 20% of the short and mixed grass plains. So the precontact plains, so empty to the European farmer's eye in the 1880s, were marked out by their varying populations of plants (250 species), and the animals that ate them; by buffalo trails and wallows; by prairie dog towns; by buffalo traps; by the rings of stones that marked the encampments of the Hidatsa or Arapahoe; by waterways. On another level they were marked by long erosional processes; by the airsheds and flows of the atmosphere that determined rainfall; by sunlight; and by the stored carbon in their soils. Settlers arriving when the animals were gone commented on their expansiveness and silence. Buffalo trails were the basis of the Wilderness Trail from Virginia through the Cumberland Gap to Kentucky; of the rail line from the Potomac River to the Ohio; of the route of the Union Pacific Railroad up the Platte River in Nebraska.

By the early nineteenth century, the prairies, oak savannahs and woodlands of the Old Northwest, part of the Louisiana Purchase, were being surveyed by teams employed by the government. Unlike the animals, or the Native Americans who depended on them, and whose stories and foot paths went together with the pastures and trails of the animals, the surveyors marked out the landscape into rectangles, a mile on a side. The survey created Jefferson's enlightenment landscape, the marketable landscape whose pale squares cover the Middle West and plains as seen from the air today. The survey was another one of those breathtakingly simple schemes by which the West has taken over the world. The survey lines ignored topography and led to many difficulties for the men. Country considered home by the natives, who knew what times of day and what seasons to use different places, and how to get around, was often described as a hell by the surveyors, who went in a straight line from sun-up to sun-down regardless of the season, or of the country's abundant insect life. By marking out the territory as saleable land, the surveyors changed the lives of large animals like the buffalo. The survey meant the plains would be settled up to the limit of rainfall necessary to grow wheat: 15-18 inches annually, or approximately to the 100th meridian; that is, a line from Bismarck, South Dakota to Amarillo, Texas; a line that divides the humid continental steppe (subtropical in some parts of Texas) from the subhumid dry one; and the so-called eastern birds and mammals from the western ones. The exact position of the line can shift a few hundred miles east or west depending on rainfall. This line is also notable now in that west of it, on the high, dry, depopulated lands between the wheat fields and the Rockies, exists the only landscape in which it is possible to imagine the return of the buffalo.

The buffalo was one of the keystone animals of the plains. Large herbivores raise the productivity of grasslands and manipulate their vegetative structure. They create a varied habitat. A prairie, like a forest, consists of a shifting mosaic of habitat patches, the result of disturbance (by grazing herbivores, by digging rodents, by fires). Many prairie birds are adapted to the different habitat patches. Buffalo grazing and trampling created bald spots, and (as the vegetation recovered) patches of herbs. By digging wallows and pawing out springs they made surface water available for other animals. Grasses produce biomass more quickly than the dead matter can decompose; herbivores and fire (the megaherbivore) increase a grassland's productivity by speeding up the recycling of nutrients. By reducing the amount of flammable material, grazing makes fire on the plains less severe. (Most lightening fires on the plains occur during the growing season and burn less than a acre.)

Another plains herbivore was the Rocky Mountain locust. Rocky Mountain locusts are probably now extinct thanks to the plowing of the high river bottomlands where the insects bred. They are now studied in part by collecting the animals that thaw out of mountain glaciers. The overall effect of the locusts isn't known, but may have been substantial. A swarm 1800 miles long and 110 miles wide, consisting of some 6000 tons of insects was reported in 1875 in Nebraska. A typical swarm ate 50 tons of vegetation a day. (Five acres of a good corn crop, if the insects ate the corn to the ground, more like 25 acres in 1880, or—since the vegetation was not totally consumed—tens to hundreds of acres of grassland.)

Another grazer, one whose range corresponds rather closely with the High Plains Aquifer and the short grass prairie, is the white-tailed prairie dog. Prairie dogs were another keystone animal of the plains. In general prairie dogs prefer sandy soils and short grass, though the black-tailed prairie dog lives amid the taller mixed-grass prairies to the east. Their strategy for avoiding predators is to keep the grass about their burrows nibbled short and thus be able to see approaching prairie falcons, golden eagles, ferruginous hawks, foxes and coyotes; short grass did not impede the black-footed ferrets, which pursued prairie dogs down into their burrows. The constantly re-growing grass was also nutritious. Nibbling prairie dogs controlled invasive shrubs, such as mesquite. Prairie dog burrows reached densities of 50 per acre in prairie dog towns. The burrows went down 10 feet.

It is thought that 100-250 million acres, one fifth of the plains, had such towns, so prairie dogs were as abundant as passenger pigeons, with something like 5 billion animals. The animals moved tens of millions of cubic yards of earth annually; this reshuffling raised soil fertility. The towns with their short grasses and burrows also served as aquifer recharge areas. Water penetrated through to underground more easily than in undisturbed grassland. The towns' nutritious grasses, fertilized by prairie dog droppings, urine, new soil and insect parts, were preferentially grazed by buffalo. Where they remain, prairie dog towns are now preferentially grazed by cattle, though cattle gain slightly less weight grazing on prairie dog towns than on adjacent pastures. Grasses growing in the towns are reduced in quantity by a third to a half by the constant grazing of the prairie dogs, but are more digestible and nutritious. The prairie dogs needed a large grazer like the buffalo to help keep the grass low. Both they and the buffalo benefited from the constantly mown, fertilized grasses of the towns. The urine-soaked grass at the entrances to the burrows held salts needed by the buffalo. Prairie dog towns were refuges from grass fires, as the short, green grasses wouldn't burn. And prairie dog towns, like burned areas, would green up early in the spring.

The twentieth century has seen prairie dogs rendered virtually extinct on the American plains. Prairie dogs were removed because they were thought to compete with cattle for grass. (That horses broke legs stumbling into their burrows was apparently a canard.) Removing the prairie dogs greatly simplified the biota of the plains, by doing away with the animals that depended on the prairie dogs. These included not only their predators, but the box turtles, cottontail rabbits, skunks, burrowing owls, snakes, insects and amphibians that used prairie dog burrows. The yearly temperature 4 feet underground in a prairie dog burrow in the Oklahoma plains ranges from 50° F. to 80° F., compared to -10° F. to 120° F. at the soil's surface. Ten species of amphibians, 15 of reptiles, 101 of birds, and 37 mammals found food or shelter in prairie dog towns. The burrows were a sort of built environment on the grassland, a constructed oasis.

C. Hart Merriam, an influential nineteenth century biologist, pointed out that 256 prairie dogs eat the same amount of grass as 1 cow. What Merriam didn't realize was that prairie dogs and buffalo grazing together result in more total biomass (more grass-fed animal and more available grass) than the buffalo or prairie dogs grazing alone. Prairie dogs reduce the forage effectively available to cattle by only 4-7%. Eliminating them was never economic. The biomass of the prairie dogs was substantial. At 1 buffalo per 30 acres, an acre of grassland supported approximately 33 pounds of buffalo. At 50 animals per acre, an acre of prairie dog town supported 100 pounds of prairie dog. By comparison, the richest parts of the Serengeti in Africa, which have more rainfall, support 100 pounds of mammalian herbivore per acre (plus 50 pounds of termites). The sensible thing in this case would have been to manage the prairie dogs for leather and meat; or for the skins of their predators (coyotes, foxes, ferrets). Prairie dogs were probably not much limited by predation but by competition with each other and by bubonic plague, a microbial predator, which was endemic in the colonies. Now of course they are limited by federally funded trapping and poisoning programs.

It is unlikely that buffalo and prairie dogs limited each other's populations. Grazers that share a landscape tend to partition the environment, concentrating on different parts of it. Large grazers can move great distances (away from drought and prairie dog towns) and tend to raise the potential productivity of grassland, partly by grazing and trampling the grass, partly by temporarily storing grassland nutrients in their bodies and recycling them in feces and urine. (The nutrients in their bodies end up stored in the bodies of their predators and scavengers, and released in their feces and urine; or leach slowly from everyone's bones into the ground.) In general, once rainfall reaches 800 millimeters a year (about 31 inches), soil fertility declines because of leaching by rainfall. On African savannahs with large grazers soil nutrients peak at rainfalls of 900-1100 millimeters per year, thanks to the grazers. So a writer has called the plains piss driven. Grazing and trampling also promote regrowth and more rapid recycling of nutrients. Moderate grazing can double the productivity of grasses and increase their nutrient content and digestibility (young growing grass tips have more nutrients and are more digestible)—which is why prairie dogs didn't limit the numbers of the buffalo. The situation is a bit more complicated than this: many plant eaters leave growth promoting substances on the cut surfaces of the plants; cow saliva contains thiamine that promotes the growth of grass, mouse saliva contains epidermal growth factor, moose browsing encourages branching in the browsed plants. Thus animals and their food plants dance around each other's abundance.

The buffalo shared the plains with several other large herbivores whose total biomass was a not-inconsiderable fraction of theirs: elk, with an estimated precontact continental population of 10 million, perhaps 2-4 million on the plains (perhaps more), and an average body weight of 700 pounds; 9 million pronghorn antelope (out of a continental population of 30-40 million), with an average weight of about 100 pounds; 2-4 million bighorn sheep (no one really knows), at 200 pounds each; 3.5 million mule deer, abundant in lowlands, rough country, and river valleys; an unknown number of white-tailed deer, an eastern species that with other eastern species like the opossum colonized the plains up the river valleys; and millions of black-tailed jackrabbits, an abundant small herbivore with a strongly cyclical population, important in some Native American diets, especially west of the plains in the Great Basin. All these population figures are rough estimates. Most of these animals were shot out relatively quickly and some (especially the sheep) were susceptible to the diseases and parasites of domestic livestock. Perhaps the biomass of the prairie dogs was 30% that of the buffalo, and that of the large grazers also 25-30% of the buffalo's biomass.

The elk and bighorn sheep may have made altitudinal migrations in summer up to the pastures of the Rockies, or into the Black Hills of South Dakota, as they do today, thus effectively expanding the acreage of the plains (modern bighorn sheep grow larger if they are able to make such migrations); or they may not have and their modern movements may be an attempt to avoid humans. Some populations of both probably stayed down on the plains in winter, browsing in bad weather in the gullies and the timbered river valleys. Browse is important in the diets of both elk and sheep. Mule deer prefer rough ground, where they can escape wolves, and white tails were restricted to the river valleys. These rivers also had beaver until the 1840s, and some of the Indian tribes netted fish when they ran upstream in the spring. Antelope are primarily browsers, eat almost no grass and thus compete little with cattle or buffalo. A large part of the diet of the other large herbivores is also browse. Buffalo and elk prefer grass, but will eat browse. (Half of elks' winter diet is browse.) Pronghorn eat sagebrush (it is the basis of their winter diet), as well as various prairie forbs.

Cattle will not eat sage, though buffalo will: sage, a common component of American dryland ecosystems, has a higher protein content and more carbohydrates and fat than alfalfa. Many smaller prairie herbivores, such as sage grouse and black-tailed jackrabbits, also eat it. With their different eating habits, migration patterns, and environmental preferences, the large herbivores partitioned the plains habitat. Some direct competition existed, for instance between elk and buffalo for the standing dead grass of late winter and early spring. Such habitat partitioning used to be taken advantage of by old-fashioned farmers, who would pasture a small herd of sheep with a herd of cattle. The sheep would clean up the growth the cattle wouldn't eat and were thus "free." This was at a time when the income from 10-20 sheep mattered.

In general, a single grazer cannot use the natural biomass of an area efficiently. For this reason, innovative grass farmers follow a herd of cattle with a flock of chickens, which eat the grass the cattle didn't, and also the insect larvae in the cattle's dung. A mix of browsers and grazers would provide a higher yield of hides and protein in the dry rangelands of the Sonoran desert than cattle; on this basis some have suggested re-introducing the elephant, Chacoan peccary and llama (as well as the panther, lion and jaguar) to the Southwest. While buffalo by themselves are not efficient users of the plains flora, they are better at it than cattle. They have a more efficient digestive system and eat a wider range of plants. Buffalo are a dryland species, like gazelle, which are 3 times more productive on African savannahs than cattle. The lack of adaptation of cattle to the climate (they withstand drought and cold poorly), together with their effect on streams, where they browse down the willows, trample the banks, and near which they leave much of their daily 20 pounds of urine and 50 pounds of manure, makes them a poor choice of animal to raise on the plains. Several writers have pointed out that without agricultural subsidies for cattle and wheat, farmers would start raising buffalo: nothing else would make them money. (In North Dakota, a cattle and wheat state, agricultural subsidies constitute 80% of farm income.) A decade or more ago, two sociologists from New Jersey proposed a communal buffalo range on the drier, poorer parts of the plains, which are now more empty than when Native Americans hunted buffalo there.

Modern examples give one an idea of what a managed buffalo range would produce. The productivity of animal populations is commonly manipulated by controlling their sex and age ratios. In Custer State Park in Montana in the 1970s a herd of 1000 buffalo (the winter count of the population, following the fall round-up and slaughter) was kept on 73,000 acres. The ratio of cows to bulls was kept at 10:1. Buffalo are serial monogamists; that is, a bull stays with a cow until she is bred and then approaches another; this takes time and sets a lower limit on the number of bulls. So 10% of the adult animals are bulls, 60% are cows of breeding age, the remaining 30% are calves, yearlings and two-year olds. The breeding cows are slaughtered at 10 years. Some breeding bulls are also kept this long and sold to hunters as trophy animals. Bulls are reliable breeders at 3 years and still edible if slaughtered at 4 years. The manager at Custer State Park tried to keep 70-80 animals in each age class up to 10 years. At this stocking rate and under this management regime, the calving rate was 90%, with some twins; some calves were born to two-year old heifers (the usual age of first calf in buffalo is 3 years). Ranchers call this a healthy herd; that is, one in which the animals are capable of rapid growth and reliable reproduction. The herd produced 500 calves annually. With a 5% death rate, and slaughter at 2 years, the herd was producing 450 marketable animals a year (90% of its yearly increase), or about 1 animal every 162 acres. (Not all the two-year olds were slaughtered, some became replacements for older animals.)

When at one point Custer State Park let its herd get to 2600 buffalo, there was some deterioration in the range and the calving rate fell to 60-80%. A calving rate of 80% isn't bad, and with so many more breeders the yearly slaughter would have been considerably higher: such numbers explain why pastures get overgrazed. With the 1000 wintering buffalo in the park are 500 elk, 150 white-tailed deer, 120 bighorn sheep, and 120 pronghorn antelope. These animals constitute a total biomass of 35-40% that of the buffalo. So if we take these numbers as a rule of thumb, the plains could produce 1 buffalo carcass for every hundred plus acres, plus an equivalent number, at something less than half the weight, of the carcasses of other large herbivores. These numbers are based on a stocking rate higher than that of Custer State Park, but Custer State Park, in south central Montana, is very dry. (The carrying capacity of a habitat varies with rainfall. The National Bison Range in Flathead, Montana, is also dry, and runs 1 buffalo for 56 acres and also has a 90% calving rate.) Current estimates for a bison population on the plains are one animal per 24-30 acres (high for the drier parts of the plains, where one animal per 50 acres is probably more reasonable), or 1 carcass for double that amount. These animals are produced with no additional feed and very little manipulation of the landscape. The sex, age structure, and size of the herd is manipulated to increase its productivity.

The animals taken from the manipulated herd amount to 45% percent of its fall count (its "population"), compared to a sustainable take of 7% of breeding females (and say 10-15% of two-year old males) in a natural herd. Most of the animals slaughtered in buffalo ranching schemes are young males. Such manipulations of wild herds can create problems. When Canadian wildlife biologists limited the take of musk ox on Baffin Island to mature bulls, the survival rate of the younger animals was reduced. The bulls were necessary for some other function than breeding, perhaps for exposing snow covered lichens for the other animals to reach, perhaps to defend the herd against wolves. The Inuit had warned the biologists that killing the mature bulls wasn't a good idea, but they didn't express themselves in a biologically acceptable way; they said the large males "encouraged" the other animals. More generally, slaughtering males before they reach breeding age is a genetically risky strategy. One tends to remove the more aggressive animals. The result is that the animals become tamer, but it isn't clear what else is being removed. More intensive management of buffalo also destroys a herd's social structure and prevents buffalo from learning to be buffalo, something buffalo calves learn from their mothers, aunts and uncles. Elk managed for antlers and meat rapidly become domesticated and exhibit physiological, morphological and behavioral changes (the "tameness") associated with domestication. Silver foxes selected for neonatal characteristics such as floppy ears and round heads, as well as for behavioral tameness, become tame within several generations.

The modern bison was probably partly created by human predation. Killing large numbers of adult animals results in populations that breed at earlier ages (when their sizes are smaller). While one reason the modern bison is smaller than the North American steppe bison of glacial times is thought to be due to human predation, another is the warmer modern climate. (Large compact animals are better at conserving heat than small angular ones.) In fact, the modern bison is more closely related to the European wisent than to the steppe bison and may have come, like the moose, from Eurasia over the Bering Steppe with human hunters.

There are modern examples of the evolutionary effect of hunting. Hunters taking trophy rams from an isolated population of bighorn sheep in Canada (57 animals over 30 years) reduced the mean body weight of four year old rams from 200 pounds to 160 pounds and their horn length from 28 to 20 inches. (Big rams are hunted for their trophy horns.) Fish are extremely susceptible to size selective predation; experimental populations can rapidly be made smaller by removing the larger breeding individuals. (This also reduces the population's total biomass.) The effect is reversed by removing the smaller breeders. The point is that human predation on wild animals differs from that of animals like wolves. Wolves are cautious and take the young, the old, the starving, the sick. Buffalo mothers whose calves were not eaten by wolves may have been more wary or more fierce, letting these traits increase in the population. Wolf predation is thought to increase the size of elk, as wolves tend to take the smaller individuals, leaving the larger ones which are more difficult for them to handle. Evolution keeps predator and prey exquisitely attuned to each other, the healthy elk just able to outrun the vigorous wolf. Golden eagles take old, young, sick, and pregnant prairie dogs (a problem for predation theory: but pregnant females are probably slightly slower), as well as males obsessed with sex. Humans take healthy animals of breeding age, which creates a different selection pressure. Ideally, culling managed herds of buffalo for food should be random. If the point is to manage the buffalo as a genetically fit wild animal, one would have to see how the herds reacted to a scheme that artificially reduced the number of bulls, that was probably biased toward removing the more agressive animals, and that eliminated old animals altogether. Older females are important in a herd's social structure. (Buffalo can live 20-30 years but are usually culled at 9-10.) Elk and buffalo bands are traditionally led by older females, who have good spatial memories for food and water. Bighorn sheep have trouble colonizing new territories without an older female to lead them on their yearly migrations. In a new territory, no one knows the way around. So sheep management schemes must allow older "boss" females to remain as leaders.

Seton deduced from fur traders' records that the buffalo herds made yearly migratory circuits of 300-400 miles; the large herds were formed of small groups of 10-20 cows, calves, heifers and bulls to 3 years of age (the age of sexual maturity). Current thinking is that many small herds were non-migratory, while large herds (tens of thousands of animals), which were also composed of small family groups, moved all year. The plains are a very variable habitat, with rainfall varying from year to year and from place to place, and movement, even a more or less regular circuit, is a way to deal with this. (African pastoralist systems in which people and animals follow the grass tend to be more productive than fenced ranches in similar dry climates.) Different environments had different uses. Places with early growth thanks to an early season fire, or many prairie dog towns, were useful in the spring. (Early spring fires were set by Indians to lure buffalo; summer fires were less desirable but usually small, especially if the area had been grazed, and fall fires were a disaster because they destroyed the winter feed.) In summer, buffalo need shade and mud, in which they roll to coat their skin against biting insects. Their wallows, located in low ground, constituted another "built environment" on the plains. They were sources of water and mud for birds and insects, and also aquifer recharge areas: breaking up the sod helped water sink into the ground. In winter, buffalo favored broken ground with gullies and brush, partly to get out of the wind, partly because slopes are likely to have their forage exposed by the wind, partly because brushy browse is more accessible than grass in winter. River valleys constituted 7% of the Plains habitat and were important for the buffalo's survival. With their massive heads and forequarters buffalo are built to move snow out of the way. Their compact bodies and thick coats help them survive cold weather. While they go into a state of reduced energy demand in cold weather, saving energy by taking advantage of what topography and weather offer improves their likelihood of survival.

So what would a buffalo ranching scheme produce? If we take a reasonable estimate of the grassland's carrying capacity, from somewhere west of Bismarck, North Dakota, to the Rockies, of, say, 1 animal (winter count) per 30 acres, one could harvest 1 two year old animal or its equivalent every 75 acres, more or less. (In drier areas the harvest would be less.) Buffalo, being scarce, are worth more than cattle nowadays, but an animal should be worth $1250 (meat, head, hide, bones). From the same acreage one could take 1 other large herbivore. Since they vary so much in size, their value is harder to estimate—perhaps $500 for an average carcass. A gross return of $23 per acre isn't far from the net for dryland wheat, in the years in which it will grow in this environment (perhaps 1 year out of 3), and compares well with the returns from cattle. Of course trophy hunters will pay more for an animal: $2500-$5000 for a large buffalo, elk, or bighorn sheep.

A semi-natural plains means giving up some control over the animals on it. Buffalo are difficult and dangerous to handle, and also likely to injure themselves in close confinement (their skin tears easily and they fight to escape). Fences that contain buffalo are expensive, costing from $5000-$20,000 a mile, as are confinement facilities such as corrals. If the animals could be marked from a distance with marks that could be recognized at a distance (paint, implanted chips), they could be slaughtered (shot) and dressed in the field. Much of the offal could be abandoned to feed wolves, foxes and other scavengers and to return some of the buffalo's nutrients to the grassland. (By comparison little is wasted from cattle; of the 35 million cattle slaughtered annually in the United States, the equivalent of 11 million animals goes into products other than human food, much of it animal feed.) Slaughter would take place over several months in the fall and early winter. The offal (like the 5% of the animals that die naturally from accidents every year) would help support populations of grizzly bears, wolves, coyotes, foxes, badgers, bobcats, skunks, all animals once abundant on the plains. Some of these animals could also be trapped or shot in a limited fashion for their fur and flesh. Letting a part of the buffalo return to the soil better mimics the natural situation and also supports less obvious cycles of dependence and opportunism, for instance, early returning mountain bluebirds that feed on flies hatching from maggots in the decaying flesh of winter killed buffalo. Prairie dogs and their predators would be left alone; perhaps lightly harvested. One might spread some mineral nutrients to replace what is removed with the buffalo and grow some irrigated hay, or cut some wild hay, and leave the large bales about in favored situations for supplemental winter feed. (Large round bales remain usable for several years.) As with any herbivore, the health and longevity of buffalo depend on the health of their teeth, and reducing the wear on their teeth from eating coarse browse can double their productive life. The productivity of the grassland, and perhaps to an extent, the movement of the animals, could be manipulated by burning. Burning every second or third year maximizes grassland productivity. Early spring, when the ground is wet, and the dead grass eaten down, is the favored time to burn.

Any attempt to maximize yields of herbivores always brings one into competition with their predators and the question of whether to harvest predator or herbivore (here, wolf skins or buffalo meat). On a buffalo plains people would replace wolves as the predator limiting buffalo numbers; and people would also try to ameliorate the worst effects of the weather. The number of wolves would have to be kept low (say 50,000-100,000 rather than 1.5 million), so as to keep calf mortality from wolf predation low (5%?). A healthy wolf population would be one reason for letting some older buffalo survive. Wolf predation in general falls on old animals, as well as on young animals, or those weakened by injury, disease, or starvation. Such predation is not "waste" but a way of maintaining the buffalo's (and the wolf's) fitness. The remains of the animal go back into the soil. Like a regenerative wheat field, a plains with buffalo is a world in which economic growth has to accommodate the productivity of the underlying ecosystem: the grass, soil, rain, the behavior of the buffalo and that of the other herbivores and their predators and parasites.

Tourism would provide another income base, comparable to the $20-$30 an acre the buffalo, elk, antelope, bighorn sheep, mule deer, prairie dogs, wolves, bears, golden eagles, prairie falcons, sage grouse and the other animals and plants would bring. The plains are now largely empty, cleared of everything but cattle, irrigated hay, and grain, and some native herbivores, whose populations are perhaps 1-10% of precontact levels, and whose behaviors have been altered by fences, loss of habitat, predator control and hunting, and some of whom, especially elk and bighorn sheep, suffer severely from the diseases and parasites spread by domestic stock. (More cattle and sheep die from the poisons spread to control predators than from the predation itself, an example of the moral idea of a proper landscape—that is, one free of predators—trumping economics or even sense.) Suppose we set a goal of $10 an acre for tourism. If the average tourist spends $100 a day for room, food, gas, and so on, then a value of $10 an acre means 10 million people a year visiting for 1 night each, or more likely 2-3 million visitors for several nights each, probably concentrated in the better weather of spring, summer, and fall. Three million people visit Yellowstone National Park in Wyoming each year. They don't all stay there; but they stay somewhere nearby. Ten million yearly visits at double occupancy, with the rooms occupied 150 days a year, would require 30,000 rooms (8000-10,000 small bed and breakfasts) on the 100 million acres of the dry plains. Something like 70,000 extra people would be visiting the plains at any one time (2 per square mile). The stress so many people would put on the landscape would depend largely on how and where they went into it. A network of seasonal roads would be necessary to bring out the slaughtered animals; such roads would also provide trails for walking or riding. If most people stayed on the roads, much of the area would remain undisturbed.

A buffalo commons was proposed as a solution for the poorest areas on the plains, whose agricultural economies and population are shrinking. While such a scheme would never cover the plains as a whole, it might work for large sections of them. The reason for putting large sections of the landscape together to raise buffalo is partly because the enclosing fences are expensive, and one perimeter fence is cheaper than several smaller ones (internal fences aren't necessary for buffalo, who practice a sort of rotational grazing on their own), partly because fences that keep buffalo in will also keep other wild animals, such as wolves, grizzlies, mule deer and elk, in (or out), and partly to let buffalo be buffalo. A small herd of buffalo leading a more or less natural life requires 4000- 5000 acres. One can raise buffalo like cattle, on 100-200 acres, and make a profit, but then one has a ranching operation with a fairly dangerous wild animal. Some people do this successfully. Raising wild buffalo along with their natural commensals (wolves, prairie dogs, elk, grizzly bears) requires tracts of several hundred thousand acres. A hundred million acres amounts to about 156,000 square miles, or a square a little under 400 miles on a side. This comes to 1600 miles of fence (actually more, because the land won't be a square), or $32,000,000 at $20,000 per mile. Thirty-two million dollars is less than a year's agricultural support payments on 100,000,000 acres of farmland. Grasslands now constitute one fifth of terrestrial environments. The advance of grasslands is a rather recent development. Grasslands, like peat bogs, store carbon; it has been suggested that the coevolution of grasslands and grazers is one reason we live in a glacial age: together they remove carbon dioxide from the atmosphere and store it in the soil, continually lowering the atmosphere's level of carbon dioxide. Undisturbed grassland soils are 10% organic matter a meter down; forest soils are that rich to 10 centimeters. Most of the biomass of perennial grasses is below ground and the annual root turnover (the death and regrowth of small roots) continuously stores carbon in the soil. Grasslands store between 0.5 and 1 ton of carbon per acre per year. Storage of a ton of carbon (paid for by power plants, auto manufacturers, and other industries fueled by combustion; or more simply, by the extractors of fossil fuels themselves) is currently valued at $10-$50 a year. (Many writers say $20 a ton; the Norwegian government in 2000 was charging its offshore oil industry $38 per metric ton of carbon dioxide released. This charge made it worthwhile to sequester the gas.) The buffalo bring about $20 an acre. So a buffalo rancher who doesn't have a bed and breakfast or a wind farm (wind power being a major potential source of energy on the plains, but one that depends on the construction of new power lines and some sort of storage facilities), might, in a regenerative world, expect another $10-$20 an acre from carbon storage, for a total return of $30-$40 an acre, a yield considerably above cattle or dryland wheat.

Perhaps in a late industrial world, such a scheme would work. Similar ones to raise wildlife, harvest sustainable timber, or collect wild fruits in Kenya, Peru or Brazil rarely work. Part of the problem is ownership—who owns which buffalo, which wolf pelt? One can see a corporation made up of current landowners managing their section of the plains. Reluctant landowners whose land lay amidst others' would have to join, or be bought out, or fenced out. Land would have to be converted to shares, goals established, compensation for labor agreed upon. The corporation could invest in tourist facilities or license that development to members. The place of federal land in the scheme, if any, would have to be negotiated. As part of modern society, the gross return from the land should be sufficient to pay for a modern life: heated houses, higher education, medical care, pickup trucks. The infrastructure demanded by such a scheme, which reflects its integration into and value to the larger economy, is small: gravel roads, mobile meat-packing units, some sort of four wheel drive vehicles to haul the field-dressed animals to the packing units, trucks to haul away the meat, freezing plants to store the surplus; tourist accommodations. While all this falls far below the capital demands of farmland, it is comparable to those of raising cattle on grass, which is considered part of a modern economy. The buffalo would have to be kept to the plains, kept out of towns and cities, off major highways, and away from grain-growing areas; this could be done with fences and cattle guards, or perhaps other, smarter barriers would work: living hedges, ditches, highway overpasses, the incorporation of natural barriers like mountains, rivers or bluffs.

With the technology of the 1860s or 1870s such a scheme was also possible. The Native Americans were already managing the buffalo, but for a low rate of return, in modern economic terms: the feeding and housing of 120,000 people, plus a few hundred dollars income per household per year in robes. (A modern buffalo scheme would support a similar number of people with a modern life, a reflection of how the value of buffalo has risen.) Sustainable management of the environment and its animals was never the point of capitalism. Nineteenth century capitalist development broke apart the more or less sustainable systems of agriculture, fishery and forestry that had supported Europe and limited European economic development. If they grow (most can't), sustainable systems grow too slowly to make sufficient profit or to meet demand. One reaches a point where one can't raise the productivity of the biological system without compromising its long-term stability. Biological systems are capable of flipping up or down, to greater or lesser levels of productivity, and then remaining stable at those levels. North American cod stocks for instance, may remain stable at their current very low levels for a long time (several decades, centuries), until something (several good years for cod recruitment, several bad years for the fish that prey on small cod, temperature changes in the ocean favorable for food resources for larval cod, ocean currents favorable for spawning) raises levels once again. Or cod may never recover; their abundance may have been due to a combination of events that happened once, and won't happen again (the climate, the ocean, the currents may be too different). Once abundant, the cod could maintain themselves, but once reduced they may stay that way. Populations of moose are also capable of flipping in this way.

Lack of sustainability is also the downside of modern agriculture: under modern management, crop yields rise, but at the expense of soils, waterways, ground water, the health of the farmers, their neighbors and the sea, the long-term productivity of the landscape. Generating electricity from wind and sun presents new possibilities for plains farmers, but any sort of human industrial use, with its associated demands for water, power lines, roads, factories and people, would have to work around the existing ecosystem in order not to destroy it; the buffalo would have to be seen as part of an ecosystem that mattered. The natural inflation in land values that accompanies industrialization would tend to drive a buffalo-raising scheme out of business, as property taxes rose to reflect the land's potential industrial value. Restricting land use so as to maintain an ecosystem is not impossible (it is done in some cases now for agricultural land), but it means seeing the value of whole ecosystems as greater than that of their destruction for economic development. (Luckily enough, in this case buffalo and prairie dogs coexist well with windmills, though birds of prey and bats do not.) The economic costs of destroying an ecosystem are always large, but are distant in time and space, and so not a current worry. In an ideal world, terrestrial landscapes (prairies, agricultural lands, forestlands) would be valued at what they can produce in a biologically appropriate way; and the amount of land developed in a non-biologically appropriate way would be limited, and the effects of its development mitigated. Such schemes would not destroy the current economic landscape but re-arrange it; and probably require compensating monies to flow in many directions. At this point our accumulation of capital and fossil-fuelled power allows us to accomplish many things; we could still create an intellectually satisfactory, sustainable, modern material life; but with the end of cheap energy and the economic disruptions caused by climate change, this opportunity is likely to close.

The end of the buffalo came swiftly. The plains tribes had lived as nomadic hunters on horseback for something over a hundred years. Many of the famous tribes were immigrants to the plains; the Cheyenne and Dakota had been corn-growers and maple syrup makers in Minnesota and Wisconsin, until pushed west by the Ojibwa, who were being pushed by other tribes, by Iroquois expansion and by the Americans. Horses, introduced by the Spanish, began to be used on the plains in the 1680s and by the late 1700s all the plains tribes had horses. Twenty-five separate Indian nations, of 7 language groups, lived on the plains in 1800, their existence there (and to an extent their culture) a creation of Euro-American settlement. The take of buffalo by the Indians before the horse has been estimated at 300,000 animals a year. Many of these animals were killed in jumps or fire drives, so little selection by age or sex was involved. (One can guess from the normal composition of the herds, that cows of all ages, calves and young bulls would have been the animals killed.) Fire drives and jumps were sometimes wasteful; not all the animals killed could be used. But all those trapped were killed, since people feared that escaped animals would warn others about the trap. The take of 300,000 animals (estimates go up to 450,000) was well within what the herds could sustain.

The acquisition of the horse changed the life of the tribes and opened up the plains away from the large river valleys for habitation. Many tribes abandoned agriculture completely. Horses were owned individually, so the social structure of the tribes became less egalitarian. Wealth in general increased. Tipis became larger. On average, 4 times the weight in possessions could be transported with horses than with dogs, the former beast of burden. At a ration of 5 pounds of meat per person per day (the Hudson's Bay ration was 7-8 pounds) the 120,000 Indians on the plains in 1800 required between 720,000 and 840,000 buffalo. Hunts now took place from horseback and the animals killed were selected. Most of those killed were young cows. This reduced the population of breeders. The robe trade with Americans traders became important in the 1840s and 1850s and probably added 100,000 buffalo to these numbers. Robes were prime in winter, when the animals had their thickest coats. Young bulls were acceptable for robes, but cows, with thicker fur and more pliable skin, were preferred. The winterkilled animals usually weren't needed for food, and the carcasses were often abandoned, except for the hump and tongue. The robe trade further changed Indian social structures. While women prepared the robes (10-20 per winter per person), men kept the proceeds. A good hunter with several wives could accumulate wealth and influence. More available carrion from the winterkill may have increased the number of wolves, increasing the pressure on buffalo calves. Indian numbers on the plains during the nineteenth century rose from immigration as more tribes were pushed west by the Americans, but fell because of disease, mostly smallpox. There was a net fall of 50% in the population by the end of the century.

During much of the nineteenth century the buffalo herds were stressed by droughts. There was a drought on the southern plains from 1822-1832, and one on the northern plains from 1836-1851. The lack of grass would have interfered with the animals congregating in late summer to breed. Cows would have been in poor condition, fewer would have become pregnant, and fewer would have carried their fetuses to term. During the droughts one can assume buffalo numbers fell. The native take of perhaps 900,000 animals, mostly cows of breeding age, would have further reduced them, and reduced their potential for renewal. At the same time buffalo had to compete with another large grazer, the horse, whose numbers were increasing on the plains. There were 2.5 million horses on the southern plains by 1850. Buffalo's winter habitat along the valley of the Missouri River was being degraded by fuel cutting for steamboats and by the Native Americans collecting cottonwood branches and bark for winter feed for their horses. The spring grass along the Platte was being eaten by the oxen and cattle of the emigrant wagon trains, whose members gathered in early spring in Missouri awaiting greenup further west. Buffalo were also susceptible to the diseases of cattle. All this, plus the shooting that accompanied the westward movement of the Americans meant that by the end of the 1850s buffalo were considerably reduced; perhaps the population on the plains amounted to 20,000,000 animals, perhaps fewer. The buffalo had once drifted east to the better watered areas in times of drought, when their favored short grasses began to compete well with the taller grasses of the eastern prairies. As settlement progressed onto the prairies, animals that followed the grass east (once an adaptive behavior) were killed.

Buffalo then became part of the industrial world. Buffalo leather was more durable than that from cattle or deer, and harder to tan, but in the late 1860s new tanning techniques made bison leather workable. The British army favored the long wearing skins for footwear. The Americans found it ideal for the belting needed to connect steam engines to the machinery in their factories. There was a growing demand for belting in the economic expansion that followed the Civil War. A drought in southern South America in the 1860s shut off the usual supply of cow hides from Argentina. Finally, as a matter of public policy, killing the buffalo was seen as a strategy that would eliminate the plains tribes once and for all. When railroads reached the plains in 1867, it became possible to ship out the hides by rail. Then the commercial hunt was inevitable. The old growth hemlock forests of the Adirondacks and northeastern Pennsylvania provided the tanbark. (The logs, too heavy to float out, were abandoned in the woods.) Only hides and tongues were removed from the dead animals. By some accounts 80% of the hides were thrown away because of poor handling. The hunt lasted about 15 years. Soon afterwards rubber from Brazil came into use for belting, and soon after that electric motors made steam engines, with their maze of connecting belts, obsolete. Less than 1000 wild buffalo were left by 1889. According to an Indian story, the buffalo in their final days, distraught at their treatment by men, went mad, and raced round and round bawling in despair and pain, before returning to the center of the earth from which they had come. In the 1890s, thousands of tons of buffalo bones were collected from the plains and sent east to be ground up for carbon black (a filtering agent, used for purifying sugar) and fertilizer. The bones (by estimates based on their weight) amounted to those of 30 million buffalo.

With the buffalo gone and the Native Americans on reservations, the plains were opened to American settlement. Rainfall on the plains tends to go in 20 year cycles. The 1820s and 1840s were times of drought. In the 1870s, with buffalo becoming scarce, ranchers began to drive herds of cattle north from Texas to take advantage of the summer grass. Wolves, which, if abundant, can make cattle ranching difficult, and sheep raising impossible, were killed for a bounty (80,000 between 1883-1918 in Montana). Good rainfall years from 1878-1887 led investors from the eastern United States and Britain to invest in cattle ranching on the northern ranges. Then the blizzard of 1886 killed most of the cattle. Cattle are not adapted to winter on the northern plains. In 1886 they froze to death in the deep snow, suffocated as it drifted over their heads in the draws. They drifted before the wind and froze standing up. Unable to eat because of the deep snow, they starved.

At the same time the open range was being broken up as homesteading followed the railroads into the Dakotas, Nebraska, Colorado and eastern Montana. Much of this arid land was fertile and would grow wheat in a wet year. Drylands accumulate nitrogen since they are little leached by rain, and their vegetation, lacking water, is unable to make use of the accumulating nutrients. The first crops of the homesteaders were good ones. Dryland wheat in Oklahoma, in a year with sufficient rainfall, yielded up to 75 bushels an acre when the sod was first broken. After 80 years of cultivation, such lands now yield 45 bushels an acre under irrigation or 8 under dryland farming. (Eight bushels an acre means a gross income of about $24, or a net of perhaps $5 an acre; but $5 an acre over 10,000 acres comes to $50,000.) Rain was said to follow the plow, but the wet years were followed by dry ones. The southern plains had a boom in wheat growing during the wet years of the late teens and twenties (blockade of the Dardanelles by the Turks during the war blocked wheat exports from Russia and greatly raised prices). Then farmers chased falling prices and deteriorating weather by plowing up more grassland to grow more wheat so by the early thirties tens of millions of acres were bare. (By then farming on the northern plains had collapsed from drought.) Dryland farming on the plains results in soil loss (up to a foot a year) through wind erosion. Plowing and dryness tend to destroy the ability of the plains soils to flocculate and absorb water; unless kept vegetated, they turn to dust and blow away. For the most part wheat farming west of the 104th meridian was a failure except in river valleys or on irrigated ground. The drier land in the Dakotas, Montana, Wyoming that didn't have sufficient water to grow crops was slowly consolidated into large holdings of several square miles, used mostly for cattle pasture, and to raise some hay, and wheat in favorable locations if the year allowed. Much of the soil in the southern plains blew away in the dust storms of the 1930s. Where available, river water was used for irrigation and farmers who lived over the High Plains Regional Aquifer began experimenting with using the water below them.

Ground water pumping had begun in the 1880s where the aquifer was near the surface, with small acreages irrigated by homemade windmills. Such windmills were common in Nebraska, under whose soil lies two-thirds of the ground water in the plains. Irrigation systems for large acreages of commodity crops like corn using wells and pumps were experimented with throughout the early part of the twentieth century but it wasn't until the late 1930s that well drilling technology, with pumps powered by automobile engines, and guaranteed government loans (a gift of the Depression), put groundwater irrigation from the Ogallala within reach of the average farmer. The invention of center pivot irrigation in the 1950s made irrigation much less laborious. A single irrigation of 150 acres by hand from a central wellhead with irrigation pipe and ditches took 2 weeks of hand labor; the center pivot system, once set up, moved continuously on its own. Center pivot systems required the removal of the windbreaks and hedgerows that had been planted with government support after the droughts and duststorms of the 1930s, when dirt from the plains had fallen on Washington and drifted over the Atlantic to France. The trees broke the plains wind and helped stop the soil (once held in place by its cover of grass) from blowing away. The equipment's high capital cost ($50,000-$60,000 for a system that would irrigate approximately 130 acres, a quarter section, the corners being left out by the circular run of the rig) also made farmers that much more vulnerable to changes in the price of grain, or in the tax system. (For a while in the 1970s, the tax laws were all that made new irrigation rigs in Nebraska profitable.) Regular irrigation also leached pollutants down into the aquifer. Such pollutants included nitrogen fertilizer, herbicides and pesticides; gasoline and diesel oil; nutrients and toxic chemicals from feedlots and septic systems. Polluted runoff seeped out of abandoned water or oil wells into the aquifer directly. Wells throughout the plains have tested above the level of nitrates toxic to babies. Nitrates in rather low amounts are also toxic for cattle and hogs. The Agricultural Extension Service recommends that farmers pumping irrigation water out of the aquifer count its nitrogen content as part of their fertilizer applications. Life on the plains depends on underground water. Aquifers like the Ogallala are essentially impossible to clean, once polluted.

But irrigation of the plains worked. The crops were profitable, and so the plains were mined for water. (The Ogallala accumulates water from rainfall at a fraction of an inch a year, while pumping depletes it by feet a year. Over the whole aquifer, use, at 22.2 cubic kilometers per year in 2000, is 3-4 times recharge, at 6-8 cubic kilometers per year.) Where there was water, the plains prospered. Towns gained population. Property values rose. Different states had different laws governing the use of groundwaters. In Texas, a landowner could pump a well until it ran dry, whether the well was drawing water from under someone's else's property or not. So a Texas farmer could sink wells along his fencelines and pump the aquifer under his and the surrounding lands dry. Pumping in much of Texas, in eastern New Mexico and in Colorado has already depleted much of the recoverable water. On the plains as a whole the irrigated area peaked in 1980, when the water pumped from underground exceeded the flow of the Colorado River.

Only about 15% of the water stored in the Ogallala is economically recoverable for agriculture. The rest is too deep, too salty, or too difficult to extract. So a great deal of water will be left when irrigated agriculture on the plains is over in 20-30 years. It will be more expensive to recover, but water for household use (and perhaps for some industrial uses) is worth orders of magnitude more than water for agriculture, which must be literally as cheap as dirt, and thus is always provided with a substantial government subsidy, either directly or through the tax system. Groundwater pumping has already dried up much of the surface water habitat on the plains. Hundreds of miles of once perennial streams are now seasonally or permanently dry; springs are gone. Recently, local water management committees were formed to regulate depletion. Their rules vary mostly in the length of time they favor depleting the resource to zero. (Unregulated ground water pumping has gone out of favor because your investment in a well might be affected by your neighbor sinking several wells nearby and pumping out all the water too quickly for you to recover your investment.) The time scales for depletion range from a 1% loss a year in Kansas to a 20 year pumpdown in Oklahoma. One Kansas committee is planning to stop use shortly before economic depletion of the water.

When the water is gone, some areas of the plains will still be suitable for dryland farming. Current operators claim the use of deep plowing, which brings up heavy clays from the subsoil (and is economic because of cheap fossil fuels) will prevent another Dust Bowl in the next droughts. Perhaps the changes in patterns of rainfall with the changing climate will bring more rain to the plains; but this is doubtful. As far as I am aware, no rules exist to require revegetation of irrigated lands as the water disappears. Revegetating the landscape with native grasses is neither simple nor cheap. The two major grass species of the High Plains do not reproduce well from seeds and require irrigation to get started. They are more difficult to start in eroded, silt depleted soils, and do most of their propagation vegetatively, which allows them to withstand grazing better than most American bunch grasses, and to expand rapidly after droughts. (An introduced Eurasian grass, crested wheatgrass, stands summer grazing better than any of the native species.)

Perennial grasses, native or not, turn the soil into a carbon store. Soils store three-quarters of the carbon in terrestrial ecosystems, more than the plants that grow on them. Cultivated grassland soils in the United States are thought to have lost 30-50% of their carbon to date. Turning the plains once again into a carbon hungry ecosystem of perennial grasses and grazers, their restoration and maintenance partly paid for by the burners of fossil fuels, is a happy possibility. And using windmills to harvest some of the constant wind doesn't interfere with uses like grazing (buffalo and cattle seem to ignore windmills). A considerable proportion of the United States' coal reserves lies under the plains and thousands of acres of regraded strip mined land will be available to site wind machines. Irrigation agriculture and strip mining currently return the greatest income per acre on the plains; sustainable uses (grass, buffalo and wind power) would return the greatest total income on the plains over time, and a better income than dryland farming. As global warming worsens, and the place of people, soils and vegetation in the global context becomes clearer, perhaps attitudes will change, and industrial combustion will begin to subsidize more biologically appropriate uses of this landscape.

Chapter 16
Chesapeake Bay

Apart from trees, and the animals that lived among them, the fishy worlds felt the effects of agricultural settlement first and strongest. More extreme environments would join them in first showing the effects of industrial chemistry, such as mountain ridges, where ultra-violet light is strong, and where industrial aerosols are filtered out of mountain fogs by tree needles and accumulate in soils. Early in the twentieth century high altitude spruce forests in parts of the southern Appalachians were killed by acid rain from the steel industry in Birmingham, Alabama. Mountaintops in Central America concentrate the pesticides used in the lowlands, which rise on the heated air and condense in fogs; levels of pesticides on mountaintops can be 10 times those in the fruit plantations below, perhaps one reason for the decline of amphibians in Central American cloud forests (the others being night time warming which favors the spread of the chytrid fungus and the rising cloud level that dries the habitat). Then there is the Arctic, where thanks to atmospheric distillation and the constant movement of heated air north, industrial chemicals tend to end up. Fish swim in the denser fluid of water, which also accumulates chemicals running off the ground, condensing out of the air, concentrating in the surface layers of lakes and oceans; fish accumulate toxins through their food and absorb them directly through their skin and gills.

Fish suffered first and most from human settlement. Starting three and a half centuries ago in the United States, mill dams stopped fish spawning runs; and stocks that had formerly spawned above the dams died out. Sawdust from sawmills built over rivers covered spawning beds. Sawdust spread out over the bottom of Lake Michigan for several miles from Green Bay, Wisconsin; removing sawdust from the mills was always a problem, which running water removed. Silt eroded from fields; and later, industrial effluent, oil, stockyard waste, and sewage made rivers less and less habitable for fish. Fires on industrial rivers were common in the early twentieth century. But dams, the muddying and warming of rivers from agricultural erosion, and overfishing, reduced fish numbers long before pollution became a problem. Dams, overfishing and agriculture—that is, the destruction of riverine habitat—ended runs of Atlantic salmon in both North America and Europe; and overfishing in the salmon's foraging areas in the seas off Greenland and in their migratory pathways off the Faeroes and Ireland finished them off. The Atlantic salmon's population is now less than 1% of historic levels. There are 900 dams on New England rivers that once held salmon.

Land and sea are connected by rivers and flowing ground waters. Chesapeake Bay illustrates the effects of intensive settlement on the ocean edge. Chesapeake Bay is the drowned channel of the Susquehanna River. Created by the rise in sea level at the end of the last ice age, the shape of the bay was originally hollowed out by a comet. The presettlement bay was one of the most productive estuaries in North America. One of its keystone species was oysters, which by filtering the water as part of their feeding behavior kept the water column clear. In 1500 oysters grew on reefs built up of oyster shells that had accumulated over the last 4500 years. Sea levels had risen 3 feet per 100 years from 11,000-8,000 years ago, then slowed to 6 inches per 100 years by 4000 years ago. The more stable estuarine conditions let the oysters and other bay flora and fauna establish themselves and provided the resident peoples (resident then for several thousand years) with abundant, dependable acquatic resources. Oyster reefs formed a rough, irregular shelf platform along the edge of the bay and its rivers and creeks, protecting the shoreline and extending out to a depth of 6 meters (about 19 feet). Six meters corresponds with the layer of rapid increase in salinity and therefore of density of the water column in spring and summer. The reefs formed up to mean low water and broke the surface at low tide. Chesapeake natives fitted their canoes with a shallow keel to protect them from the sharp shells. Water draining off the reefs may have increased the vertical mixing of the water column in spring and summer, oxygenating the deeper waters more completely and reducing the vertical variation in salinity, letting the oysters extend their domain to greater depths. (Oysters favor brackish water.) Native Americans began intensive use of the biotic resources of the bay about 5000 years ago. Oyster harvesting increased 3000 years ago. Signs of large scale drying of fish and shellfish appear 1000 years ago. Biologically speaking, the bay in 1500 was new.

Oyster reefs make life easier for young oysters. After oysters spawn, and the fertilized eggs hatch, their larvae float about in the sea until it is time for them to attach and grow into adults. Attachment is the one big thing a larval oyster has to do. An oyster shell offers an oyster larvae a firm place to attach; and the young oyster uses the dissolving calcium in its ancestor's shell to build its own. Oyster reefs are home to hundreds of other species of estuarine organisms, including mussels and sponges. Many of these species are fish, food for fish, or food for organisms that are food for fish. In the 1600s, European ships entering the bay navigated around oyster reefs, pods of whales and rafts of sea turtles. Sturgeon leapt from the water, endangering small boats. The abundance of fish and marine mammals reported by Europeans in North American waters was once thought to be exaggerated, but paleoecological studies of Indian middens have shown the reports were probably true (skeletons in the middens indicated fish were abundant and large, though oyster shells at the bottoms of the middens were larger than those on top, and the fish caught shifted somewhat from predatory to herbivorous, showing that harvest by the native Indians over time had some effect). Later studies have shown that European riverine fisheries were as abundant as aboriginal North American ones up to about the year 1000, sea fisheries to 1300. The oyster beds of Chesapeake Bay were once capable of filtering the whole water column of the bay every 3-7 days. Chesapeake Bay averages 21 feet deep (20% of the bay's bottom is less than 6 feet deep) and the water cleared of algae and other small plankton by the filter feeders (oysters, mussels, clams, worms) let in the sun and allowed the growth of sea grasses and rooted underwater algae. Several species of aquatic grasses covered perhaps half the bay's bottom, that is 500,000- 600,000 acres, and helped oxygenate the bay during the summer months. Summer anoxia in the deeper parts of Chesapeake Bay is normal. It is caused by cold saline seawater pooling below the warm fresh waters from the rivers that flow into the bay. It was not a serious condition in the precontact bay. Since the bay is so shallow, winds and tides would break it up and move it around. However 15 million people now live in the watershed of the bay, and the equivalent of 4 million more are found in the 500 million broilers raised annually on the Delmarva Peninsula, which forms the bay's eastern shore. The result is that the waters of the bay are tremendously overfertilized. (If they still existed, the seagrass meadows would absorb about half the nutrients in the sewage effluent now entering the bay.)

Spring along the east coast of North America is accompanied by an algal bloom. The vegetation of the tidal marshes has been ground up by ice shelves working back and forth over them; and recycled by feeding waterfowl. The rivers bring down fresh sediment and nutrients with the spring thaw. As temperatures rise, the organic material decays, feeding the growth of algae, which respond to the rising temperatures by dividing more quickly. The nutrients brought down by the rivers include silicon (from sand), needed by diatoms, the major component of a healthy bloom off these coasts, to form their skeletons. The growth of the algae may start as early as February and peaks sometime in June.

A working modern ecosystem (modern in a biological sense) is capable of turning the nutrients transformed into algae during the bloom into the tissue of higher plants and animals. Filter feeders like sponges, clams and oysters filter algae and bacteria out of the water; zooplanckton feed on the algae, small fish feed on the zooplanckton, larger fish eat the small fish, and so on. Beds of annelid worms, living in the bottom mud, their filter feeding parts protruding (nipped off by feeding fish, they regenerate) take in the living and dead material that filters down. Some of the nutrients are used by the underwater grasses. Tides wash the bay's water back and forth over tidal marshes, whose plants are now growing and also absorb nutrients. By the time summer anoxia sets in and makes the bay more vulnerable to oxygen depletion, the nutrients have been absorbed by the system. Any excess accumulates in the bottom muds, where they are recycled into the water column over the year.

Cores from Chesapeake Bay show changes in its diatom population to species that tolerate increased turbidity as early as 1700, a hundred years after settlement began. The increased turbidity was caused by Europeans growing corn and tobacco in the watershed. Both tobacco and corn are demanding crops and required the constant clearing of new forestland. Trees would be girdled and tobacco or corn planted for 3 years; after that wheat might be grown for 2 years; then the land would be abandoned. Recovery took about 30 years. Tobacco planters sent their product out by ship (most plantations had a wharf on a river or the bay itself) and so were sensitive to the dangers of siltation caused by agricultural erosion. At first, tobacco, the cash crop, was grown by slaves or indentured servants using hoes on only the best, least erosive land. As more and more forestland in the bay's watershed was cleared for agriculture, and the bay's wetlands were drained for agriculture and housing, the load of nutrients entering the bay rose and became continuous year-round, not a springtime pulse. Growing wheat in Maryland and Virginia after the 1730s meant plows, draft animals, permanent pastures and fields, the spreading of manure, more land under cultivation. Eroding agricultural land sent more and more sediment into rivers. Sedimentation was at least 4 times pre-colonial levels from 1760 to 1860 and caused extensive shoaling in the upper parts of the estuary. (The Potomac River was 42 feet deep at Alexandria, Virginia, in 1794 and 18 feet deep—it would have been shallower except for dredging—in 1974.) Cutting the forest raised stream temperatures in summer and lowered summer water levels, both making life more difficult for anadromous fish (who in general need clear, cold water), and increased spring runoff by 25-30% over that of a forested watershed. In general, the higher the spring water flow down the Susquehanna, the greater the summer anoxia in the bay. More fresh water means more nutrients, more silt and more salinity stratification (warm fresh water sitting over cold salt water). Summer anoxia probably became a regular feature of the bay in the 1760s.

The growth of a market economy for fish in the early 1800s meant stocks of herring and shad were overfished. In 1833, 750 million river herring and 25 million shad were caught and salted from the bay; this number was reduced by 99% by 1878, mostly through overfishing; the fishery later recovered somewhat, but the condition of the rivers was too poor to admit much recovery. Toward the middle of the century, railroads and ice (for fresh oysters), and canneries, made it profitable to overfish oysters. Sail dredgers removed layer after layer from the reefs. Dredging went on day and night. Oyster fisheries in New York State and Connecticut, nearer large markets, were in decline by 1808, when they began to import spat (young oysters) from the Chesapeake to replenish their beds. (In 1600 the estuary of the lower Hudson, including New York harbor, had 350 square miles of oyster beds and perhaps half the world's oysters.) Chesapeake Bay produced an average of 15 million bushels of oysters a year throughout the 1800s, demonstrating the ecosystem's tremendous resilience, which was thanks in part to the habitat developed over 4500 years of occupation by oysters.

However production began to fall in the twentieth century with the physical removal of the reefs and oyster production is now 80,000 bushels a year.

Farming and dams also destroyed fish habitat. Dams held back sand, its silica needed by the diatoms (the base of the spring bloom), which was heavy and settled out. They blocked fish runs, preventing anadromous fish like shad and river herring from reaching their spawning grounds. Sawdust and silt covered fish spawning beds. During the nineteenth century chemicals from tanneries, paper mills, steel mills and chemical industries, along with sewage and silt, polluted rivers. Sublethal toxicity from pesticides, herbicides, various hydrocarbons, oil, heavy metals (arsenic, copper and lead were used as pesticides in the 1800s) affected the seagrassses and the fish and invertebrates, especially the animals' larvae. Tides resuspended the pollutants and moved them up and down the bay. (Like spring high water, tidal movement was probably an advantage when all the water was moving was nutrients, in a nutrient-limited system.) Some oyster reefs were dredged for shipping channels. Removal of the oyster reefs removed places for oyster larvae to settle and also shelter for other invertebrates and fish. The vertical structure of the reefs had increased water turbidity over the oysters, reducing their exposure to low oxygen levels, enhancing their feeding, and preventing them from being smothered by silt. Overfishing had eliminated the natural oyster fishery by early in the twentieth century. Artificial propagation kept the oyster fishery alive, at a much lower level, until a disease was introduced with asian oysters in the 1960s. With the oysters essentially gone, their ecological work of filtering the waters of the bay also ended and the abundant algal growth made the bay much more turbid. One of the remaining forage fish, menhaden, a plankton eater, whose feeding might have helped control the algal blooms, was fished down for animal feed (those chickens of the eastern shore). The menhaden were fished down to such an extent that when striped bass, a popular sport and culinary fish, began to recover after closure of the bass fishery in the 1980s and 1990s, the growing population of bass starved, and died of opportunistic infections.

By the 1950s the systems that had maintained Chesapeake Bay had started to collapse. In the 1980s nitrogen entering the bay from its rivers amounted to about 220 pounds per acre of bottom (a heavy dose for farmland; several tens of pounds per acre also entered from the atmosphere). Phosphorus amounts (partly from detergents, partly carried with eroded sediments) were also substantial. By then the oyster fishery and the eel grass meadows had collapsed and many fish (shad, striped bass, river herring, yellow perch) and shellfish (including the oysters) were down to 5-10% of aboriginal levels. In general the water quality in the rivers was too compromised, their summer flow too low, and dams too many, to support anadromous fish. The bay's food web started to shift from bottom dwelling organisms in a clear water environment (oysters, herring and their relatives, striped bass, eel grass) to a web dominated by microscopic algae, bacteria, and other single-celled organisms that live suspended in turbid water (a floating ecosystem), eaten by jellyfish. The normal way the ecosystem had dealt with the spring algal bloom no longer worked. The algae now grew until they ran out of nutrients, then died and sank to the bottom of the bay, where their decay used up the oxygen in the water's lower levels, always in short supply in summer. The lack of oxygen smothered the grasses on the bay's shallow shelves, already suffering from siltation and from a lack of light from the algal turbidity; fixed animals like oysters and mussels also suffocated. Much of the primary productivity of the presettlement bay was channeled through the seagrass meadows, which removed nutrients, provided shelter and breeding habitat for fish and invertebrates, oxygenated sediments and the water, served as food for tens of millions of overwintering waterfowl (those flocks of canvasbacks, redheads, black ducks, shot commercially from 1870-1910, whose droppings helped fertilize the bay). When the meadows were gone, primary productivity shifted to algae. The reduced tidal marshlands absorbed less of the silt and nutrients that washed with the tides back and forth across them. Eroding riverbanks sent down more silt. Mudflats as a habitat increased. More nutrients accumulated in the bay's bottom muds, were recycled back into the water column, and fed more algae. The siltation and the increasing summer anoxia, along with commercial trawling, destroyed the bay's benthos below about 22 feet (half the bay's bottom) so what was a garden of sea pens, sea anemones, sponges, corals, clusters of worms, rooted algae and other bottom-dwelling organisms, that oxygenate sediments, recycle nutrients, and provide places to feed and hide for invertebrates and fish, became bare anoxic mud. Toxic pollution was also now a problem but nutrient pollution, overfishing and habitat destruction, which had been problems all along, were probably worse problems.

Changes like this are now universal in lakes and shallow seas around the world. One reason the Atlantic right whale population is failing to recover is suspected to be excess nitrogen in Cape Cod Bay. The bay is a major winter feeding ground for the whales. (Sailing into the bay, the Pilgrims remarked on them.) It is thought microscopic plant growth in the bay overwhelms the zooplankton and other grazers on the plants (they can't keep up) and the resulting mix of plankton is no longer rich enough to allow these whales (filter feeders like the oysters) to gain sufficient weight to reproduce: the whales have been turned into vegetarians. A similar explanation has been put forward to help explain the collapse of fur seals in the Bering Sea: as the sea has warmed, fatty cold water fish have been replaced by leaner, warmer water fish; the new diet lacks sufficient calories for the animals (who, like the whales, spend much of their time in water near freezing).

Overfishing may also be a problem in Cape Cod Bay: algal blooms can be enhanced by changes in the abundance of different species of fish; that is, by fishing pressure. Small plankton-eating fish preferentially feed on the larger zooplankton that eat (more of) the algae; if the numbers of the larger fishes that eat the small fish are reduced by fishing, the small fish increase, and eat more of the zooplankton, and it is easier for the algal population to get out of control. So we have a new estuarine world, perhaps harvestable for algal broth (and whose overall productivity, boosted by fertile runoff, may be greater, but of less use to us). Jellyfish also harvest algae (and, preferentially, the zooplankton that eat the algae), and some shrimpers off the coast of Georgia have begun harvesting jellyfish rather than shrimp, which are depleted there; there is a market for jellyfish in Southeast Asia. Seas whose biomass consists largely of algae and jellyfish (the Black Sea is an example) are what we would call destroyed ecosystems, though several hundred million years ago they would have been normal.

Of course the usual modern changes also occurred in Chesapeake Bay's watershed. Most rivers along the east coast of North America carry substantial amounts of oil (poured down drains, washed off roads and parking lots). Polycyclic aromatic hydrocarbons in the oil and in the fallout from power stations and automobile combustion cause deformities in developing fish embryos that kill many of them before they reach adulthood: a tax levied by industrialization. (Birds pay a similar tax levied by lighted windows, cats, and cell phone towers, amphibians one from agricultural chemicals in surface waters.) Estrogen in sewage effluent changes developing male fish into hermaphrodites. Passage through electricity generating turbines turns migrating fish into stunned food for ospreys and eagles. Dams cut down on the supply of silicon needed by diatoms, a major part of the original algal system, to build their shells. This together with the excess nitrogen has shifted the population of Chesapeake Bay's planktonic algae toward the dinoflagellates, some species of which are poisonous and form so called red tides. Riverine sand and silt, like nitrogen and phosphorus, are both nutrients and pollutants; it's the timing and the amount that matter.

Part III
Future Landscapes: Atmospheric Agriculture and the Upland Ocean

Chapter 17

Fields

Farmers and loggers are the landscapers of the continent. The 900 million acres of the United States (about half the land area) that is cropland, rangeland and pasture is managed by less than 2% of the population. Farmers and loggers shape the larger landscape and determine its ecological effects: its vegetation, its rates of erosion, its downstream, downwind, down through time effects. Industrial chemistry and introduced species also shape the landscape, though more subtly, at least to the casual observer: these changes show up in changing patterns of abundance of plants and animals. For the most part, as some species of birds and trees disappear, others take their place; the forest (perhaps less stable, or less useful) still echoes with birdsong and still looks green. Developers also do their part, especially near rivers and coasts, where most people live. Changing patterns of terrestrial land-use propagate downstream through riverine marshes and wetlands and into the oceans, influencing the productivity of bays and estuaries, and thus of marine and estuarine fisheries. Seventy percent of marine fishes spawn in estuaries; and 80-90% of fish catches occur on continental shelves. Because of this, intensive crop production in upstream river valleys, with the resulting erosion and leaching of nutrients, can reduce the total amount of food a landscape produces, through agriculture's effect on associated riverine and marine fisheries.

Much of the American landscape is settled in an anachronistic pattern. Buildings and industrial installations are set close to surface waters; or to ocean harbors. Roads follow river terraces, the paths of least resistance; these landscapes were carved out by glacial meltwaters and leveled by subsequent downcutting by the river and accumulation of streamside sediment: roadbeds prepared by river geology and time. Rivers were also useful for floating logs, for running water-powered mills, as sources of water for people and animals. Some rivers floated boats, which were smaller in earlier centuries than now; large birch canoes (the "North Canoes" of the Hudson's Bay Company) held several thousand pounds of furs, worth paddling halfway across a continent. Heavy loads could be floated more easily than carried by wagon, horse or foot. Later, canals followed rivers, but by the middle of the nineteenth century, railroads had made river and canal transportation obsolete. Only public funding keeps river transport alive.

But labor, whether human or mechanical, lets one ignore topography. The cut of a nineteenth century railroad drives west through the Pennsylvania countryside, rising slowly above sealevel, its right of way dug out by hand. It runs almost entirely below the surface of the ground: the point was speed. The towns along the little Adirondack river where I live cluster along the water, about 6 miles apart. Three miles is an easy trot by horse, a moderate walk for a man. The towns were once trading settlements for farmers in the floodplain and up the side valleys, now for the most part they are simply sites of human habitation, property whose value is found in its remoteness, as automobiles and modern roads (artifacts of machine labor) have made most of the previous economic infrastructure obsolete. People shop in supermarkets 30-40 miles away, at stores whose headquarters are in Los Angeles or Holland, and children are bused to schools located sometimes in a village, sometimes merely centrally, that is, socially and economically speaking, nowhere. Sometime in the twentieth century trucks made log drives on the local river obsolete (good news for the fish); while the rich farmland of Ohio, and the irrigated vegetable fields of California (their water provided at 2% of its real cost by the government), together with refrigeration and railroads (also government supported), made much of northeastern agriculture obsolete.

We live in a new world: still with abundant energy from fossil fuels, and still with remnants of the original, preagricultural ecosystems that once covered the planet. Such ecosystems are relatively recent as wholes; especially in the temperate zone, they are post-glacial, post-modern-climate, 6-15 thousand years old, depending on the place. These ecosystems were assembled quickly because they were assembled out of much older pieces. Some of these pieces now don't work very well; for instance the North Temperate trees that now occupy the Mediterranean basin shed their leaves during the mild and damp Mediterranean winter and then leaf out and grow during the Mediterranean summer drought. These trees colonized the area as the glaciers retreated; and as the climate became more and more dry, they became less and less well adapted. They still work, but would be out-competed by better adapted vegetation, transplanted from someplace else. Similarly the American Southwest lacks species of browsers that could make use of its difficult vegetation; and forests throughout the Americas lack large mammalian fruit-eaters and tree-breakers (the elephants and their relatives) and must rely on birds to eat their large fruits and fire to renew the forests. A diverse mix of browsers and grazers would produce more animal flesh per acre in the dry rangelands of the Southwest than cattle: should the elephant, the camel, the Chacoan peccary, the llama, panther, lion and jaguar (all once resident here) be reintroduced to North America? Instead we introduce into such landscapes the crop plants and animals of the ancient Middle East, including our house cats: the domesticates of our current civilization.

Roads link the industrial landscape, which is our landscape. Through roads it penetrates the countryside. Every road has its effects: a minor one is that songbirds avoid road noise; and those that nest near roads have smaller clutches. (Like people, in whom road noise causes sleeplessness, compromised immune systems, and a slightly elevated risk of heart attack, they find road noise stressful.) Chemicals that end up on roads include rubber dust that contains polycyclic aromatic hydrocarbons, zinc, and cadmium; brake-lining dust containing copper, nickel, chromium and manganese; lubricants containing cadmium, copper, molybdenum, vanadium, zinc; motor oil; brake fluid; antifreeze and unburned hydrocarbons from fuel (the last two, like cadmium and the aromatic hydrocarbons, are cancer causers and mutagens). Phosphorus and nitrogen also run off roads, in surprising amounts. Nitrogen levels in national forestland near Denver are 10 times those in more isolated forests. Much of the nitrogen is thought to come from automobile traffic, that is, from the burning of gasoline. Much material in road runoff could be contained in properly maintained stormwater ponds, or vegetated swales, though proper maintenance would be labor-intensive and expensive. (Recycling the metals and nutrients might help pay the costs: for instance, the concentration of platinum from catalytic converters in road dusts would make mining them profitable.)

Roads interfere with natural drainage patterns, flooding areas above them; they interrupt subsurface flows and lower water tables (roads are an efficient drainage mechanism); they change subsurface flows into connected streams; they isolate animal populations. A population of bank voles in southern Germany separated by a fourlane highway cross it so seldom that the animals on each side are becoming genetically distinct. Grizzly bears, especially females, also hesitate to cross roads, with the same effect. But roads also provide habitat: butterflies sun on them; moths fly along them; bats hawk for insects above them in the evening; migrating woodcock and robins search for earthworms in thawed roadsides in spring; northern flickers forage for ants; ravens cruise for roadkill; winter finches pick up grit; young loggerhead shrikes seek beetles near their edges and are run over by cars (one reason the species is in such steep decline). Turkey hens bring their poults to forage for grasshoppers in roadside edges; singing amphibians are crushed on wet nights (some populations forced to migrate across roads to breed have been exterpated, others provided with tunnels); alien plants and animals follow roads long distances to new environments. The drying effect of roads extends a hundred yards into the forest, as does the effect of salt spray. Invertebrate variety is reduced in pools near northern roads, probably because of the use of road salt. A million vertebrate animals a day may die on North American roads (or 10 times that), and untold numbers of insects: a writer has compared cars to filter feeders cruising through the insect world. Thus roads extend the industrial world into the countryside.

Despite what another writer has said, we have not reached the end of nature. We have discovered that the world turns out to have larger limits. (It is not after all so subduable.) It has limits in terms of the gases its atmosphere can absorb, such as the excesses of the greenhouse gases carbon dioxide, nitrous oxide and methane. These gases are produced by the combustion of fossil fuels, the burning of forests and grasslands, agriculture, domestic animals, land development, oil drilling and coal mining, as well as by lightning, beaver wetlands, lakes and digesting termites. It has limits in terms of the compounds of chlorine and bromine that interfere with the stratospheric ozone layer. (Nitrous oxide from agricultural fertilizer also has a place here.) This layer shields the green world from ultraviolet light from the sun. Chlorine compounds as a whole have increased in the stratosphere by 10 times since the 1950s; one reason the ozone hole keeps growing despite the phasing out of the worst of the offending chemicals is that new, useful chlorines and bromines are constantly invented; and people resist giving up old ones, such as the agricultural fumigant methyl bromide, used to sterilize soils in which crops like strawberries are grown. (In most such cases, substitutes exist, such as soil sterilization by sunlight or crop rotation in the case of strawberries.) There are limits to what the green world can absorb in atmospheric aerosols of sulfur and nitrogen, in heavy metals, in hormones or hormone mimics in sewage water, in mutagens, or chemicals that attach to DNA; limits to the amount of carbon dioxide the oceans can absorb without reducing the ability of their plants and animals to make calcium carbonate shells (necessary for organisms like corals, clams, some algae); limits in what new plants, animals, insects, fungi can be spread among the different continents and seas without seriously simplifying existing ecosystems; limits on how much an ecosystem can be broken up and still function.

We will not reach the end of nature until we are all dead. But the natural world to which we are attached is threatened. Many of its larger animals are going to go extinct, and populations of its less charismatic members continue to collapse (such as the amphibians, whose bodies form the base of many terrestrial food webs). The natural world our economic system is constructing will be less abundant, less joyful and work less well. The costs of its decline, in terms of ecosystem degradation and climate change, may amount to 20% or more of world economic production. We can probably deal with this. Much will depend on how we handle the transition to a poorer earth. Our success as an animal (biological success is measured in numbers, and humans are the most abundant mammal in their size class ever to inhabit the planet) has brought our influence on the planet to levels of long-term geological processes, of climate, of cataclysmic events, of the processes dominated by microorganisms.

But microorganisms are still in charge. The mass of microorganisms is probably greater than that of all other organisms put together. The domination of the earthly landscape by microorganisms ("Slimeworld") lasted from about 4 billion years ago to 500 million years ago, when the metazoans (multi-cellular animals) evolved and ate up all the available slime. Slimeworld retreated to underground rocks, to deep ocean muds, anoxic wetlands, water surfaces, to soils and the atmosphere, and remained greatly reduced on the earth's surface, which was taken over by green plants. But the archaea and bacteria of slimeworld still play the defining role in making the earth habitable: in maintaining oxygen levels; in manufacturing cloud-seeding chemicals; in fixing and unfixing nitrogen gas into forms plants and animals can use. Modern trawls of genetic material of the ocean's surface indicate there are thousands of species of unknown bacteria present in very small numbers. Such organisms, now just hanging on, are ready to expand in numbers when conditions change. Their ability to swap genetic material among species lets them evolve quickly. Interferring with processes on this level is more risky for us; at the least, real estate values fall as lakes overfertilized by septic tank effluent or agricultural runoff are taken over by algae and cyanobacteria, eutrophicate, and stink.

———————

Farming is a complete break with the earlier landscape. Most native plants and animals are eliminated. As a rule of thumb, modern settlement reduces native flowering plants and vertebrate animals by 90-95%. With clearing for agriculture, water runoff increases: flood levels rise, but summer flows are lower. The water table falls. Soil and nutrients run off with the water and also blow away. Wind speeds near the ground are higher. In temperate regions the ground is left bare for 7 or more months a year. On bare ground the snow melts more quickly. The ground also dries out and warms up faster in the spring (a help to farmers). Soil microbial activity and nutrient turn-over and loss are higher in the warmer soils. Leaching of soil nutrients by rain increases. Cultivated soils also lose carbon, which adds to the carbon dioxide to the atmosphere. After a few decades of cultivation in temperate regions, soil organic matter falls to half its level in unfarmed soils. Levels of soil organic matter fall more quickly in the tropics. This makes the soil more erodable. When bare ground is covered with snow, its reflectivity is high, compared with forestland, so the ground gets colder and the frost goes deeper. Where I live water pipes should be put 4 feet below open ground to avoid freezing, deeper if the ground is compacted (as under a road). But frost in the woods is rarely a foot deep, probably partly because the winter water table in the woods is so high, partly because the trees shelter the ground from the wind and from radiant losses on clear cold nights.

The ecological effect of agriculture on watercourses, or on the plants and animals of a landscape depends on its extent. Limited fields, in a larger natural landscape, like those of the Native Americans or of the early period of English settlement in New England, create a more varied environment, with more edge habitat, and more game animals, which are for the most part creatures of the edge. The typical eighteenth century North American subsistence farm, with its weedy shrubby fencerows, different crops in different small fields, uncultivated woodlots, marshes and steep or rocky areas, provided habitat diversity. Much of the burning of Native American peoples was intended to create the berries and browse of early successional landscapes that game animals favor. Forests of nut-bearing trees were created by fire, and people harvested the nuts, and the animals that ate them. Slash and burn agriculturalists in tropical forests over time modify much of their forestland whose soils are suitable for agriculture; but slowly, with long rotation times between clearing a plot again. A few years after clearing crop yields fall dramatically as the nutrients in the soil and in the ash from the burned vegetation are used up. (In Indonesia rice yields fall 80% between the first and second crop.) Native plants (weeds) take over and the grain, tubers and vegetables in a garden are replaced by planted food trees. Those are replaced 20-40 years later by forest trees; so much of the forest seems like primary forest to the untrained eye. Some of it would be, since not all soils were cultivable. But in the Amazon, for instance, much more forestland was probably cultivated forest (including planted trees) than first thought; man-made *terra prieta* soils may occupy 10% of the landscape and forests with a high proportion of useful trees, many planted, another 20%. The Iroquois of central New York State practised a temperate variety of swidden agriculture. With perhaps 10,000 acres under cultivation yearly, they might have cultivated a considerable percentage of the light upland soils they considered usable over the thousand or so years of their occupation (the fields also had to be near good village sites and sufficient supplies of firewood); but because of the time involved, their overall effect on the larger environment was small (though local effects may have been considerable). Fish, whose abundance is a sign of healthy waterways, provided a major part of their food. Slash and burn systems, like any agricultural system, break down when the human population gets too high. Fifty people per square kilometer—which seems a very high number—has been called the limit in Indonesia. Since the population of the Iroquois was small (12,000-20,000 people on several million acres), their overall effect was little different from the effects of the animals and plants which whom they shared the landscape. The feeding habits of elephants and moose also alter landscapes;

so do earthmovers like prairie dogs. Elephants destroy acacia forests in Africa and leave behind them grassland, habitat for grazers (including cattle), which over time succeeds back to acacia forest.

In such earlier worlds, with much smaller human populations, the influence of people on landscapes was more like that of other beings. In general, plants control the distribution of resources across a landscape; nutrients accumulate under shrubs in deserts, making life difficult for grasses growing in between; most of the nutrients in tropical forests are found in the vegetation. Plants control fire cycles (some encourage them, some discourage them) and the height of water tables; in Australia, eucalyptus trees, by taking up rainfall, prevent salt from deposits lower in the soil from migrating upward and make life in the upper levels of the soil more favorable, and surface runoff into rivers less salty. In the northeastern United States nestholes dug out by yellow-bellied sapsuckers are made possible by fungi feeding on aspen trees. The nestholes are subsequently used as winter shelter by chickadees, nuthatches and white-footed mice. Beavers turn small streams into a series of shallow ponds, transforming the watery habitat into a teeming paradise of freshwater invertebrates, fish and amphibians. The ponds neutralize acid surface waters, trap sediment and agricultural nutrients, and improve the habitat for songbirds and wolves. Deer and meadow voles influence forest succession by browsing some tree seedlings preferentially; the seed-eating white-footed mice prefer the large seeds of oak and pine (but probably little influence those trees' abundance). Squirrels may affect the evolution of nut trees by selecting, burying and abandoning the largest, thinnest shelled nuts. Kangaroo rats in the Chihuahua Desert of Arizona suppress large seeded grasses by eating their seeds; their feeding helps maintain the shrub dominated desert. Plants set the scene: plants change levels of atmospheric gases; modify water and nutrient cycles; engage in chemical warfare; promote or suppress forest fires; provide shade; and alter wind speeds, relative humidity, the penetration of frost and a landscape's albedo.

Agricultural people manipulate the landscape by changing the plant cover, by hunting grazing animals (who also affect plant cover), and by introducing grazing animals of their own. Where agriculture becomes extensive, the natural habitat is reduced and simplified, and agriculture becomes the dominant influence on the environment. Erosion rates on cultivated crops are 10-100 times those in forests. Numbers vary. A study of Devonshire clay loams found that heavily grazed permanent pasture would absorb 0.1 millimeter of rain an hour (very little: a quarter inch a day), bare ground 4 millimeters, freshly seeded ground 11 millimeters (a heavy rain, about 0.4 inch an hour), freshly plowed, uncompacted soil 50 millimeters, and woodland soils 180 millimeters an hour—this is a flood, 7 inches an hour; the figure is hard to believe. The pasture must have been very compacted because generally land in pasture or cover crops absorbs 2-4 times the rainfall of land in cereal grains.

Measured erosion rates from the United States range from about 60 pounds of soil per acre per year for undisturbed forest (that is, almost nothing, 4 bucketsfull), to 600 pounds per acre per year for grassland (still extremely low), 7 tons (14,000 pounds) per acre for cropland, about 12.5 tons per acre for logged forest (compared with 60 pounds per acre for unlogged), to hundreds of tons per acre for an active surface mine or a construction site. The maximum soil losses compatible with sustainable agriculture (where soil loss equals the rate of soil formation) are site specific, and vary from 0.4-0.8 tons per acre to 8 tons per acre; but generally losses greater than 4 tons per acre result in a loss of topsoil and thus also of fertility (some writers put this number higher, up to 10 tons per acre, or one large dump truck load of dirt). Four-tenths of a ton per acre (one ton per hectare) means loss of an inch of soil every 250 years, while an erosion rate of 4 tons per acre means loss of an inch of soil every 30 years. The yearly cost to the farmer of a loss of 4 tons of topsoil per acre is small (40-50 pounds of nitrogen, 10 pounds of phosphorus, plus a small loss of soil rooting depth and water holding capacity). The farmer's cost in replacing those nutrients is probably smaller than the cost to the municipality or utility in whose reservoir the soil ends up. But the farmer's cumulative cost (the costs rise every year) is high.

Erosion rates vary with rainfall and land use. Topsoil is maintained by a balance between erosion and the disintegration of the underlying rock. A soil's nutrients come from weathering of the rock, the decay of plant and animal material, atmospheric fallout and biological fixation. Rocks are dissolved ("weathered") by the carbonic acid in rainfall; and by the acids secreted by plant roots, soil fungi and microorganisms. They are broken up by frost. Small bits of rock are reduced further by the acid in the guts of worms and by the grinding action of earthworm gizzards. As Darwin pointed out, where worms are abundant they shape landscapes, raising the soil level, burying surface features, and moving soil downhill. The topsoil in abandoned English fields consists of worm casts and rock fragments. If more soil is produced than erodes, the soil thickens and eventually buries the underlying rock beneath the reach of active soil forming processes. Then soil formation stops. So the thickness of a soil is characteristic of a certain landscape (its underlying rocks, its slope, rainfall, plants, animals, microorganisms). Some soils are deposited by wind, such as the loess soils of the north temperate regions, formed during the last glaciation, which created the great grain-growing regions of North America and Eurasia. Soils, landscape, and plant cover evolve together, often with characteristic animals and fish. A soil's living organisms affect its structure, color, chemistry, the way water flows through it. The phosphorus and sulfur in soils comes from weathering of rocks, the nitrogen from biological fixation of gaseous nitrogen in the atmosphere. The slow accumulation of organic matter in soils binds them together and makes them harder to erode. Rocks rebound as the weight over them lessens (so hills remain much the same height as their soil erodes) but erosion slowly shapes landscapes. Once a landscape has become heavily agricultural, the continuing effect of agriculture on it depends on what crops are grown, how the crops are cultivated, and where natural land is allowed to remain.

From the 1850s to the 1990s about 2.5 billion acres of forests, grasslands and wetlands were converted to farmland worldwide, with tremendous effects on natural habitats and watercourses and also effects on global climate. The land conversion produced 30% of the anthropogenic carbon dioxide added to the atmosphere during that time. Currently agriculture contributes about 25% of the gases causing global warming. These include nitrous oxide from fertilizer, methane from rice paddies and the digestive tracts of cattle, sheep and goats, carbon dioxide from cultivated soils. While little new land has been cleared in the United States since the 1950s, cultivated American landscapes have been more and more cleaned up. Since colonial times, every increase in cash or animal power on the farm has meant more changes in the original landscape: wet spots in fields, or small marshes, drained; woodlots (no longer needed for timber or fuel) cultivated. With the agricultural prosperity that followed the Second World War, the agricultural landscape became much less diverse. Home orchards and woodlots were converted to cropland, hedgerows between fields removed. The loss of woodland was hard on birds. Many year round resident birds of central Iowa shun blocks of habitat less than 35 acres and need 150 acres for successful breeding. Neotropical migrants that breed in Iowa need several thousand contiguous acres if the area is to be capable of producing a surplus of birds and not be just a sink for surplus populations from elsewhere. In woodlots under 500 acres in Iowa exterpation and recolonization by migrating songbirds is common (over 75% of the ovenbirds in woodlots less than 360 acres are unmated). So the birds come and go.

The average square mile of farmland in Ohio and Illinois in the 1940s contained over 10 miles of hedgerow habitat (most of Ohio and about 40% of Illinois was forested when settled). Hedgerow trees were animal and wind planted. Their trees included black cherry, elm, hickory, oaks and honey locust. Chokecherry bushes and native shrubby dogwoods lapped up against the trees. The hedgerows slowed winds and sheltered birds, small mammals, and pollinating and predatory insects; they protected small drainages. Along those former hedgerows, fiberglass and steel have now replaced the rotting wooden fence posts, in which red-headed woodpeckers once hollowed out nest holes. The holes were occupied the next season by bluebirds, who raised their broods on leaf-eating worms scavenged from the corn (far enough from the barn to escape the attentions of the aggressive English sparrows). Pollinators no longer fly out to pollinate seed crops, leaf eating caterpillars no longer feed on weeds, predatory insects no longer capture aphids, parastic wasps no longer lay their eggs in the bodies of various beetles and caterpillars. As crop prices stayed low or fell, while the costs of farming rose, small farmers failed and were bought out by more prosperous farmers, banks or insurance companies. Farms became larger and more capital intensive. Most crops need attention at the same time of year and larger fields were faster to plow and cultivate. So more hedgerows were removed and small in-field wetlands were drained or filled, removing more habitat and also the ecosystem services of aquifer recharge and nutrient removal. Streams were rerouted to square off fields. With no vegetation to hold their banks, they turned into silty drains. The silt ended up in downstream waterways. (Much of Iowa drains through channelized streams and ditches.)

As agriculture became more commercial, cultivation also became more intensive. Perhaps half the agricultural landscape of the Middle West was in sod or sod crops from the 1860s-1950s, some to feed farm animals, some to sell; as tractors replaced horses in the 1950s, soybeans replaced sod crops and pasture. Crop rotations were reduced or eliminated as fertilizers and pesticides let farms specialize in row crops. (Row crops are crops seeded in straight rows, the rows spaced so the plants can be cultivated with a tractor.) That it takes the manure of 1 cow to fertilize 2 acres, or the manure of 80 cattle to farm a quarter section, 160 acres, put a certain limit on farm size. Farmers with the machinery to cultivate more land needed more cattle or synthetic fertilizer. As farming became non-organic, farm animals went elsewhere. From the prairies of the Corn Belt, cattle moved west to feedlots on the Great Plains and were fed irrigated corn; chickens went to the Southeast to live in huge chicken houses, some hogs stayed in the Middle West, but were now raised in confinement. (Many also went to the more moderate climate of the Southeast.) In the 1990s, 75% of Nebraska cropland was in continuous corn, much of it irrigated with water pumped from the Ogallala aquifer. Thirty percent of cropland in Iowa and Illinois is in continuous corn, 60% in two years of corn, followed by one year of soybeans. Much of the farmland in both states is drained. Drainage greatly improves farmland, resulting in better soil aeration, faster decomposition of organic matter, deeper rooting of crops, a warmer soil. It also increases rates of water and nitrogen runoff by several times. The nitrogen comes both from applied manure and fertilizer and from the stored nitrogen in the soil; losses of stored nutrients in drained soils take decades to deplete. Both corn and soybeans are very erodable row crops; and corn rotated with soybeans can result in much higher rates of erosion than continuous corn. Such cropping is only possible with large applications of fertilizers and pesticides. Fertilizer use was given a boost by the Second World War. During the war hydropower plants along the Tennessee River were used to manufacture ammonia for explosives; after the war the ammonia, now surplus, was marketed as fertilizer. Agricultural ammonia is currently synthesized from natural gas and shipped as a liquid by pipeline from plants in Texas, Oklahoma and California. Factories that had been making tanks were retooled for new markets in cars and farm tractors; pesticides were derived from nerve gases developed for war; and the industrial foods of field rations (dehydrated potatoes, processed cheese, powdered orange juice and coffee) entered the modern diet.

Corn and soybeans replaced the mixed farm with its cattle and hogs. Soybeans fix nitrogen but only enough for the crop itself. The traditional four field crop rotation of the American Middle West consisted of legume hays, corn, oats and winter wheat. Approximately 25% of the land at any one time was in sod (including small grains as sods brings the number to over 50%). The animals (including the work horses) were fed the hay, and some of the grain; some grain was sold. Feeding grain (converting it into animal flesh) was a way, through an additional investment of capital and labor, of increasing its value. The corn and oats were manured; and livestock grazed the new growth of the winter wheat in the fall and the stubble of the harvested grains. With the animals gone and the land in row crops, erosion and soil degradation have increased, with negative effects on rivers, lakes and connected ocean waters. Soil fertility declines under synthetic fertilizers unless crop straw or manure (some sort of carbonaceous material) is added to the soil. However, keeping animals and rotating crops into legume hays reduces the amount of land in the (easily marketable) cash crop. With rotations into sods cropping systems become more complex (and more thoughtful and interesting); more animals and farm equipment may be required. Marketing new crops and animals presents new challenges. Cultivation of corn and soybeans was intensified in an attempt to increase profits; a capitalist farmer and his banker try to maximize the returns on capital investment: in this case, land. (But until recently—2008—commodity crops did not return much income per acre.)

More intensive cultivation has many effects. Hayfields, which once provided breeding grounds for grassland birds, are now mowed too frequently for the birds to finish a reproductive cycle; so meadowlarks, bobolinks, upland plovers, savannah sparrows cannot breed in modern hayfields. Partly as a result, grassland birds are among the most endangered in the United States today. Pollinators such as bees, moths and butterflies (already deprived of hedgerow habitat) cannot feed on a flowering crop, since most recommendations are for cutting at 1% bloom. Isolated patches of prairie orchids are pollinated by moths that must cross miles of chemically treated fields to reach them, so many go unpollinated. The situation could be improved by timing mowings so as to give the birds a sufficient period to breed; this means a lower forage quality in the first cut, which in my area would have to be harvested about two weeks later. (I'm not sure what the cost in lost quality would amount to, perhaps \$5-\$10 per acre, that is, \$5000-\$10,000 over 1000 acres.) Some nearby habitat, such as the grass waterways meant to capture water and nutrients running off the field would have to be left unmowed until mid-summer to provide cover for the young birds. Since birds return to specific nesting sites each year tricks like this might work; though overflow populations that tried to breed in adjoining, conventionally managed fields would be unsuccessful.

Modern commercial agriculture depends on applied chemistry, applied genetics and machine power: the use of artificial fertilizers, pesticides, irrigation and crops bred to produce large yields with large inputs of water and nutrients. So-called organic or regenerative agriculture depends on applied biology. In any agriculture, but especially in an organic agriculture, soil microbes (the mostly unknown archaea and bacteria) and soil fungi are the basis of productivity. A small part of soil organic matter, soil microbes hold its largest pool of labile nutrients. Many crop plants have connections with mycorrhizal fungi, which supply them with water and micronutrients, in return for sugars. This connection is suppressed by the use of nitrogen fertilizer, under whose influence the plants withhold their secretions from the fungal hyphae. Thus artificially fertilized plants are more suscepible to micronutrient deficiencies and to drought. Many crop plants also have associations with nitrogen-fixing bacteria, which supply them with nitrogen in return for root secretions. (Plant breeding could increase this connection.) In an unfertilized soil the nutrients contained in dying soil microbes feed plants. The microbes also recycle nutrients in dead (harvested) and dying crop plants and make them available to the new crop. (In general we use only 4% of the mass of a crop plant; the rest—stems, leaves, roots—is wasted or recycled in the soil.)

The soil ecosystem as a whole (fertilized or not) depends on microbes and organic matter, its source of nutrients. Less organic matter in a soil means less activity among the soil biota, less nutrient cycling, and less vigorous growth of crop plants. Organic matter helps soils retain moisture and improves their tilth (polysaccharides secreted from soil fungi glue soil particles into clumps); the chemistry of soil organic matter helps liberate nutrient particles from clays. Both crop rotations and manure raise soil organic matter. Manure also raises nitrogen levels in soils. Much of its nitrogen is temporarily immobilized by soil microbes, while synthetic nitrogen fertilizer tends to be quickly lost from the soil. Approximately half of applied nitrogen fertilizer is used by crop plants, the rest is lost to surface runoff, ground water and the atmosphere. A 15 year study of growing corn and soybeans at the Rodale Institute in Pennsylvania found no significant difference in crop yields among plots fertilized by rotation into legume sods, those fertilized with manure, or with synthetic fertilizer. The soil carbon content of the manured plots increased three times, and of the plots rotated into legumes five times, over the carbon levels in the conventionally farmed plots. Other studies have found organic agriculture producing yields comparable to conventional agriculture in normal years, but 30% higher in dry years, and greatly reduced levels of erosion in plots where the farming practice included rotation into legume sods. In two wheat farms in the Palouse region of Washington State, one organic since 1909, one plowed in 1908, but farmed conventionally since 1948, wheat yields were similar. The organic farm grew a green manure crop (usually alfalfa) every third year. The conventional farm lost six inches of topsoil from 1948 to 1985. The organic farm retained its topsoil, which had twice as much organic matter, more moisture holding capacity, more available nitrogen and potassium and a greater density of microbes than the conventional farm. Of course, because the conventional farm didn't rotate its fields, it produced more total wheat over the period, but at the cost of its long-term viability.

One innovative farming scheme, so-called rotational grazing, lets animals rather than people harvest the grass. Grazing in this case mimics what happens on a natural grassland: less hay is cut for the animals, few if any row crops are grown, and more land goes into pasture. The animals are kept on pasture for as long as it is available (7-12 months, depending on the climate). Each pasture is grazed intensively for several days and then left for approximately a month to regrow. Birds use the pastures, which also have a much more varied invertebrate and plant life, and probably function better as whole ecosystems than mown hayfields. (Function may partly depend on the plants and animals: the structure and water holding capacity of the natural prairie established within the circle of the Fermilab Nuclear Accelerator near Chicago is dramatically better than that of nearby planted pasturelands. Cattle graze the pasturelands, buffalo the prairie.) Rotational grazing schemes produce meat and milk from grass, which cattle are adapted to eat, rather than from corn and soybeans, which in large quantities make cattle sick. The animals' manure and urine fertilize the pasture and add to their organic matter; the grazing raises the productivity of the grasses. In some schemes, chickens are brought in a few days after the cows leave. The chickens eat insects and some of the remaining grass and scratch apart the cow patties to get at the hatching fly larvae. So the chickens spread the manure and keep the fly population down and the farmer harvests eggs and chicken flesh from his pasture. Joel Salatin's Virginia farm produces 35,000 dozen eggs, 10,000 broiler hens, 40,000 pounds of beef, 30,000 pounds of pork, 1200 turkeys and 1000 rabbits on 100 acres of grass and 450 acres of forest and a small amount of purchased corn. The farm was eroding and worn out when Salatin's father bought it several decades before. Now erosion from the grassland is essentially zero. (Loss of nitrogen from pastureland is approximately 35 times less than that from fertilized row crops.) In a nice twist, Salatin periodically scatters corn over the surface of the manure that builds up in his cow barn over the winter. In the spring pigs are let root in the manure for the fermenting grain. The pigs have a good time, and in doing so turn the manure and speed up the composting process.

Organic farmers are interested in keeping nutrients, including carbon, in their soils. Expensive to add, soil nutrients provide their livelihood. Some writers claim organic agriculture could absorb 8-17% of U.S. carbon emissions. On the other hand, farms that don't rotate crops or spread manure must use chemicals. Manufactured nitrogen fertilizers raised farm output substantially when they came into common use. Much of the doubling of the world's output of food from 1950-1980 is thought to have come from the application of artificial fertilizers. Fertilisers both increased yields and enabled the growing of continuous crops of grain on existing farmland. But perhaps half the increase in output came from increased irrigation, and some from bringing another billion acres of land into cultivation from 1920-1980 (mostly in South America). Plant breeding also raised output. American corn yields rose 72% from 1975-2005, mostly thanks to the development of higher yielding varieties, many of which needed large amounts of nitrogen. The American corn crop has grown steadily at 2% a year for some time, while the land under cultivation doesn't change much. Nitrogen fertilisation raises the nitrogen content of leaves and makes them more attractive to leaf eating insects, which become more vigorous and fecund on their rich diet. Fertilisation also seems to lower the production of phenolic compounds in the leaves. The phenols act as insect repellants and, paradoxically, as anti-oxidants, beneficial to us, when we eat the plants.

Growing the same crop in the same field year after year lets populations of pests and weeds build up, sometimes catastrophically. In the Red River Valley of North Dakota, growing continuous crops of wheat during the twentieth century resulted in outbreaks of fungus disease so uncontrollable that wheat had be be abandoned as a crop. In general, continuous grain requires pesticides and herbicides. In the 1990s American farmers used almost a billion pounds of pesticides on their crops. Farm chemicals have effects on farmers (farm families have among the highest rates of cancer in the United States) and on wild animals; they kill non-target insects and invertebrates, including pollinators, and insects used as food by birds and other animals. Increased use of agricultural chemicals including dieldrin, DDT, malathion and atrazine have reduced populations of the northern leopard frog in the United States by 90%. The chlorinated hydrocarbons reduce the frogs' immune systems almost to zero; thus they are easily parasitized; this results in the large population of deformed animals. Atrazine also interferes with the frogs' sexual development. Dieldrin and DDT are now banned in the United States and Canada, but still manufactured and exported, and still arriving on the air from Africa, Asia and South America: freshwater bogs in the eastern United States are still accumulating DDT from the atmosphere.

The tidying up of the countryside (plowing to field edges, spraying field margins for weeds, draining wet pastures) that machine power and farm chemicals made possible eliminated much of the habitat of the harvest mouse in Britain between 1970 and 1990; the animal's food plants and insect prey were gone. (So the mammals and birds that fed on the mouse also decreased.) Pollinating insects and the insect predators and parasitoids of crop pests require a complex architecture of hedgerows, wildlands, crop mixes and weedy field edges to survive in useful numbers. They need food and habitat apart from that provided by the crop plants. Evergreen blackberry plants at the edge of California vinyards provide shelter and prey for a wasp parasite of vine leafhoppers and keep their numbers high enough so that when winter is over and the leafhoppers once again become a problem in the vinyard, there are enough wasps to make a difference. In fruit orchards of the Pacific Northwest, strawberry plants and wild roses provide winter habitat for wasp parasites of leaf rollers. A few Canada thistles along the edge of an Iowa field provide places for the diseases that attack its leaves and roots to overwinter; and a place for the painted lady butterfly (whose larvae eat thistle blossoms) to breed. A field without insect pests is also a field without their predators, a situation knowledgeable farmers find alarming. One needs both pests and predators. It's a question of relative numbers: of competitive adjustment.

In the Middle West, river floodplains and wet upland prairies were settled last; the forested lands on the slopes were thought to have the best soils. In an ideal world, the floodplains would never have been settled. While they are usually good agricultural soils, they are also places where nitrogen is removed by denitrifying bacteria from surface or ground water. Carbon rich, damp sites at the junction of aerobic and anaerobic conditions are major sites for denitrification. Such sites are usually found along the riverbank but may be located all over a floodplain, and throughout upland fields. Floodplains also remove topsoil and nutrients from water flowing over them (so-called "overland flow"); floodplain forests shade and cool streams (and the groundwater that flows beneath them into streams). Sufficient forestland along streams will convert much of the overland flow into subsurface flow and remove most of the nutrients in the water; so the water entering the stream will be cleaner and cooler and its movement into the stream not so rapid—that is, more like the original stream conditions. Estimates of the width of the strip required to perform such services vary from 60-200 feet. The strip should contain any ponds or wet depressions connected to the river. Studies in Tennessee show that 6% of an agricultural watershed in strategically planted forest (most modern temperate landscapes were once forested) will convert half of the agricultural runoff into subsurface flow and remove a considerable proportion of the excess nutrients. Studies in North Dakota show that 15% of the landscape in wild land will support a population of wildlife sufficient to allow hunting and thus bring in additional income. Such lands also support insect and bird life helpful to the farmer: Eugene Odum recommended 40% of any landscape be left undisturbed; some organic vinyards in California leave half their land in wild land. (Such vinyards also use falcons to scare away starlings, golden retrievers to sniff out female vine mealybugs, and chickens in mobile coops to control cutworms and weeds.) Unplowed grasslands along streambanks also help slow water flow and remove nutrients; so do grass waterways that carry surface flow from fields; both can be mowed or lightly grazed and maintain their value as nutrient traps.

All the same, some nutrient pollution from farmland is unavoidable. Heavy rains (especially in winter, when the ground is frozen) tend to channelize the flow of water, reducing its contact with fallen branches, leaf litter, logs and grass stems, and thus send more runoff, with its nitrogen and phosphorus, directly into streams. Some of the nutrients and water may be stored and removed by downstream wetlands connected to the river—another reason to preserve large parts of the natural river. River bottomlands are important habitat for wildlife and also serve as corridors for animals and plants moving among different habitats. The species diversity of plants is greater in wild lands connected by corridors, probably because animals that spread seeds can move more easily among them.

Such numbers give us a means of designing new landscapes. Restoring semi-natural landscapes along streams is probably essential for stream health. Such landscapes are also useful for the farmer (as nutrient traps, sources of pollinating and predatory insects, habitat for birds and animals that control plant-eating insects and rodents). They are not extensive enough to restore the plants and animals of the original environment; but this is not an argument against them: like current farmland, these are new landscapes. For instance, the neotropical birds that once nested along northeastern and midwestern rivers need thousands of acres of unbroken forest for successful nesting, partly because of nest predators that prefer edge environments (skunks, feral cats, opposums, raccoons, crows, jays—all more abundant because of the absence of large predators), partly because the forest near an edge is drier. One study found that the mass of soil insects and invertebrates in the interior of large blocks of woodland in ovenbird habitat in southern Ontario was 10-36 times greater than that near the woodlands' edge. So apart from losses caused by nest predators and nest parasites like the brown headed cowbird (also birds of fragmented habitat), ovenbirds find better foraging and raise more young in large blocks of forest.

Since the effects of the edge (changes in air temperature and humidity, in ground moisure levels) penetrate from 100-200 yards into the woods (some scientists now say up to 1000 yards, which makes all but enormous blocks of woodland edge environments, but the effect runs along a continuum), bands of woodland 50-200 feet wide along a stream will not provide good reproductive habitat for many neotropical migrants (the most common and abundant birds of the eastern forest), and thus the habitat will lose the ecosystem services they, if abundant, would provide. A full compliment of birds in the eastern deciduous forest, by helping to control grazing insects, is calculated to be worth about $7 an acre, or $5,000 per square mile of forest per year. (Probably an underestimate; but using those figures songbirds are worth about $1.5 billion per year to the forests of the eastern United States.) Self-maintaining populations of large animals like elk, wolves and mountain lions would probably need more habitat than protected riverine areas would provide. (While Eurasian wolves survive in rural Tuscany and mountain lions on the outskirts of Berkeley, California, North American wolves in lightly settled parts of Canada move out when the human population rises above three per square mile.) But protected stream corridors would provide habitat for many songbirds, small mammals, amphibians, predatory and pollinating insects, as well as other invertebrates and some larger mammals and predatory birds. Owls (nesting in owl boxes) and raptors (nesting in tall forest trees) might become a natural part of the farm scene (preying, say, on English sparrows, starlings, rats and feral pigeons).

While the purpose of the streamside strips is to control the temperature and flow of water into the stream, remove excess nutrients, and to provide habitat for animals and insects useful to the farmer, riverside wildlands may also provide income from hunters and trappers and from municipal water systems wanting to reduce the flow of nutrients and silt into reservoirs; they also give the landscape connectivity. For all these purposes, the wider the strip, the better. The nutrient removing function of riverine lands is largely accomplished by bacteria and depends on soil temperature, soil carbon levels, and the degree of aeration, as well as on uptake by plants; strips should be calculated so as to capture most of the nutrients flowing off a farm. Riverside wetlands are capable of removing 90% of the nitrogen and 80% of the phosphorus in groundwaters flowing through them. (Most of the phosphorus is not dissolved in groundwater but attached to fine clay particles and must be filtered out of overland flow by the surface of the ground.) In an ideal farm landscape, all lands functioning as nutrient traps would not be along streams; some would occupy usable farmland. Some farm crops, like hazelnuts grown for oil in Minnesota, can also function as nutrient traps; so can tree crops, such as pines or poplars grown for pulpwood.

Farming systems, whether conventional or organic, matter in stream health. Natural streamside filter strips can be overwhelmed by large amounts of nutrients (especially phosphorus), and even if they remove 90% of the nitrogen that moves through them, 10% of a very large number is still a large number. So controlling the nutrient levels in rivers and lakes also means modifying agricultural processes. Erosion under cultivated annual crops (corn, soybeans, lettuce) on gently sloping land (slopes of less than 5%) may be 10 times that in forests, several times that in grassland; on steeper slopes, erosion under row crops is several hundred times that under forests. Erosion of soil, and of the nutrients and pesticides carried off with it, is a concern for rivers and lakes; for water supply reservoirs, which must supply high quality drinking water at a reasonable price and whose lifetimes are measured in the amount of silt that enters them; for estuaries, such as Chesapeake Bay and San Francisco Bay, whose marshes, underwater meadows, and beds of worms and molluscs are susceptible to siltation, and whose waters now receive several times the nitrogen of an average fertilized farm field. Erosion is also a concern for the sustainability of agriculture at a site. Fertility declines about 6% with each inch of topsoil. Topsoil is being lost by American agriculture at about 6 inches per century. Iowa loses 0.5 inch a year. In Georgia and Tennessee, losing 6 inches of topsoil reduced crop yields by 50% (somewhat more than expected). Fertilizer also runs off fields: a 40% reduction in the nitrogen now carried by the Mississippi is necessary to shrink the dead zone in the Gulf of Mexico. The cost of reservoir siltation alone likely exceeds the cost of agricultural practices that reduce erosion. But farmers don't pay that cost, or—except on Conservation Reserve land—receive compensation for using practices that reduce siltation. Generally in the United States, a dollar invested in soil conservation yields $3 in increased crop yields and saves $5-$10 in the external costs of farmland erosion, such as dredging rivers, building levees, replacing dams and other flood control works.

Erosion limiting agricultural practices include crop rotation; terracing and contour planting; strip cropping (strips of row crops alternating with strips of sod crops); minimum tillage (seeding into the stubble of last year's crop); stubble mulching (not plowing fields in the fall, but leaving the stubble, preferably with a cover crop seeded into it, until spring); restoration of hedgerows (which break the wind); plowing of grain straw back into the soil (incorporation of straw in temperate climates leads to a 40-50% increase in the soil biomass of carbon and nitrogen over 20 years and much improved soil tilth). In dryland farming, the appropriate degree of tillage helps control erosion, maintain soil porosity (and thus the ability of water to infiltrate the soil) and increases the root density of the crop plants. Soil conservation trials in Texas, Missouri and Illinois slowed soil loss by 2-1000 (!) times and increased yields of cotton, corn, soybeans and wheat by up to 25%. The costs of soil conservation are paid back by increased crop yields, while over the long term erosion reduces the farmer's income. A small soil loss, 6 tons per acre, means a loss of 110-130 pounds of nitrogen and 11-18 pounds of phosphorus per year, and the loss of an inch of topsoil in 20-25 years. The missing nutrients will have to be supplied by the farmer. One-half of the fertilizer used in the United States replaces nutrients lost by previous erosion. Erosion over a site specific amount also results in a loss in the soil's rooting depth, in its water holding capacity, its degree of aeration. Erosion of farmland soils is now 10-40 times the rate of soil formation. Such rates of erosion have meant the end of former civilizations. It seems reasonable to think that agricultural activity should not generate soil movement beyond that for which cultivation of the land can compensate. In the end, soil fertility is a matter of public interest. (But people have been saying this for 3000-4000 years.)

The movement of soil downhill is ineluctable; what matters is its speed of motion. Soils exist in a dynamic balance between depletion (through leaching, erosion, gaseous losses, removal of nutrients by plants and animals) and renewal (through weathering of rock, decomposition, nitrogen fixation, inputs from the atmosphere and from plants and animals). Natural soils vary in their success at maintaining this balance. (Thus soils are thicker or thinner.) Farmland topsoil is re-created by incorporating crop residues and manures, by catching the dust that falls out of the air, by adding rock powders (ground limestone, rock phosphate), by plowing in sod crops, by slowly incorporating the upper layer of the subsoil (itself renewed from the rock below).

The sediment, phosphorus and nitrogen that enters streams in hilly prairie landscapes can be reduced by 80% by crop rotation that incorporates hay crops and small grains (wheat, oats and barley function as annual sods, though with runoff rates up to 8 times that of a prairie); by adding 300 feet of grassland buffer along streams (not a minor amount, but the land can be grazed or used for a late hay crop); and by feeding animals on pasture rather than in feedlots. Legume hays not only reduce erosion; they also restore soil tilth and water-holding capacity and increase the soil's pool of available nutrients. Wheat and corn rotated with clover hays yield 30-50% more per acre. The clover, or clover-grass mixture, is pastured, or harvested as hay. On flatter prairie landscapes, the same practices also reduce nutrient runoff and the buffer strip along the streams can be narrower.

Rotating land into legume sods reduces the land in the cash crop (usually soybeans or corn). Hay crops bring animals back to the farm. Corn and soybeans are for the most part grown to feed animals. About 65% of the American corn crop goes for animal feed, while about 40% of grains and soybeans grown worldwide goes to feed animals. Another way to use hilly prairie landscapes is to raise cattle (beef cattle or dairy cattle) on grass, by rotational grazing. Rotational grazing divides up the land into paddocks, which are grazed for short periods (three days to a week), and then rested (for perhaps a month); the manure and urine from the animals provides fertilisation, and the resting period lets the grasses regrow. Each paddock requires a supply of water for the animals. Paddocks are periodically rested for longer periods and not grazed until the end of July; this lets them reseed, the roots of the grasses develop, and results in better long-term productivity. It also lets some native bunch grasses survive and makes better habitat for grassland birds. Surface water ponds in the prairie pothole region of Iowa, Minnesota, and South Dakota once provided 30 million acre-feet of water storage per year, water that now flows through drains into streams and rivers feeding the Mississippi. The region was a major North American habitat for breeding waterfowl, a "duck factory." Much of the region has been drained. In this region, blocking the water flow at the bottom of a depression (blocking the drain) provides a pool of water for the cattle. Paddocks cluster around the restored pothole. The wetland edge is then used by ducks, grassland birds and cattle (which tend to graze around the duck nests). Pintail ducks and geese seem to prefer the grazed vegetation (perhaps as food, perhaps because they can see predators at a distance). This mimics the natural situation: elk and buffalo once grazed these landscapes. Rotational grazing becomes a functional replacement for corn and soybeans: one ends up with the same crop (beef; plus ducks, geese and bobolinks; perhaps chickens), minus the row crops. The agro-ecosystem better retains its nutrients. One may need some corn and soybeans to "finish" the animals before slaughter. Raising buffalo on the same land would obviate the need for internal fencing, since buffalo, unlike cattle, space themselves out and rotationally graze a landscape all by themselves.

Organic agriculture doesn't use artificial fertilizers or pesticides; so-called sustainable or regenerative agriculture focusses on the whole environment of the farm; and its place in the larger landscape. There are two long-term sinks for carbon in agro-ecosystems: woody crops and perennial grassland or prairie. Besides storing carbon in their tissues, both trees and grasses transfer it to the soil, where it accumulates. Cattle, sheep, goats and hogs can be raised on grass (the first three are adapted to eat grass, which we cannot). Chickens also can be raised on permanent pasture, in small flocks in bottomless cages that are moved daily; the birds eat grasses, soil insects, and invertebrates. Such birds lead good chicken lives, taste better and sell for double the price of battery raised hens. Their eggs contain less chloresterol. In Minnesota, hazelnuts (a perennial woody crop) yield a similar volume of nuts per acre as soybeans. The nuts are a source of proteins and oil and can be managed (pruned, harvested) mechanically. Like pastureland, hazelnuts add to the soil's carbon stores. They hold soils. Mixtures of native prairie plants can be harvested for their edible or oily seeds, though yields are still below those of annual grains. The forage that remains after the seed harvest can be fed to animals. From the perspective of the long term, oak woodland in the hillier parts of the Middle West, interplanted with black walnuts, will slow down soil movement essentially to zero and yield as much net income as any other crop; but unless one seeds the young woodland with morel mushrooms or truffles, or cultivates portobello mushrooms in stacks of thinned oak logs, or grows berries as a understory crop, the harvest won't come in until one is dead. After the first century or so, the land could be sustainably logged for a net income equivalent to that from row crops ($55 an acre in 1996; so at current prices, one tree per acre every 20-40 years). Such long-term conversions would have to be supported by government policy. Perhaps loans based on the value of the future crop; annual payments based on the value of the land in the water cycle; annual payments (subsidized by a tax on fossil fuels) of $20-$50 a ton for carbon storage.

Young, growing temperate forests store carbon at a rate of half a ton to a ton per acre per year in their stems, and old forests, where growth in the stems has slowed, continue to store carbon in their soils. Borders of forest or grassland along streams, as well as permanent prairie or hayland, also accumulate carbon. So do soils under minimum tillage. Soils hold considerably more carbon than the atmosphere. At $50 a ton, carbon storage could be worth a billion dollars a year to American farmers. Evergreens grown for paper pulp, as on the prairies of Minnesota, would provide a faster return than nut trees or natural forest, but since they are clearcut, and carbon neutral over a rotation would not be available for carbon payments; however they also restore soils, slow soil movement, provide habitat for birds, mammals and amphibians, and return surface water to underground flow.

Prairie potholes are easily restored wetlands, functionally speaking (one plugs the drain), though it takes a very long time for their vegetation to return to something like the original condition (if ever: this may or may not matter much to the ducks, amphibians and songbirds of the original landscape, most of which soon recolonize the pond). Pothole wetlands also can be harvested in late June, after the ducks have done nesting, for wild hay. Because the yields in the wet soils are so high (3 times normal, up to 8 tons per acre), and more reliable year to year than grain, the return from wild hays over the long term is much the same as corn. (One needs a market for the hay.) The wild ducks that once nested in the prairie pothole region are now probably worth more (net) than the grain the land produces, but since the ducks are hunted elsewhere, wild duck farmers don't receive a return for their crop. In an ideal world, grain growers in the prairie pothole region would combine their farms into larger entities and raise free-range buffalo and elk instead of soybeans and corn, receive a duck subsidy from the sale of shotgun shells, a carbon subsidy for their carbon accumulating soils, a payment for keeping excess water and nutrients out of streams, and live comfortably. Forestland and restored prairie pothole habitat is of course land taken out of intensive food production, though both landscapes still produce food (the forestland mushrooms, a high protein food).

The matter of sustainable agriculture brings up three issues: feeding a world population of 8-10 billion people (that is, yield per acre); profit to the landowner (net income per acre); and the effects of an agricultural practice on its soils and the environment downstream (its biological effects and its sustainablity). The ecological value of undisturbed uplands has been put at $200-$400 per acre per year. While such figures are suspect (and probably much too low), they provide a basis for discussion. We can say few people are going to pay a farmer that for leaving his land undeveloped, for instance. We can also say farming should try not to extinguish that value. There are few trade offs in modern human use of the land: only losses. Considering agricultural use a trade off in which the return from agriculture replaces the land's ecological services is part of the capitalist rationalization that supports the methodical extinguishing of natural landscapes in favor of human economic ones. The loss of ecological services will be felt economically somewhere sometime (usually downstream, several months to several centuries later).

Regenerative agriculture is a step beyond organic agriculture, in that it attempts to regenerate healthy soils and function within a whole working ecosystem. It regards soils as the basis of agriculture and not something that holds plants up. In nature, soils are formed out of their parent material (rock, glacial till) by climate, topography, biotic processes and time. Half the volume of a healthy soil is living material (mostly plant roots but also microorganisms). The main biogeochemical transformers in the soil are the soil archaea, bacteria and fungi, which together constitute 1-4% of soil organic matter but its largest pool of rapidly recyclable nutrients (carbon, nitrogen, phosphorus, sulfur). They provide the nutrients for plant growth, partly by growing and dying, partly by breaking down decomposed plant material (including the recalcitrant humus), partly by leaching new nutrients out of the soil's parent material. But most species of soil organisms and their functions are unknown. As in a forest, soil fungi expand the root area of crop plants and connect the roots of different plants and species of plants: soil fungi are thought to transfer nutrients from the roots of dying crop plants to those of the germinating new crop, helping the new crop get off to a rapid start. Cultivated plants were once wild plants and regenerative agriculture treats crop plants as though they were wild plants adapted to live in mutualistic relationships with soil microorganisms and to grow successfully on the pools of organic nutrients in the soil..

Soil nutrients come from incorporating crop residues, legume cover crops and manures. On a regenerative farm, rotating land from row crops into legume sods, incorporating crop residues, rotating grain crops, and adding composted manure, through work is done by soil microorganisms, fungi and invertebrates, provide sufficient nutrients for a crop slightly less or equal to that of one's high tech neighbors in a good growing year, and considerably more in a drought, when regeneratively farmed soils perform better (probably because their higher fungal populations provide more mycorrhizal connections with the roots of crop plants). Good regenerative practices raise levels of soil carbon; expand the labile pool of soil nutrients; enhance nitrogen fixation by soil bacteria; reduce soil erosion; and improve soil tilth, porosity and water-holding capacity. Soil tilth—its crumb structure, which strongly affects its water holding capacity—is created by gluey polysaccharides secreted by mycorrhizal fungi. The ground freezes less deeply under fields in better tilth, letting the farmer plant earlier in the spring. The water that flows off the land is cleaner. Some added soil amendments, such as ground limestone, are usually necessary for long-term agriculture on a site. Sometimes amendments create new agricultural soils: small applications of lime and phosphorus, together with plant breeding that produced varieties of soybeans that would grow in the tropics, opened up Brazil's Cerrado, the savannah south of the Amazon, to commercial agriculture, with the result that land values there rose 10 times from 1997-2004, and that very little of it is left in a wild state. Sugar cane, the world's largest crop, can be successfully grown without any nitrogen fertilizer if the soil is inoculated with the right species of endophytic bacteria.

Most soils show microbial and fungal activity and have plant roots down 5-8 meters, that is well into the soil's parent material. Alfalfa draws up nitrate from 2 meters down the first year, greater than 5 meters in subsequent years and thus (besides its work in providing a home for nitrogen-fixing bacteria) increases nitrogen levels at the soil surface. (The nitrogen at deeper levels of the soil comes from leaching and will be replaced by leaching from above.) The amount of carbon in a soil is a good measure of its fertility. Studies have shown a 20% increase in potential yield with every 1% rise in soil carbon. A rise of 1% is however an enormous rise: the carbon content of undisturbed soils is about 2%. This falls to 1.5% during the early stages of cultivation and to 0.5% in badly damaged agricultural soils. Soil carbon largely comes from decomposed plants and in temperate regions incorporating grain straw will increase the carbon and nitrogen in the soil 40-50% percent over 20 years (that is, result in about 0.5% increase in soil carbon). Topsoil can be increased under appropriate agricultures at about an inch a decade (raising productivity about 6%, that is, almost doubling it in a century), while topsoil production under natural conditions is about an inch every 250-1000 years. Productivity declines caused by soil loss and degradation in conventionally farmed soils have been masked over the last 75 years by planting higher yielding cultivars and by increasing rates of fertilization and irrigation. Breeding can produce plants that perform well under high nitrogen regimes; or plants that are more efficient users of nutrients and water, and thus suitable for regenerative farming schemes.

The nitrogen that flows off farm fields comes largely from artificial nitrogen fertilizer, but also from manure, from nitrogen fixed by legumes and free-living soil bacteria, and from that which falls from the air (fixed by lightning, forest fires, and by industrial and vehicular combustion). The nitrogen fixed by people is now about equal to that fixed naturally. Its effects on the natural world are largely unknown. Nitrogen falling from the air encourages weedy annual grasses in prairies of native perennial grasses in Minnesota (reducing the soil's storage of carbon) and speeds up turnover of tree litter in urban forests (so one sees few dead leaves on the surface). Streams in unpolluted drainages in South America cycle mostly organic nitrogen, while for the last century the natural nitrogen cycle in most north temperate ecosystems has probably been overwhelmed by inorganic nitrogen from combustion and fertilization.

At any rate, about half of the nitrogen supplied by fertilizers to crops is used by them. Ten or fifteen percent leaches out with rainfall or irrigation water; some blows away with the soil; up to half (the number depends on the climate, the time and method of application, the type of fertilizer used) is denitrified by soil bacteria and returned to the atmosphere as nitrogen gas or nitrous oxide. Nitrous oxide is a greenhouse gas that now contributes about 6% of the anthropogenic greenhouse effect; it also depletes the stratospheric ozone layer. A doubling of the amount of nitrous oxide in the atmosphere, which would be a disaster, would lead to a 10% decrease in the ozone shield. So conventional agricultural soils lose not only carbon (as carbon dioxide) but also nitrous oxide, and global agriculture's atmospheric connections affect both climate and the ozone layer. Agricultural activity is thought to produce 15-25% of current global warming, some from its production of gases like carbon dioxide, nitrous oxide and methane, some from dust and soot (from land clearing or from burning grainfields or grasslands), which, falling on Arctic ice and snow, increase its absorption of heat and help it melt. Some agricultural chemicals, like the soil fumigant bromine, now banned, but used by California strawberry farmers under special license, also destroy stratospheric ozone. Soils under no till agriculture (which reduces or eliminates plowing), store carbon instead of losing it; soils cultivated this way might be able to sequester 8-20% of the carbon emissions in the United States. No till agriculture also reduces soil erosion and slows water runoff. Conventional no till agriculture uses chemicals for weed control but since water runoff is less (keeping the chemicals in place), degradation of the chemicals is thought to be more complete.

The nitrogen that runs off fields enters surface waters, or sinks into groundwater. Some of it is removed by denitrifying bacteria in the damp anoxic lowlands near watercourses, and also in the river waters themselves. These bacteria return it to the atmosphere as nitrogen gas or nitrous oxide. Continental ground water constitutes 40% of global fresh water, rivers constitute something like 0.005%. Rivers are in contact with groundwaters many times their size and exchange soluble nutrients with these waters. The water in a river spirals downstream, sometimes in the channel sometimes not; the water one sees in a river is a small part of the river/groundwater flow. Riverine invertebrates have been collected from 10 meters down in wells in the floodplain of the Flathead River in Montana up to a mile from the river channel. Such groundwaters act as both a source and a sink for nutrients, and nutrients in the river water can be removed all along its course in riverside wetlands and low banks (as well as by free floating denitrifying bacteria); more reasons why floodplains and wetlands and aquifer recharge areas all along the courses of rivers should be protected. Since most rivers in industrial countries are overwhelmed with nitrogen, and much of the capacity of riverine wetlands to remove nutrients has been destroyed, most of these nutrients now end up in the sea. The load of nitrogen and phosphorus in British estuaries from farms, industrial combustion, human urine, from fertilizer spread on lawns, from pet feces, and from land development (losses of nitrogen and phosphorus are speeded up in disturbed soils) is 100 times the level needed to cause eutrophication of those waters.

Nitrogen compounds from farms and industrial combustion also fall out of the air. In large parts of the northeastern United States nitrogen rains out of the atmosphere at the average rate of its application to American farmland. Such excess nitrogen over fertilizes and acidifies forests that are naturally nitrogen limited, causing a weakening of the mycorrhizal connection between fungi and tree roots, and a reduction in fine root mass (both are adaptations of trees to nutrient limited lives); the loss of such connectivities makes the trees more vulnerable to droughts and micronutrient deficiencies. The nitrogen also reduces the competitiveness of nitrogen fixing species of bacteria, including those in lichens. So lichens become less abundant. (Nitrogen fixation by lichens is important in older natural forests.) Calcium in acidified soils leaches out and is replaced by (toxic) aluminum. In native grasslands, the excess nitrogen encourages weedy (often introduced) annual grasses in their competition with the deep rooted perennial grasses and thus reduces the landscapes' carbon storing capacity, which is mediated through the huge root masses of those perennial grasses (ninety percent of the mass of native perennial grasses is underground and one third of their roots die and regrow every year, adding huge amounts of carbon to the soil). Careful hosts in farm country give you coffee that has been made with filtered water, since the ground water (their well water) contains so much nitrogen (as well as various mutagenic and hormone mimicking pesticides) that it is not safe to drink. Iowa lakes and rivers have some of the highest levels of nitrogen and dissolved phosphorus in the world.

A study in England listed the externalities of modern agriculture as the costs to water companies of removing nitrates, pesticides and farm pathogens from drinking water; the costs to companies of workers absent with food poisoning from eating mass produced animals contaminated with salmonella; the costs of restoring damaged habitats; the costs of compensating for the air pollution and greenhouse gas production from agriculture; the costs of soil erosion; and the costs of the damage to farmers' health from farm chemicals. (I find it doubtful the list is complete: habitat fragmentation, introduction of exotic plants, animals, insects, genes and diseases, and elimination of large predators come to mind.) Total costs per acre per year were calculated at $120. This cost amounts to a hidden subsidy to modern farming. A hundred dollars an acre applied to all the world's cropland would make the hidden subsidies to agriculture about the same as the direct subsidies paid to agriculture in the developed world (about a billion dollars a day; $2.20 per cow per day in the European Union).

A cost that study seemed to ignore is that of antibiotic resistance caused by feeding farm animals prophylactic antibiotics. Antibiotic supplements help animals gain weight and deal with the stress of living in crowded conditions, with no outlet for their instinctual behaviors (rooting, nest-building, scratching, the formation of social hierarchies, social rubbing, pecking). In the case of cattle raised in feedlots, the antibiotics also help them deal with the consumption of a feed (maize) to which their stomachs are not well adapted. Cattle are adapted to eat grass, which we cannot: one good biological reason to eat meat. The acid rumens of corn fed cattle also seem to encourage the growth of the deadly 0157:H7 strain of *E. coli* bacteria, which is common in the manure of cattle fed on corn. (*E. coli* 0157:H7 appeared in 1982 and is thought to have evolved in the guts of feedlot cattle.) The numbers of 0157:H7 *E. coli* in the manure fall by 1000 times after a week of feeding cattle hay. Genes for antibiotic resistance are common in bacteria and bacteria commonly swap genes among and between species. (Some archaea share genes so often that species distinctions become questionable; this makes their evolution in the face of new environmental challenges extremely rapid and helps explain why archaea and bacteria run the planet.) So bacteria exposed to continual low doses of antibiotics quickly develop antibiotic resistance. Antibiotic resistant bacteria are found in soils affected by feedlot runoff, in manure from dairy cattle, in the air around farms, in wild animals, and in retail meat and poultry. Antibiotic resistance can easily jump to bacteria that affect humans. While it is difficult to trace resistant bacteria in humans to the use of antibiotics in farm animals, antibiotic resistance in bacteria that cause disease in humans is a growing problem. Feeding prophylactic antibiotics to farm animals is foolish, on a level with the feeding of processed cattle carcasses to cattle. Such industrial cannabilism gave us mad cow disease; and probably wasting disease in wild elk and deer (which are also fed food of questionable origin by humans). Such results might have been predicted, since the ritual eating of partly cooked human brains produces a similar neuro-degenerative disease in humans. The eventual economic costs of such capitalistic practices are likely to dwarf the benefits.

Runoff waters from farms and feedlots contain not only nutrients and antibiotic resistant bacteria but diseases and parasites from the animals. Animals produce about a billion tons of manure per year in the United States (several times the manure of people) and most of this material is not handled well. The disease-causing organisms end up in rivers and eventually in the marine environment. Some of them end up in drinking water and cause outbreaks of disease. Oocysts of *Cryptosporidium*, *Giardia*, and *Cyclospora* (all of which cause human disease) have been found in high levels in bivalves on the east coast. Cysts of *Toxoplasma gondii*, a parasite of cats, are washed with cat feces into the ocean off California. The cysts are concentrated by shellfish. The shellfish are eaten by sea otters, in which the parasite causes severe mortality.

Uptake of nitrogen fertilizer by crops could probably be improved by about 10%, through proper timing of fertilisation, and proper balance of phosphorus and potassium applied with the nitrogen. One problem is that the nitrogen in artificial fertilizers is extremely available—to plants, to leaching by rainfall, to denitrifying bacteria. It is meant to be available. Anhydrous ammonia becomes available as nitrate at 50° F., before most crop plants can use it. So-called organic nutrients, derived from bacterial decomposition of material in the soil, a biological process that is also temperature dependent, become available somewhat more slowly and at temperatures more commensurate with the ability of the plants to absorb them. Some fertilizers (especially the ammonias) tend to volatilize rather than leach, so more of them is lost directly to the atmosphere than to surface and ground water. If denitrified to nitrous oxide before volatilization, they contribute to global warming; if volatilized as ammonia, they contribute to acid rain and end up in watercourses by a different route. (Some of the ammonia oxidizes to nitrogen gas and is harmless.) About 50% the nitrogen entering the Gulf of Mexico comes from commercial fertilizer, 15% from manure (mostly from feedlot runoff). Many modern high yielding crop varieties need excess nitrogen to grow well. Under good management 35-40% of that nitrogen is going to run off. Some of it will sink directly to ground water, some of the rest will be trapped by grasslands, wetlands and woodlands downstream of the field, but the more nitrogen that runs off, the more land is needed to soak up the excess, and the less effective that land will be. Land used to catch nitrogen is also land taken out of agricultural production (though it may produce hay and therefore meat or milk).

Thus fertilizer use comes up against limits. By ruining riverine and ocean fisheries, intensive commercial agriculture may in fact reduce the total amount of food a watershed produces. One can't maximize grain yields and have a healthy aquatic environment; but one can have lesser amounts of grain, together with grass and hay (that produce milk, eggs and meat), some timberland (that produces logs, mushrooms and game animals), and riverine and marine fish. Such changes may imply a smaller human population.

Organic agriculture takes land out of grain production through its rotations into legume sods. (So an organic agriculture will always produce less total grain.) Modern crop plants have a much greater proportion of the plant in seeds, rather than in leaves and stalks, compared with traditional varieties. (This makes them high yielding.) Thus recycling their residues replaces less of the nitrogen and phosphorus lost in the crop than formerly. One writer claims that a complete recycling of all the organic matter from currently harvested land and from all confined animals would not replace the nutrients lost in crops under modern high yield agriculture. Composting food waste, spreading manure produced by humans and capturing the nitrogen in human urine would make up most of that. (Freezing sewage effluent is one of the more off the wall suggestions on how to separate out urine: since urine freezes at a lower temperature than water, the excess water can be removed as ice. Methane from the sewage would power the process.) Adding rock powders and rotating land into sods also help. Legumes (alfalfa, vetch, clover) fix considerable amounts of nitrogen (300-500 pounds per acre), but only 5-10% of this is available for the following crop. The rest adds to the soil pool and slowly becomes available over time. Crop breeding has raised corn yields by 31% since 1995, 76% since 1975. These crops were bred to be fertilized with artificial nitrogen. Crops could also be bred to take advantage of symbiotic soil bacteria and fungi, that is for high yields in an organic agriculture.

Use of manures and sludges (human manure plus other things that go down the drain) have their own problems. Both animal manures and sewage sludges contain heavy metals (including cadmium, arsenic, chromium, lead, mercury, nickel and vanadium). Those in animal manure originate partly in synthetic fertilizers (especially phosphate fertilizer). Some of these could be eliminated. Other metals come from plants. Some plants have an affinity for metals, which can make them useful in cleaning up contaminated ground. Alfalfa plants for instance, will sequester gold, while aspen trees take up mercury and volatilize it from their leaves (thus distributing it more widely to the atmosphere: how much this helps is another matter). Plants with metals in them (perhaps metals falling out of the sky from coal burning power plants) are eaten by animals and the metals come out in the manure. Sewage sludges contain metals from the plants and animals we eat, as well as from roadways, rooftops, factories, and labs (such as metal plating or film developing facilities). So, if they are to be used as a soil amendment, sewage sludges must be cleaned up; and matched to the soils they are spread on; and the rate of application of both sludges and manures limited, so as to avoid a build up of heavy metals in the soil. Cleaning up urban sludges is partly a matter of controlling what goes down the drain, that is, capturing metals at the source. This is often possible with a small capital investment that pays for itself (in the reclaimed metal) over a few years. There is no question that human and animal manures can be used safely as a fertilizer: some Chinese landscapes have been farmed for 7000 years using human manure.

Some manures are contaminated with drugs and pesticides fed the animals. The chicken feed used to raise chickens (one billion a year) on the Eastern Shore of Maryland contains arsenic and anti-biotics. The arsenic is a biocide used to control parasites in the chickens, the antibiotics keep the chickens healthy and make them grow faster (it improves their absorption of nutrients from the feed). Excrement from the chickens is used to grow corn and soybeans for chicken feed. So arsenic now contaminates surface and ground waters along the Eastern Shore and the soils contain antibiotic resistant bacteria. Eliminating arsenic and antibiotics from feed would mean changing the way chickens are raised.

Manures and sludges are heavy and expensive to move, but using them on farms is undoubtedly the best way to use them. Large animal facilities worsen the disposal problem. A dairy farm with 1000 cows, whose animals produce feces equivalent to a village of 10,000 people, needs 2000 acres to spread the material; somewhat less if the farmer puts it through a methane digester or composts it. (Both these processes will reduce its pathogens virtually to zero.) Most confined animal operations store their manures in open lagoons, where organisms that cause human disease are not appreciably reduced and where anaerobic decomposition produces considerably more carbon dioxide and methane than aerobic composting. During heavy rains such lagoons often fail, and send their nutrients, in toxic amounts, into rivers. Most confined animal operations have far too little acreage to spread their manure. Their investment is in their facilities. Phosphorus especially is a problem. Phosphorus is often the limiting nutrient in rivers and lakes and adding it causes algal blooms. If the ration for the animals in a confinement barn or a feedlot comes from grain raised a hundred or a thousand miles away, the manure may be difficult to dispose of. Thus there are projects to recycle it into animal feed.

In general, while organic, soil regenerating farming produces yields slightly lower than modern chemical agriculture, it yields higher returns to the farmer. This is partly because the price the farmer receives is higher, partly because purchased supplies are less. But yields per acre are not only dependent on the method of fertilization; they also depend on general farm practices. Planting spring wheat, corn or soybeans two weeks earlier in the short growing season of Manitoba boosts yields from 10-40%. While good management of crop residues helps prepare the land for early planting, reliable early planting usually means drainage (and then appropriate handling of the water that comes out of the drains). Careful management of crop residues enhances nitrogen fixation (providing nitrogen for the next crop), reduces compaction and erosion, and improves the soil's water holding capacity—so the soil freezes less deeply and can be worked earlier. Such organic techniques might be adopted by conventional farmers. In California's Central Valley a study compared tomato yields with organic and conventional methods. Without the use of conventional pesticides and fungicides, yields were predicted to fall 36%. But chemicals turned out to be less important than inherent soil fertility, water availability, the cultivars grown, and management practices (such as transplanting time, which strongly influenced the degree of insect damage). The density of herbivorous insects was only 5% greater on organic farms, but the herbivores were considerably more diverse. The predators and parasitoids of the herbivores were also much more diverse and much more abundant on the organic farms.

In some cases, pesticides reduce total yields of food. Farmers in Bangladesh who stopped putting pesticide on their rice so they could grow fish in their paddies, found their yields of rice rose 25%; and they also were able to harvest the fish. One reason for the increased rice yields may have been a speedier cycling of organic nutrients by the fish; part of a generally more efficient food web in the paddy that had been suppressed by the pesticides. In China, rice yields have been increased by draining the fields several times during the growing season. This saves water and also reduces methane production in the rice fields by 40%. (Of course, such draining eliminates the possibility of raising fish.) Some rice farmers in the Sacramento Delta of California now flood their rice fields after harvest instead of burning the rice straw; the straw decomposes in the water, improving the organic content of the soils, and migratory waterfowl use the fields as a substitute for natural wetlands. The waterfowl recycle nutrients in the fields and add more nutrients from their feeding in upland habitats. Such cycling can be substantial. It is thought that nesting seabirds contributed two million tons of guano annually to New Zealand coastal forests, a thousand pounds an acre, before Pacific rats brought by the Maori destroyed their nesting colonies. Some of the California rice fields are leased to hunters, producing additional income, but overall such management is a good deal for the ducks and geese, whose winter numbers in California have fallen from 100 million to perhaps 6 million over the last century, as 90% of California's wetlands were drained. (Large parts of the Central Valley were marshy in winter and spring; and winter wheat can be grown there without irrigation.) Rice farmers in Louisiana grow crawfish, and also provide habitat for native waterfowl, in their paddies during the off season; the crawfish eat the vegetation that grows in the paddies.

In Chinese cities urban gardens provide over 85% of the fresh vegetables used in the city, plus some meats, grains and tree crops. On the outskirts of Italian cities one sees gardens on the slopes of drainage canals. Such hand agricultures are always more productive than extensive cultivation. Farmers in the United States now need 30,000-45,000 square feet to feed a person on a diet high in beef, considerably less if more of the animal products are from dairy cattle, poultry, or pigs (which are more efficient converters of grain to meat, eggs or milk), 10,000 square feet for a vegetarian diet. Such numbers vary; these figures are for single-cropped, chemically farmed cropland, and include the large amount of waste currently in the American food supply (about half the food produced is wasted at the retail level). Such farming is energy intensive; on average in the United States, food contains a tenth of the calories in the fossil fuels used to grow and fertilize it. On the other hand, a bio-intensive gardener needs 2000-4000 square feet (20-40% of the land of the commercial farmer), seeds and hand tools, to feed a person on a vegetarian diet, with a small amount of chicken or pork. This includes land for compost crops. In this case the food contains several times more calories than the work that went into it. Among the most productive hand agricultures in the world are the dike-pond units in the Pearl River delta of China's Guandong Province (now disappearing under urban sprawl). Irrigated agriculture and aquaculture are integrated. The nutrient rich runoff from the fields fills fish ponds, fertilizing algae that are eaten by the fish, which are also given crop wastes. Several different carp species with different feeding niches are raised: surface feeders, middle layer feeders, fish that eat plants and organic wastes, bottom feeders. Ducks also feed on the plants and insects that grow in the ponds. Periodically the fish are harvested and the ponds are drained to grow crops. Crops include cows, pigs, vegetables, mulberries, sugarcane, ducks and fish. This organic agriculture yields 30-50 tonnes of fish and crops per hectare, or enough to feed 50 or more people, while intensive rice farming in the same region feeds 11 (as would American commodity agriculture). Figures like this have led a geographer to quip that to improve agricultural yields we should suburbanize the landscape, and let people grow their food in gardens. In cities with appropriate climates and sufficient capital, rooftop gardens under greenhouses made of photo-voltaic panels (some are now transparent), could produce electricity and grow vegetables, fish, chickens and rabbits, a variation on the rural schemes of the far away New Alchemists.

Feeding the human world, even a world of 10 billion people (though that number now seems less likely), is less a matter of improving the yields of farmland (though in some cases that would help) and more a matter of focus on the problem of population, diet, agriculture and landscape as one whole. In 1948, 50 million pounds of pesticide were used in the United States and 7% of crops were lost to pests, preharvest. In the 1990s, 20 times that weight of pesticide was used and 13% of preharvest crops were lost. Some of this is due to the fact that plants have been bred for yield rather than for insect residence (sprays take care of the insects), some from the planting of the same cultivars on hundreds of thousands of contiguous acres. With their short generation times, insects and plant pathogens adapt to changing conditions much faster than plants, so such plantings are vulnerable to small adaptations on the part of the pests. Some of the increase may have been due to the growing size of fields, the increasing use of the whole habitat for crops, and spraying for weed control, all of which made the environment less favorable for insect predators of pests. Most agricultural chemicals are bad for people and animals, and make fields and orchards dangerous places for wildlife and farmers. They kill insects that birds and mammals need for food. (Insects are the means by which many animals access the energy stored in plants.) In sufficient doses pesticides kill birds, mammals and other vertebrates directly (60-70 million birds a year in the United States). When predatory animals eat the dead insects, the dose is concentrated in them, so orchards are not healthy places for predatory insects, or for small falcons that eat large insects, or for red-tailed hawks that eat mice sprayed with organophosphate pesticides. In very low doses many agricultural chemicals weaken the immune systems of amphibians and mammals, promote cancers, feminize developing embryos, cause genetic mutations.

While spraying is sometimes necessary, there are also other ways of controlling pests. Codling moth is a major pest of pears, apples and walnuts in California. Pear orchards suffer 5% damage or less when a bat colony is less than a mile away; and 60% damage if the colony is over 2 miles away. Large bat houses were once built in the American South as a method of mosquito control. Building bat houses in orchards (or installing boards for bats to roost under along the sides of barns) would be much cheaper than spraying. And one could spread the guano. An organic farmer in Oregon controlled the insects in his cherry orchard by hanging nesting boxes for swallows on wires among the trees; the birds ate the insects, and when he cleaned out the nest boxes in the fall, he used the nesting material to fertilize the orchard. Swallows and bats are limited more by roosting and nesting habitat than by food supply, so populations can be increased (at least locally) by providing nesting and roosting sites. Similarly, shrike populations can be increased by providing perching posts. Shrikes eat small birds, large insects and mice. Citrus growers in China place colonies of predatory ants on their trees, and dig moats about the trunk to keep them there; such trees have 60% less insect damage than sprayed trees: the ants do a better job than the spray. (Ant colonies of certain species inhabiting acacia trees in Africa swarm to protect them from browsing by elephants and giraffes—so some trees are left after the elephants have moved through.) Owl nesting boxes and hawk perches are used to attract predatory birds to control mice in organic vinyards in California. Some organic orchardists graze pigs among their apple trees. The pigs keep the grass and weeds down and eat the fruit from the June drop, killing the larvae of the plum curculio, a major pest of apples, thus preventing them from migrating into the soil and developing into adults. Beds of alyssum in lettuce fields (5-10% of the field) support hoverflies ; the hoverflies lay their eggs on lettuce leaves; their larvae eat aphids. Beds of alfalfa in strawberry fields attract the lygus bugs whose feeding deforms strawberries. The bugs can be vacuumed off the alfalfa.

Modifying the structure of the agricultural landscape provides habitat for predatory insects and rotating crops breaks population cycles of pests. Here we are looking at things like smaller fields (pollinators are more effective within 100 yards of their nests, predatory insects are more abundant at field edges); replanted hedgerows; in-field wetlands; strips of crops of different heights and leaf characteristics (providing a more varied habitat for predatory and prey insects); several varieties of the same crop in a field (not one cultivar); regular crop rotations; very little spraying: all this against a background of a semi-natural environment (a later successional ecosystem) that occupies 15-40% of the agricultural area (including some prime land) and provides natural habitat for insects, birds, mammals, amphibians and invertebrates (many useful to the farmer) and also serves to regulate the quality and flow of runoff water. Agriculture is always full of disasters, climatic and otherwise; different crops provide insurance.

By some estimates there are 5 billion acres of degraded and abandoned agricultural land worldwide (current cropland is about 3.75 billion acres, with 7.5 billion acres in pasture and rangeland). With sufficient investment (say, $100-$1000 an acre), this land could be rehabilitated. Successful rehabilitation under regenerative agriculture would produce food, oils, fiber and habitat; the soils might absorb half or more of the 6-7 billion tons of carbon the human world emits annually. (And continue to absorb the carbon for some centuries: soil carbon levels seem to rise linearly with the amount of recycled crop residue added to them, and continue to absorb carbon indefinitely.)

———————

We cannot feed the present world population on a modern western diet. But while probably 15% (about one billion people) of the global population is currently underfed, we produce more than enough food to feed them. Farming is a business. Farmers must be paid. Animals eat 35-45% of the grain grown worldwide, though the grain would yield more calories and protein to people if people ate it directly (it's more profitable to feed it to animals). Quite a high percentage is wasted before it reaches the consumer. About 10% of grain is wasted in developed countries, up to 20% in underdeveloped countries. Spilled grain in the United States supports

Canada geese about Chesapeake Bay and other waterfowl in the rice fields of California and Louisiana. Grain provides 20-30% of energy intake in affluent countries, compared with 60-80% in poorer economies. The richer the population, the more meat it eats: so bread consumption in France went from 600 grams per person per day in 1880 to 170 grams per person per day in 1980, the difference being made up by grain processed through animals—eggs, cheese, meat. Global milk production is now about 80 liters per person per year. Traditional milking societies in northern Europe consumed the equivalent of about 100 liters per person per year. Much of the world population is lactose intolerant and can't digest milk, but lactose intolerant people can digest yogurt and cheese. Global meat production is about 75 pounds per person per year. This is also enough: a reasonable annual meat consumption, based on an evolutionary understanding of people's needs, is about 40 pounds, or what the Iroquois ate. Western societies now consume the equivalent of 300 liters of milk and 150-240 pounds of meat per capita, probably several times that needed for people to reach their full genetic potential for growth (a common measure of dietary sufficiency). In foraging societies, one of our better benchmarks for diet, plant and animal protein provided about one third of food energy. People got more vitamins and minerals, including calcium, from wild fruits, nuts, roots, tubers and legumes than currently recommended and much more fiber (fiberless beet and cane sugar provides 20% of the energy in a modern diet). Foragers also took in much less sodium and much more potassium than sodium, perhaps a factor in maintaining healthy blood pressure. Finally, in the West, about 40% of food is wasted after being purchased; some waste of course is unavoidable. Many writers claim that an organic, soil improving agriculture, located within a living natural landscape, cannot feed the existing world. The necessity to rotate grains into sod crops (thus reducing total grain production) makes this so; but such calculations ignore all the degraded land that could be brought into production. At any rate, at this point, human hunger is essentially a choice (largely a choice made by the rich nations): like the Irish potato famine, partly a matter of income, partly a matter of eating habits, partly made for us by the invisible hand of the market. The current problem is less food supply than food distribution: the poor (in global terms) can't afford food; and their countries either can't afford to subsidize their agricultures or don't choose to.

Feeding more people is also a matter of diet. It is much more efficient for people to eat grains, than for them to feed the grain to animals, and eat the meat. Grains are virtually equal in energy density to meat (they have 5 times the energy density of tubers like yams or potatoes), and are not that low in protein: rice is 7% protein, durum wheat 15%. Mixing them with legumes makes a diet balanced in protein; and eating fruits and vegetables provides vitamins and minerals. However there are good arguments for including some meat in a diet. People have a much shorter gut than our close relatives among the primates. Some writers have proposed that (energetically speaking) reducing the size of the human gut made our large brains possible. Cooking, which doubles the carbohydrates available in plant foods, would have been the adaptation that made a reduction in the size of the gut possible (smashing foods before eating also helps make them more digestible). One writer proposes that the adoption of fire and cooking by protohumans long ago led to animals with smaller guts and teeth, bigger brains, more running and hunting, longer lives, more bonding between males and females. At any rate our short guts have adapted us to foods rich in energy: nuts, seeds, meat. If we ate leaves like gorillas or chimpanzees, we would have to spend much of the day eating and digesting, the fate of the once carnivorous pandas that turned vegetarian and now spend most of their days eating the tips and leaves of bamboo. Meat is not necessary in a human diet. But herbivorous animals, thanks to the microbial symbionts in their guts, can eat foods that we cannot (grasses, twigs, leaves: black bears eat deer droppings and thus acquire their bacteria, which lets them digest grass); and soil regenerating agriculture is not possible without manuring; or rotating crops with legume hays. The system works better both economically and biologically if something eats those hays, converting them to meat and manure, which can be spread on the grainland in the years after the rotation from hay. Some areas, like the North American plains, are adapted to grass and grazers. Grasslands constitute 20% of the modern global landscape and, helped by the grazers, store carbon in their soils. Using (some of) them to produce meat is far better than using them to produce grain.

No terrestrial agriculture will be able to feed a population that grows forever. Earlier peoples, without our notions of progress, limited their populations, or starved. One could argue that a population reaches its limits when it no longer fits within a working natural landscape. If so, we reached our limits several thousand years ago, when there were perhaps 100 million people worldwide and the gases released by agricultural began to moderate global climate. (And Deep Ecologists, who say 500 million people are enough, would be right.) However that first mild warming was hugely advantageous to us, and more or less a toss-up to the rest of the biological world. Such matters are fluid. We could also say we have passed population limits when we can no longer feed our population on a biologically appropriate agriculture in a biologically functioning environment. The falling birthrates of the West (the U.S. population would be falling too without immigration) are regarded as a disaster, but if we could figure out how to get through the demographic squeeze (an economic problem), the falling populations would result in a much healthier landscape. If every fertile woman restricted herself to one child, population would fall to 1.6 billion by 2100, a level that would allow a modern life life for everyone. At two children per woman, population would still fall, but more slowly. It was long thought that development (a certain level of affluence) was necessary to bring population growth under control. But what seems to matter more is improving the survival of children and improving the status of women. If women are educated, can support themselves, and see that their children are going to survive, they—if they can—will limit their fertility. Populations then stabilize at much lower levels of income. All this of course is in direct conflict with the capitalist doctrine of continuous growth in population and income, one constantly driving the other. (And with the religious ideas of many fundamentalist Christians, Muslims and Jews.)

A sustainable life for a large human population in a working biological environment means dealing appropriately with nutrient flows, whether this is "economic" or not. In Dutch agricultural districts, where large numbers of animals have been raised for decades on imported grain (the Dutch invented raising animals in confinement), the animals' manure is now dried and turned into a bagged fertilizer that is exported from the region. This is to keep the nitrogen levels in groundwaters, already high, from rising further. In broad economic terms, it is certainly cheaper to subsidize the costs of dealing with the waste from the animals than to deal with groundwater too toxic for people or animals to drink, so all water has to be filtered, or with the eutrophic bays, rivers and lakes into which the water drains. Of course not all nitrogen fertilisation is harmful. There are degraded soils such as those of the dry Ethiopian highlands where some artificial nitrogen fertilization is biologically helpful. Fertilisation doubles or triples yields of wheat and tef, so more land can go into cover crops and trees; thus improving the habitat and reducing soil erosion. In the dry wheatfields of the Dakotas and Saskatchawan little nitrogen leaching from rain occurs, so nitrogen fertilisation is less harmful to watercourses, but nitrogen will be lost to the atmosphere as nitrous oxide, and along with soil and phosphorus, to the wind. The point is to farm, and apply nutrients, appropriately for the place.

Not including ecological costs in economic costs subsidizes the destruction of the natural environment. Environmental protection may seem expensive; but reducing the nitrogen load in rivers by upgrading sewage treatment plants that empty into them will also raise property values along those rivers. Upgrading sewage treatment plants and subsidizing the spreading of New York City sewage sludge on Texas cotton fields (where its heavy metal content doesn't matter), or on strip mined lands in Pennsylvania (where sludges help, and turn wastelands into carbon storing ecosystems) make economic sense in the long run. (For one thing, abandoned lands become productive.) Appropriate use of sludges probably requires subsidies. How to do this fairly is a difficult question, since the benefits of cleaning up, say, Long Island Sound, accrue to more people than those in the sound's sewage districts. And not just sewage influences the health of the sound; all the runoff in its watershed does, including that from septic systems, roads, dairy cattle, factories, dogs and cats. We probably should subsidize the restoration of natural wetlands (or the construction of artificial ones) to filter and slow the water that runs off the land.

And to filter and slow the water that runs off the 50 million acres in the Corn Belt that have been artificially drained. Drained land constitutes about 25% of that landscape. Letting 15-25% of it revert to natural lands, by planting it to oakwoods or prairie, by blocking the drains in prairie sloughs, or by opening levees to let rivers occupy old wetlands, would restore much ecological process to the landscape. New drainage structures that allow the farmer to control the level of water flowing from individual fields may reduce the size of the filtering wetlands needed and also, by allowing the farmer to control subsoil moisture, raise yields of crops. Lands in the Conservation Reserve Program peaked at 36.8 million acres in 2007, that is, 8% of the agricultural landscape, at a cost of $1.8 billion (about $20 per acre). Putting 25% of the agricultural landscape in permanent conservation lands would cost perhaps $6 billion a year. Payments for conservation lands would replace price supports for grain. In the coming world natural lands that store carbon will be eligible for other payments. Replanting marsh grasses and cattails on farmed peatlands in San Francisco Bay creates wetlands that store 24 tons of carbon per year. Continuing to farm these lands releases about 7 tons per year (oxidation of the peat has lowered peat islands 15-20 feet below water level). At $30-$50 dollars per ton, farmers could net around $1000 per acre by planting cattails. Returns from carbon storage in the Corn Belt would be much less ($30-$50 per acre), those in the sugar cane fields of the Florida Everglades comparable to those of the peatlands of the west coast.

The runoff of water and nutrients from farmland is at least partly a matter of economic attitude. In the Pearl River Delta of China, nutrient rich runoff water is regarded as useful, and economically productive ponds, marshes and woodlands are used to filter it. (Until recently in rural China, someone using a neighbor's outhouse would weigh his contribution so he could demand a similar one in return.) In the Middle West, in a better managed landscape, the excess nutrients would grow timber, marsh plants and fish. But much water pollution in rich countries is biologically pointless. Lawns occupy 40 million acres in the United States. The fertilizer and pesticide used on lawns and gardens amounts to about half that used in the country, and virtually none of it is necessary, except economically to the fertilizer companies: compost, produced from yard waste at home or at the local landfill, will leach fewer nutrients and produce a healthier lawn and garden. Use here is a matter of economics (fertilizer is cheaper for lawn care companies to spread), esthetics and attitude (moral notions), which come together in property values: a green dandelion free lawn is considered the "right" setting for a house, and fertilizer and pesticides are "clean" materials (lending their own chemical whiff of rectitude to the suburban environment).

Land uses based on moral attitudes are hard to deal with and demand persuasion to change. "Picking up" the environment is another one: natural environments appear messy and chaotic; but the tangle of branches and trees lying on the floor of a natural forest—the coarse woody debris of the foresters—retains moisture during dry periods, provides sites for nitrogen fixing bacteria, and represents a pool of nutrients for the forest. Its removal helps explain forest decline in central and eastern Europe. The fallen logs and branches are an important part of small mammal habitat, providing dens, cover and food plants like mushrooms. They help protect tree and shrub seedlings from browsing deer. Healthy small mammal populations help maintain the ecological processes of which they are a part; forest voles disperse the spores of mushrooms, whose mycelia help maintain the trees; voles are also food for predators and—predators themselves—help control defoliating insects.)

Even without fertilization, the water that flows from suburban lawns will be polluted by the nutrients and parasites in animal feces, by metals and by motor oil. Great rivers of underground water flow below most agricultural and urban areas, all susceptible to pollution by leaching from above. The water in the Ogallalla Aquifer under the Great Plains moves at a rate of a few feet a year from the Rockies toward the Mississippi Valley. One could argue that that movement, which brings it across private property and state boundary lines, makes it public property. Its waters also feed surface waters, which, if connected to navigable waterways, come under federal regulation. Several states have declared groundwaters a public trust and thus subject to regulation. Underground waters are far too essential for human life (and for the biological world) to be polluted; or to be depleted beyond their natural rates of recharge. Once polluted, they are astronomically expensive to clean up; once depleted they are gone: where it was least abundant, the economically extractable part of the Ogallala water is already gone.

When the snow melts in the spring, one sees in the bare fields the shadows of its former hydrology: damp spots that were once depressions in the forest or prairie, the dark lines of temporary streams. Such surface hydrology, with its plants and animals and their water cleansing effects, is what the farmer eliminates. To a certain extent, the release of agricultural nutrients to streams and the atmosphere, can be controlled by agricultural practice: what fertilizers are used and how they are applied; how much manure is applied and how it has been handled; what crops are grown; how fields and hedgerows are laid out; how crops are rotated. Physical constructions also help and amount to a sort of reconstruction of the hydrological landscape: vegetated corridors along streams, constructed wetlands for drainage water, grassed drainage channels from fields, permanently vegetated aquifer recharge areas. Some of this can be profitable: one farmer in the prairie pothole region who restored his wetlands under an agreement with a conservation organization (and then kept them restored because he liked the ducks), found that cutting the slough hay when the ducks were done breeding produced about as much net income as growing corn, partly because he didn't have to seed the hay and partly because the hay yield was more dependable than the corn crop. (He fed the hay to his young cattle.) Dabbling ducks like mallards, pintails and teal prefer shallow seasonal wetlands, with their abundant aquatic life, in early spring. These can be temporary pools in a thawing field. They need somewhat deeper ponds later in the season for nesting. Such ponds provide cover for the ducklings and an abundance of the small aquatic invertebrates the ducklings need to grow. (In modern agriculture, pesticide drift can reduce these invertebrates by 90%, reducing the value of the habitat to zero.) If a farm landscape is varied and productive enough, with small grains and hayfields where a nesting can be completed, game bird populations allow a farmer to lease hunting rights. In the Midwest and Rocky Mountain states most of these birds will be introduced ones (ring-necked pheasants, chukar partridge, hungarian partridge): the native prairie chicken of the prairies tolerates agricultural landscapes poorly, the sage grouse (once a keystone species of Great Basin sagelands) hardly at all. But the new birds become part of the new landscape; goshawks, peregrines and foxes, as well as hunters, will pursue chukars and ringnecked pheasants.

In the end, controlling nitrogen in streams and estuaries means controlling its use. It is virtually impossible to control all the nutrient runoff from commercial cropland: the agroecosystem doesn't work that well and the surrounding nutrient sinks can only trap so much. Levels of nitrogen in Iowa's Des Moines River were already high under the traditional crop rotations before the Second World War. Plowing, and cultivating for weeds (up to 7 times a season), were causing considerable soil erosion. To reduce nitrogen in streams one has to use less of it. One way to control nitrogen use is by taxing it. Nitrogen currently represents about 10% of a farmer's cost of producing a bushel of corn. Regenerative farmers that rely on tight nutrient cycling for fertility have an incentive to keep nitrogen out of streams and in the field (through the use of cover crops and crop rotations, through no-till farming); raising the price of nitrogen gives the conventional farmer the same incentive. Reducing the dead zone in the Gulf of Mexico requires a reduction in nitrogen runoff of 40%. So application rates must be reduced by something like 40%. Doubling the cost of nitrogen would go a long way toward doing this: one could raise taxes on nitrogen by 5% a year for 20 years, with the taxes paid by the manufacturing companies (effectively, the price of fertilizer would rise). The monies collected would be used to restore riverine wetlands or construct new wetlands to trap nitrogen from nearby farmland.

A market approach offers a payment. Currently western governments spend over $100 an acre on agricultural subsidies; net profit for the American farmer is about $55 an acre; the ecological value of farmed uplands is $200-$400 an acre. Suppose subsidies were cut in half and farmers received a payment of $50 an acre for the ecological value—the public goods—of their lands. Most of this (say 75%) would go for the water that runs off the land: its total amount, the rate at which it runs off, the amount of soil it carries with it, its chemistry (including its load of nutrients and pesticides). A smaller part of the payment would credit the land as habitat: the size of its fields, the length and width of its hedgrows, the variety of crops grown, the connection of its natural lands with neighboring natural lands. To receive the full per acre payment for water, the water runoff and chemistry would have to approach the original condition: rates of runoff, temperature, and chemistry would have to be within (say) 20-30% of presettlement levels. The farmer would decide how to do this.

Of course, if one is attempting to lower nitrogen use, one should stop subsidizing its use. Current price supports support nitrogen use by rewarding production alone. The government essentially guarantees it will buy all the grain that is produced at a set price, that is, government payments will make up any difference between the world price for grain and what the government calculates it costs the farmer to produce. Such costs are based on those of modern chemical agriculture. The purpose of price supports is a political one, to provide cheap food. (Price supports neither support small farms nor protect the environment.) Price supports on grain hurt farmers that grow grain to feed their animals by making grain artifically cheap. (Since the costs of raising grain are greater than its market price, which is driven down by oversupply, it's cheaper to buy grain than to raise it). Thus price supports help feedlots, which are major polluters, by making it cheaper to buy grain (grown on farms without animals, fertilized with artificial nitrogen) than to raise it. An unintended consequence of the boom in ethanol made from corn—another bad but politically irresistible idea—is that the price of corn is rising sufficiently to make the cost of beef raised on grass look better.

Easily erodable farmland (sometimes streamside or riverside land) is now voluntarily taken out of production through the Conservation Reserve Program. It is cheaper to buy such land outright (usually 10 years or less of payments amount to the price of the land); but outright purchase raises the questions of trespassing (people able to walk into a farmer's property through public lands) and of local property taxes (which farmers pay on such lands: essentially Conservation Reserve payments shift the cost of property taxes to the federal government). Outright purchase also reduces the farmer's income. Another method is to purchase permanent conservation easements; in many ways this is equivalent to buying the land, but avoids the problem of trespassing and the matter of property taxes (which however may be reduced on such lands). While underfunded (conservation programs amounted to 8% of federal farm outlays in 2000), probably because it helps small rather than corporate farmers, the Conservation Reserve Program has been a great success, both in reducing soil erosion and in restoring habitat for native animals. (Soil erosion was down almost 40% from 1978 to 1997, although 29% of fields were still excessively eroding. While some of this was due to land in the Conservation Reserve Program, farmers since the early 1980s have also been required to adopt some soil conservation measures to remain eligible for commodity payments.) When the prairies were first settled, the populations of birds like prairie chickens increased (the spilled grain was a resource, as may have been the insects and weed seeds in the fields); as agriculture occupied more and more of the landscape, eliminating their breeding and wintering areas, the birds declined. In hilly parts of the Palouse wheat growing area of the state of

Washington, whose deep, light soils are easily eroded, lands in Conservation Reserve (up to 25% of lands in some counties) have slowed erosion and led to a major comeback of the sharp-tailed grouse, a game bird whose subspecies in the Palouse was nearing extinction.

———————

Modern agricultural landscape are new landscapes. They haven't existed before. Eugene Odum recommended keeping 40% of any ecosystem in its original state (let's say, in a biologically working state; since the original state was always evolving and changing). If farm fields were tucked into a larger landscape of forests or prairie their leakiness would make much less difference. The nutrients seeping from them would be turned into trees, deer, prairie grasses, game birds, or recycled in riverbank wetlands into bacteria, plants and invertebrates; some would feed fish. But in most of the world, turning the landscape and its resources into commodities has destroyed the ability of the pre-existing systems to work, biologically speaking; streams are drains; the natural environment is pulled apart; and farmers focus on their fields and crop yields alone. Our current landscape has been created by capital and government policy, with farmers as more or less willing accomplices. A major part of farm income (half, or $28 billion, in 2000) comes from price supports for commodity crops. Modern price support policy favors unlimited production in order to keep food prices low. The money is recycled to producers of seeds, chemicals, fuel and farm machinery, who, like consumers, have an interest in maintaining the status quo. Export enhancement programs help large grain companies and food processors market their products abroad. Shippers benefit from tax-supported highways and locks and dams on rivers. Seed companies, chemical companies, meat producers and the government sponsor research on increasing production at land grant colleges.

But policy—through programs like the Conservation Reserve Program, through looking at an agricultural landscape as a whole—can also create new landscapes. What would an agricultural landscape that produced grain, meat, habitat for native plants and animals, rivers with fish, jobs, healthy estuaries, be like? In heavily agricultural areas, 15-30% of the landscape would be uncultivated land. Its purpose would be to absorb the nutrients and slow the water running off the fields. It would also provide habitat for native wildlife, much of which is useful to the farmer. (The annual value of pollination by wild insects and of insect and rodent control by wildlife in the United States is calculated at about $50 billion, close to net farm income: half that number is a large number.) Foxes, hawks and owls control rodents, and need protected lounging and breeding habitat. In the irrigated alfalfa and vegetable fields of California's Imperial Valley cattle egrets, white faced ibises and curlews eat Mormon crickets and cutworms and barn owls eat the gophers whose burrows undermine irrigation ditches. Such birds also need other habitats. Pollinators need to be near the crops they pollinate, as do the insect predators of crop pests. Pollination is most efficient when the insects have to travel less than a hundred yards, so fields that require pollination should be no more than two hundred yards wide, bordered by wide hedgerows or other later successional landscapes that will support a healthy population of pollinators. In Costa Rica coffee yields are 20% higher in plantings within a thousand yards of a national forest. Much of the protected land (habitat for predators and pollinators) would be along watercourses, some of it would be hayed once a year, some lightly logged, some of it would be in constructed wetlands meant to filter and slow field drainage (some of these lands could also be hayed late in the summer). Studies in South Dakota found that leaving 15% of an agricultural area in undisturbed land (in this case, Conservation Reserve Land) gave the most economic benefit in terms of wildlife (here mostly considered as game animals); less land provided too little space for the animals, and the economic benefits per dollar fell with more wild land (the ecological benefits would have risen, however). Such natural areas should go up to 30-40% of the landscape and should connect. Not all land is now cultivated even in the Corn Belt (though 89% of Illinois is cropland). Some of the land that would be used to protect streams is already steep and eroding and many farmers would be happy to see it protected, especially if someone else paid for the grading and revegetation. Wild landscapes function better if the agricultural landscape itself is more benign—with mixed crops, smaller fields, less use of pesticides and herbicides, more rotational grazing (more animal flesh produced from grass), fields cut or harvested at times that allow birds to finish breeding. Then the wild and the cultivated mix.

What would this landscape produce? Land taken out of production to reduce nutrient runoff and surround farms with a working biological landscape (producing public goods) would cut the current corn crop by several percent, say 5-10%. Rotating row crops (corn, soybeans) with hay and small grains (wheat, oats, barley, which function to a certain extent as sods) cuts the corn crop by probably 30%; the small grain crop on the other hand would rise. Organic agriculture, or a limited use of nitrogen fertilizer, cuts the corn crop by another 5%. So the annual corn harvest is now down by 40-45% (from 9 billion bushels to perhaps 5-6: the crop 10 years ago). One result will be to raise the price of corn and make growing it profitable again. Modern genetics has been so successful in raising corn yields that in 2006 even agricultural economists said the best thing that could happen to farmers would be for the corn crop to fall by 15%. The size of the corn crop led to federal support for the conversion of grain to ethanol, a motor fuel, in order to absorb some of that corn; and that, combined with rising demand from the developing world, has currently—2008—raised the price of corn to profitable levels. A 45% drop is a big drop, but it won't happen all at once and further progress in plant breeding is likely to raise yields further. Since most of the United States corn crop goes to feed animals, and since, in a regenerative agriculture, animals would be raised on hay (and some grain) produced on the farm, and since feeding cattle corn and soybeans in feed-lots is tremendously wasteful of feed, as well as harmful to the animals, the agricultural productivity of the landscape (the calories it produces in grain, meat, milk) should not fall by anything like the reduction in the corn crop. The land taken out of cultivation will reduce commodity crops (but help fisheries, game birds and animals and produce timber, mushrooms and hay). Farm income in general under the new regime should be higher, since costs of production will be less and prices for crops higher.

My figures are of course as suspect as anyone else's. About 30% of the calories fed hogs and cattle currently come from recycled animal fats. Another 20% of animal feed comes from the by-products of food processing, that is, from the cake left by oil seed pressing, from rice bran, peanut shells and skins, distillery mash, citrus pulp, other grain milling wastes. Cattle may also be fed the high fiber wastes—processed feces—of other domestic animals. Because of mad cow disease, most countries no longer allow cows to be fed cow remains. The United States has been slow to admit the danger to its food supply and here calves are still raised on cow blood and fat. Pigs and chickens are still fed animal fats. Such practices are extremely dangerous. Bags of feed can get mixed up (so pigs eat pigs and cows eat cows). Moreover pesticides and other cancer promoting and hormone mimicking chemicals to which farm animals are exposed accumulate in their fat. So feeding recycled animal fats to animals (which are slaughtered and which we then eat) sets up a chain of bioaccumulation of these chemicals, as the fat is recycled over and over. Cutting this material out of the food chain (and composting it, digesting it for methane, or rendering it into biodiesel) would further lower agricultural productivity. Fat is also a food—one should eat it, just not too much of it. (Composted animal remains are a soil amendment, as they become on Joel Saladin's Virginia farm. In the same way, the bodies of wild animals return to the soil.)

———

Modern capitalist landscapes tend to be extremely simplified; this is their capitalist advantage and fundamental problem: thousands of acres of one cultivar of one grain (corn in much of the Middle West); forests of even aged plantations of Douglas firs or southern pines; suburbs of similar buildings as far as the eye can see. Such simplification leads to other problems. Planted forests of Douglas fir avoid the natural succession into deciduous shrubs and trees, mostly short lived, that fix the nitrogen on which the young firs grow until, 150-200 years later, they reach maturity and acquire a new source of nitrogen fixed by lichens. The increasing size of the American corn crop and the falling price of corn syrup has led to Americans consuming 120 pounds of sugar a year, as bottles of sugared soda got bigger and bigger while the price remained the same; the increased sugar consumption has led to increases in obesity and diabetes (the curves in the production of high fructose corn syrup and the incidence of type 2 diabetes almost perfectly match). While corn and its products got cheaper, fruit and vegetables got more expensive. Extensive monocultures are the result of economies of scale: maximizing profit per man-hour of labor. Smaller scale landscapes are more productive per acre but not per man-hour, because more labor is involved in them. Smaller scales and more labor allow more innovative solutions to agricultural problems. In East African corn fields planting napier grass in the row attracts stem borers away from the corn; the borers prefer the grass, but a sticky exudate from the grass traps the larvae. So spraying is not necessary. Also in Africa, planting a plant called *Demodium* keeps *Striga*, a parasitic pest of the corn plant, from growing in the field. *Striga* is otherwise uncontrollable. Small fields in parts of Africa are bordered with fast-growing leguminous trees. The branches of the trees are used as a green mulch to raise yields, and the stems yield building material or firewood.

Lines of ponds dug along the valleys of natural drainages in monsoon India capture the monsoon rains and let them seep into the ground. This keeps water tables high and within reach of shallow wells, while modern methods of irrigation using deep wells and pumps lowers water tables, putting water beyond the reach of all but the well capitalized. The stored water is used for irrigation and also provides some fertilisation. In the state of Tamil Nadhu 40,000 such ponds irrigate 2,500,000 acres. (Ponds like this, fed by winter runoff and ground water, are also used by nurseries and dairy farms in the northeastern United States and Canada.) In the Sahel, low stone walls (one stone high) laid across hillsides capture sheet flow, increasing the yield of sorghum and wheat by 70%. Both techniques could be used in the U.S. Southwest instead of river irrigation for small-scale agriculture. Capturing sheet flow with lines of stones is already used there to restore dry (overgrazed) desert landscapes.

Over the last 30 years farmers in Niger have allowed acacia trees to grow on their lands at a density of 20-40 an acre. The desert was expanding and their crops were being buried by wind-blown sand so the farmers decided to protect the trees, which germinated naturally, and allow them to grow to shelter their fields. This was an old technique which western agronomists had discouraged. As a writer has remarked, the trees form a virtuous circle which made this southern edge of the Sahara productive. They provide income from the sale of firewood. They shelter crops and prevent wind erosion. Their litterfall and roots enrich the soil. They provide fodder for animals, which leads to more animals and more manure, better crops, and more trees (whose seeds sprout from the manure). The trees are thought to have increased rainfall by 10-20% locally and have reversed desertification on 7 million acres and brought into cultivation 600,000 acres of cropland. Few modern farmers allow trees in their fields, but in California cattle pastures, cattle graze preferentially under blue oaks. The grass there is also more productive (15-100% more forage is produced under blue oaks than in the open). The cows also eat acorns and oak seedlings, as do rodents attracted to the seeds of annual grasses in the pasture, so some young oaks must be protected if the farmer wants to maintain the oaks on his grazing land.

In the United States, a field planted to a mix of corn, beans and squash yields 75% more food than one planted to corn alone, probably because the mixed planting makes more efficient use of water, light and nutrients. Mechanical harvesting of the mix, except for cattle feed, however would not be possible. Similarly, growing corn, beans and cassava together in Cuba doubles yields. Rice yields in Madagascar were raised 3-12 tonnes per hectare by transplanting the seedlings from smaller bundles (so they were handled more carefully, and more plants survived), keeping the paddies unflooded for much of the growing period (which saves water and reduces methane production), and using compost rather than fertilizer. This system has also been tried successfully in China and India. If adopted worldwide, the method would constitute a new green revolution. Japanese rice growers once used green fertilizer, a mulch of leaves and small branches collected from the woods, on their paddies. Several acres of woodland were required for each acre of rice: a riff on using the production of the wasteland for fertilization.

On a slightly different note, clever Middle Western farmers grow cover crops with a mix of up to ten species of grains and legumes to feed their animals; depending on the year, some of the plants do better than others; so yields are somewhat weather-resistant. And 30% of Argentina's farms are no longer plowed. The farmers plant winter crops on their harvested fields (such as black oats) and spray glyphosate (an herbicide that, if it doesn't leach into the soil, breaks down over several days into carbon dioxide and water) to kill competing weeds and grasses before planting in the spring. Grain yields are higher, soils lose fewer nutrients and continue to accumulate carbon. Using special planters, organic farmers grow such no-till crops without herbicides.

Crops must be adapted to the place: a truism, with implications. Winter crops can be grown without irrigation in the Central Valley of California but most summer crops cannot; some crops currently grown there (alfalfa, cotton, sugar beets) are very water demanding. About 80% of California's water goes for irrigation. For rivers to function properly, no more than 25% of their flow should be withdrawn. More than 60% of the water that flows toward San Francisco Bay is withdrawn. So, in general, (though the San Joaquin is much more exploited than the Sacramento) use should be reduced by 35%. Crops with a high demand for water use 20% of the flow. Grapes are often grown in California without irrigation, though they need irrigation to become established. By shifting to more appropriate crops, California growers could reduce their water use considerably, without reducing the value of what they grow. John Muir, who grew irrigated plums at the northern edge of the valley, suggested (half seriously) that the best, most profitable use of the wildflower plains of the Central Valley would be to produce honey. Some areas are simply lucky. Apples grown on certain ridgelines in Mendocino County of northern California, 1200 feet above the ocean (originally the habitat of redwoods and Douglas fir) don't need irrigation because of the fog coming in off the ocean. But the trees have more sun, and the apples ripen better, than in locations near the coast. The area is just south of the apple maggot line and west of the codling moths of the Central Valley; essentially the trees need no irrigation, spraying, fertilizing or cultivation; they need to be pruned and harvested.

Animals are producers of milk and meat in industrial agriculture. The air in confinement barns (another economy of scale) smells of ammonia and hydrogen sulfide (both poisons; people who work in such places can be recognized by their coughs); the animals are tense, crowded and denied their instinctual behaviors. (One senses the miasmic stench of cattle feedlots miles away.) The main job in raising chickens in confinement barns, apart from making sure the machines continue to provide feed and water, is removing dead birds. Pigs are a traditional animal of the Corn Belt, a means of marketing the corn crop as meat. Pigs, like chickens, are much more efficient converters of grain to meat than cattle (though the conversion efficiency of milk is good). Pigs were once raised more or less outside in much of the Corn Belt; now they are raised in confinement facilities, often under corporate sponsorship (corporations provide the facility and the grain; and buy the pork; the farmer provides the land and labor). Confinement facilities have their problems of odor; outdoor manure lagoons that stink and fail in rainstorms; fighting and cannabalism among the animals;

up to 20% of the pork of too poor quality to sell as fresh meat. Some argue there is a moral cost to treating animals so badly (as I would argue there is one in abusing a natural landscape). An alternative to confinement buildings are movable fabric tunnels. The animals are bedded over a deep layer of litter (hay or straw), in which they can root and build nests. They are also free to move around and socialize. The composting of the litter layer (one to two feet deep) helps keep the animals warm in winter and eliminates much of the indoor odor of ammonia and hydrogen sulfide. Allowed to be pigs, the animals are much less stressed; costs are 60% less than in confinement barns. After the pigs in a tunnel are slaughtered (perhaps one day this will take place on the farm, in familiar surroundings, without the stress of that final journey), the tunnel is taken down and moved, the bedding scraped up in a pile to finish composting, then spread on the fields. In a world more attuned to local markets, the ground under the tunnel, cleared of sod by the animals and saturated with manure and urine would be used to grow vegetables. The tunnels are large; one can also rotate them over the fields, using the land underneath to grow corn. In the warm season the hogs are raised outside in pastures (after a season's use, those pastures could also be planted to crops; first cover crops, then vegetables or grain).

The English landscape before World War I developed under the twin pressures of capitalist agriculture and the land's use for hunting. Many common lands of England became aristocratic lands after the 1200s, with the extinction of the hunting rights of commoners. (This was the time of Robin Hood.) Private lands without heirs passed to the local noblemen, who accumulated much land during the 1400s and 1500s, the centuries of the Black Death. During the 1500s common lands were enclosed by large landowners, to increase the national prosperity by raising sheep for wool. Later, as the value of manure on cropland became clearer (one acre of grain needed the manure of half a cow), Parliamentary enclosure laws allowed rich landowners to enclose the remaining common lands for pasture for their cattle. In return tenant farmers gained heritable rights to their farms. But English estates were managed for both agriculture and game. Hunting, once training for warfare, retained its grip on the rural and noble mind. Entailment of the estate to the oldest son meant long-term development of the land was possible. Labor in England was cheap. So the fields were bordered with hedgerows that bloomed with wildflowers in the spring and provided habitat for some game birds, songbirds and animals, while rough woodlands (copses) were maintained for pheasants for the fall hunt. (Red grouse moors were found on the aristocratic estates in Scotland.) The weedy edges of meadows helped shelter Hungarian partridge, while the meadows themselves yielded pasture and hay. Predatory birds and mammals were trapped and shot. Wildlife was abundant but not diverse, with few wide ranging herbivores or carnivores. The owners of the estates knew what they wanted, whether one agrees with their goals or not, and ran their farms both to make a profit and to enjoy their landscape. Maintaining the hunt was regarded as a social imperative. After the 1950s the profit motive took over and the old landscape started to disappear. By contrast, in the United States farming became commercialized almost upon arrival. Speculation soon broke up the town plans of the Puritans, which allocated to each family the amount of land it could use—even here, the rich, who could hire labor, got more. In 1910 in the United States, half the farmers had been on their farms for less than five years, and half the land in the Corn Belt was rented.

I once saw a drawing by a Mexican artist of his home. The small houses of the village were on a terrace above the river, among fields of corn, the river below. In the background was a forested mountain with an eagle perched on top. Agricultural landscapes like this are also new landscapes, but ones that better fit into the biologically functioning habitat of the earth, with its rivers and seas, eagles, migrating butterflies, geese, fish. Agricultural landscapes in this sense become the second nature of the garden; maintained not only for profit, but for the way they fit into a whole landscape. Farming becomes much more of an intellectual game. In this new landscape large trees in the gallery forests along rivers attract breeding ospreys; hollow trees shelter breeding owls or kestrels, Cooper's hawks nest in dense clumps of backyard evergreens, merlins and sharp-shins ambush blue jays at the feeder: such birds also hunt mice, rats, grasshoppers, English sparrows, starlings, rock doves and other endemics of the agricultural landscape. Behind the evergreens in the yard are beehives. The absence of large carnivores (wolves, mountain lions, bears) means more abundant mid-size carnivores (skunks, foxes, raccoons, oppossums) that eat birds and smaller animals (amphibians, Norway rats, rabbits, mice; coyotes also eat deer, especially the fawns, and in the East seem to be evolving into a larger animal, in order to be a more efficient predator on them). Cliff swallows nest on the sides of barns, barn swallows under roof overhangs, bats roost under the eves. Mushrooms grow in the woods and perhaps are cultivated by the farmer. Unplowed springtime pools in the fields provide habitat for migrating ducks and shorebirds and for breeding amphibians (these are also aquifer recharge areas that filter runoff from the fields); in the Southeast, such damp, thorny tangles are habitat for quail. If grassed, aquifer recharge areas can be mown late in the season, or grazed (along with the cover crop); if wooded, lightly logged. The fields themselves grow mixtures of crops; wide hedgerows (50-200 feet, so called ecological edges) break the wind, shelter helpful insects and birds, sloughs, ephemeral streams. Agricultural support payments no longer support crop prices but compensate farmers for maintaining a working biological landscape, with healthy streams. How the farmer maintains a healthy landscape is up to him. Not all biodiversity is functional, but all of it is beautiful. Thus one creates a new landscape.

Chapter 18
Forests

Like fields, forests affect atmospheric chemistry, the chemistry and flow of local streams, local and global precipitation, climate. Root growth accounts for 50% of the net primary productivity in forests (the net yearly production), leaves for 25-35%, growth of wood the rest. About 20% of the mass of forest trees is roots, most in the top few inches of soil. The growth and death of roots delivers carbon to the soil, removing it from the atmosphere. Thus old growth forests continue to store carbon, even after growth of their trunks has slowed. Storage varies with the type of forest. It falls precipitously in the stems of planted southern pines after 25 years, when their growth begins to slow, but probably continues underground. The dry Ponderosa forests of the Rockies, with their frequent low level fires, may be carbon neutral, while the mature rain forests of the Pacific Northwest continue to store considerable carbon. In general, forests sequester between 0.5-1.0 ton of carbon per acre per year. The alluvial forests of the Mississippi Valley are capable of storing 2 tons of carbon per acre per year—and so, at $20-$50 a ton, or $40-$100 an acre, are worth in carbon storage what they are as farmland. Tropical peat swamp forests in Indonesia sequester 200 tons of carbon per acre in their soils and standing biomass, an argument for paying Indonesian landowners to not cut them down. Draining and logging tropical peat swamp forests—for instance, for palm oil plantations to produce green diesel fuel for Europe—takes a century or so to create a carbon benefit and currently (2008) produces more carbon dioxide than the burning of fossil fuels in China.

Evergreen forests release large amounts of hydrocarbons in the summer, as part of their cooling mechanism; these chemicals become part of global chemistry and in the presence of nitrogen oxides (products of combustion that are always present in the modern atmosphere) and sunlight, produce photochemical smogs. (So President Reagan's comment that forests produce smog was partly correct, though cutting down forests to prevent smog would be nuts.) At least 120 chemical compounds are present in

Sierra Nevada mountain air. These are insect deterrents and thermo-regulating chemicals. The monoterpenes, which prevent and cure cancer, are among the most abundant. Monoterpenes enter the bloodstream through the lungs and the limbic system through the olfactory nerves. So a breath of mountain air may be healthy.

A forested landscape has a lower albedo (absorbs more of the sun's heat). It is warmer in winter; the sheltering trees also make the temperature of the ground much less variable than that of bare ground or grassland. In summer, a forested landscape tends to be cooler because of transpiration by the trees. (On Vancouver Island, in a relatively cool climate, cutting the forests has produced a permanent temperature increase of 1°-2° C.) Over large areas, the albedo of the ground affects temperature, precipitation and the speed of jet stream winds. Because of their absorption of sunlight, evergreen forests in snowy regions probably have a net warming effect on the climate, despite their taking up carbon dioxide from the atmosphere. Snow melts some weeks later in forests (especially evergreen forests) and so spring runoff is later; and slower. A later runoff makes (for instance) more water available over a longer period in California rivers. Fog-catching trees, like redwoods, increase streamflow (2-3 times more water reaches the ground under redwoods than not); fog-catching forests on the Canary Islands recharge the islands' aquifers, which dry up when the forests are cut (and cannot regrow because the ground is too warm for condensation to occur); droplets of water from fog-catching shrubs on the Galapagos Islands feed herbaceous plants and collect into pools, providing water for giant tortoises. In general however, forests lower runoff by intercepting and evaporating 35-50% of the precipitation that falls on them. They thus reduce peak flows (floods) and, by storing more water in their soils and recharging aquifers, increase the low flows of summer. The water that flows from forests is cooler. In humid climates such regulation of streamflow is generally desirable; in drier climates, forests, especially introduced forests (eucalyptus in California, saltcedar or *Tamarix* in the American Southwest, Mediterranean pines in the fynbos shrublands of South Africa) may reduce surface water flow considerably (to zero in small streams).

Forests control stream temperatures. Summer water temperatures are 10 degrees lower in shaded reaches of streams. Streams that flow through old growth forest accumulate fallen logs at a rate of about 1 every 10 feet.

The trees help stabilize the streams, creating pools. Their wood is turned by detritus feeders into nutrients for freshwater organisms, which become food for fish; the nutrients are retained in the swirling pools. Salmon reproduce and survive better in log choked old growth streams, with their shade, slower currents, more complex habitats of pools, more complete nutrient capture by stream organisms, than in the more or less channelized streams of commercial tree plantations. Undeveloped lakes in Minnesota have about 500 fallen logs per kilometer, or about 1 every 6 feet; while developed lakes have about half that, and near cabin sites have only 1 log every 50-60 feet. (People remove them and cut the forest, the source of new logs, along the shore.) In lakes, logs provide habitat and shelter for fish and detritus for invertebrates. The decaying logs become part of the lake's food chain. (Log free parts of lakes have fewer fish. Bass, for instance are 5-60 times more abundant near brush shelters.) In rivers, grounded logs catch other logs, form jams, sometimes accumulate sediment and become islands, forcing the river about them and forming side channels. The slowly flowing side channels become habitat for ducks and juvenile fish. Fallen logs along rivers also protect the shoreline, create pools, slow water flow, enlarge (with partial dams) the floodplain and thus the riverine habitat, and provide detritus-based nutrients. Sailing out to sea, rafts of logs, hung with barnacles and other creatures, increase the size of the ocean edge (several hundred thousand per year may have sailed out from precontact North America); and perhaps introduce alien organisms to new environments (a job done much more efficiently now by the ballast water in ships). For unknown reasons, floating logs in the ocean attract schools of fish. Floating man-made constructions with homing beacons are used as lures in some West African fisheries. A floating log in the eastern tropical Pacific, fished for tuna, yielded 240 tons the first day, and then 40 tons a day for 5 days. In West Africa many logs from illegal cutting are lost in river drives and float out to sea, ending up on beaches in numbers that interfere with nesting sea turtles.

In aboriginal forests much of the forest was old growth, or so-called primary forest. How much was old growth varied with the type of forest and the site. The mixed deciduous and conifer forests of Maine are thought to have been about 30% old growth. Records from early surveys in Maine indicate that 2% of stands were recently burned, 14% were birch and aspen (short-lived, early successional trees), 25% were young forest (75-150 years old), 32% were mature forest (150-300 years old), and 27% were old growth (greater than 300 years old). So about 60% were what we would call old forest. Natural forests are subject to many types of disturbance, which range from the deaths and collapse of individual trees to hurricanes or tornadoes blowing down whole tracts of forest. The type of destruction varies with the location of the forest. Natural fires in the northern hardwood-hemlock forests of Maine occur every 800-1400 years, an interval much longer than the maximum ages of the trees (250-500 years). Icestorms in interior New England occur perhaps twice a decade and serious hurricanes once a century. Windthrow, along with ice storms, heavy early snows, pathogens and fire (spreading from the more fire-prone coastal and northern forests) likely kept Maine's forests young. In the Midwest the interval between disturbances is also much greater than the lifetimes of the trees (800 years for straight line winds from thunderstorms, 1000 years or more for tornadoes). The natural fire interval is shorter than in the east (summers are hotter and drier) but about 85% of the forests of the Upper Midwest were mature and old growth when first seen by Europeans. Catastrophic crown fires occur every 150-500 years years in mesic stands of Douglas fir on the western slopes of the Cascades in Washington. Less destructive ground fires are more common. The trees are somewhat fire-resistant and also have adaptations to fire. Once the lower branches of a Douglas fir become shaded and begin using more resources than they produce, they lose their needles, die and fall off; this removes the fuel ladder that lets ground fires reach up into the crowns of the trees. Huge floods occur perhaps once a century on creek bottoms inhabited by coastal redwoods; some trees are felled by the floods. New shoots sprout from their trunks and broken-off stumps. Trees still standing send out adventitious roots into the deepened muck, and benefit from the new soil. (Fifteen such floods have raised the floodplain along Bull Creek, a tributary of the Eel River of northern California, by 30 feet over 1000 years.) Such adaptive behavior lets the trees live for two millennia and outcompete the Douglas firs and California bays on rich alluvial flats. Stand and soil-removing fires follow insect infestation every 50-200 years in boreal spruce-fir forests, resulting in complete stand replacement and temporary soil impoverishment (many nutrients are vaporized).

North America now lacks large tree breaking animals like the elephants that renew acacia woodlands in Africa, but insects like spruce budworm kill stands of balsam or spruce, which may then be replaced by birch or aspen (which are periodically defoliated by tent caterpillars). Budworm outbreaks, like that from 1950-1954 in Atlantic Canada, kill whole forests, which then burn and regrow. During a budworm outbreak, budworm larvae increase from 1000 per acre to 8 million per acre, and the populations of wood warblers that eat budworms increase by 10 times or more; but not enough to contain the insects. (A virus or the death of the forest does that.) By allowing the (less dominant) aspen and birch to replace the (more dominant) spruce and fir, budworms renew old or stressed conifer forests. Conifer productivity on a site reaches its maximun extent at relatively low levels of nitrogen. Hard to break down, conifer litterfall ties up more nitrogen, while the nitrogen in the leaves and branches of the more demanding hardwoods is quickly recycled. So patches of aspen and birch among the conifers raise the productivity of the forest and the invasion of aspen and birch after a budworm outbreak (and fire) help renew the soil. (Similarly, nitrogen availablility under an intact stand of hemlocks—another conifer—is low, but equals that of a hardwood forest when small groups of sugar maple share the canopy with the hemlocks.) Some writers claim hardwood trees and shrubs have a keystone function in conifer forests. They provide habitat for butterflies and moths, which attract birds, which eat the caterpillars that eat conifer foliage; they provide habitat for parasitoids that also help control conifer-eating insects; and their sprouting ability stabilizes soils after disturbances like treefall or fire. Under capitalist management the spruce-fir forests of Atlantic Canada, where budworms and hardwoods are controlled by spraying, and the proportion of spruce is falling thanks to heavy cutting (the forests are being turned into monocultures of the faster growing balsam fir), are becoming less productive.

Beaver dams in northeastern conifer swamps also kill conifers, which are replaced by damp-tolerant aspen and alder. The trees enrich the habitat with their litterfall, as well as provide food for beaver. The turnover of the plants and animals in the ponds (which store soil and nutrients) increases the nutrients available for the trees.

Browsing by mammals also shapes forests; an increase in browsing by deer, elk, moose, snowshoe rabbits, or meadow voles, by decreasing the survival of the more palatable species (oak, white pine, eastern hemlock, Canada yew, northern white cedar), will change the future composition of a forest, and reduce a mixed herbaceous understory to ferns and grass. Heavy browsing by elk in Yellowstone Park tends to eliminate streamside willows and reduce beaver wetlands on small and medium-size streams; the streams straighten out, become more erosive, cut more deeply into their beds, lowering water tables. Browsing by elk also prevents reproduction by aspen (another beaver food). The reintroduction of the wolf to Yellowstone let both types of tree come back, partly by reducing the elk herd (wolves seem to keep the herd 20-30% below what weather and vegetation would allow), but mostly by changing elk behavior: elk are frightened of spending too much time in dense cover where they might be ambushed by wolves. With the return of willow and aspen, beaver returned, trout became more abundant in the deeper, slower streams, and a greater variety of songbirds bred in the streamside vegetation.

People influence forests. Throughout the temperate zone, people began to shape forests long before they had reached equilibrium with their post glacial environments (3000-6000 years ago in much of North America). Anthropogenic fire, such as the cool ground fires set in early spring or fall by Native Americans in the oak and pine woods of southern New England, produced an open parklike wood of large nut bearing trees, mixed with areas of younger vegetation, amidst much larger areas (such as white cedar swamps and spruce-fir forests) that were never (or rarely) burnt. Anthropogenic fire probably created the open oak and hickory forests of the Middle West. Without fire, they are now returning to the more shade-tolerant beech and maple. Primary forests have large trees, with a standing biomass 3-6 times that of second growth (and several times the board footage, which explains the size of the early cuts). The most productive old growth forests in the eastern United States carried 125-250 tons of wood per acre. In old photographs the tall trunks and root buttresses of Middle Western oaks and sycamores dwarf the nineteenth century biologists with their high laced boots and long shotguns standing beside them. The large old trees in these forests produced enormous amounts of nuts and seeds, or mast.

Once they reach a certain age all the trees in a forest produce mast. Most trees produce some seeds after a few decades, but become large producers as they mature in size and have spreading branches and root systems capable of supporting large numbers of seeds. (The story is the same for fish, where large females produce many times the eggs of smaller animals.) Mast production is related to crown development, so burning, by thinning the forest, increases mast production; the more rapid turnover of nutrients in a burned forest may also help produce more seeds. In the cool North-east, a red oak begins producing acorns at 25-50 years, and then continues production (which would reach a peak at perhaps 150 years) for 100-300 years. White oaks and sugar maples, longer lived trees, produce large crops of mast for longer periods. The nuts and seeds of the trees, large and small, are food for birds and rodents. The several varieties of red crossbills, a finch of the boreal forests, specialize in the seeds of specific species of pine and spruce, to which the shapes of their beaks and tongues (the seed-extracting devices) are adapted. Seed production by conifers and birches (an alter-nate food) tends to be synchronous over a large area, but irregular, so the birds move long distances (hundreds to thousands of miles) over the boreal forest in their search for food, and adjust their breeding schedules to the supply of seeds. With sufficient food, they will breed in mid-winter, with temperatures -30° F. or below. Thus the populations of many northern seedeaters (crossbills, redpolls, pine siskins, pine and evening grosbeaks) are not regional but continental. The rodents that eat the tree seeds sup-port populations of predatory birds and animals (weasels, minks, martens, owls), and recycle (like the birds) many of the nutrients in the seeds back to the trees through their droppings. The larger tree seeds (beechnuts, acorns, hickory nuts, pecans, formerly chestnuts) are eaten by both the rodents and the larger animals (deer, elk, turkeys, bear, probably buffalo). Such animals and the nuts themselves were once eaten by people; aboriginal people also ate the small rodents (mice, squirrels), which are extremely abundant in mast producing forests. (About 222,000 mice and other small rodents live in a typical 10 square miles of eastern forest, which explains the presence of mouse skeletons in fossilized human dung.) Recently 3-5 billion passenger pigeons supported themselves on the mast of the Eastern forests (squirrels may have exceeded that number), nesting where the acorn crop from the last year remained on the ground in the spring. Their habit of migrating over wide fronts in parallel flocks has been interpreted as an adaptation to locating areas of sufficient food to nest. Much of the forestland that supported those birds has been converted to cornfields; corn, cattle and pigs have replaced the passenger pigeons and beechnuts.

Trees reach economic maturity long before they reach biological maturity and so, under modern management, are cut down soon after they become large producers of mast. This changes the nutritional relationship of the forest with its inhabitants, reducing the numbers of seed eating animals and their fur bearing predators, while increasing the numbers of animals that eat browse. (White-tailed deer, cottontail rabbits and grouse seem to do well, while turkeys, squirrels and bear are probably reduced. But many animals, including deer, turkeys and bear, eat both browse and mast). It eliminates the tall, old trees raptors use for nest sites, from whose dead lower limbs flycatchers hunt, and the old rotten trees in which woodpeckers and chickadees excavate nest-holes, and in whose large hollows bears and raccoons den. Economic maturity is determined by a change in the rate of increase in the mass of the tree. As trees approach their final heights, the yearly increase in new wood fiber slows. In long-lived woody plants, maintenance respiration increases with age; so there is a necessary reduction in net annual production of fiber, that is, of growth. Risk of loss of the tree from windthrow or disease also rises with age; so the trees are cut. Douglas firs, trees that reach ages of 500-700 years, that stand a century or two as dead snags, and whose fallen logs take up to 300 years to rot, and so whose influence on the forest spans a millennium, are cut down soon after they reach the majority of their height growth at 80 years. Redwoods that once grew for 2 millennia or more are also cut at 80 years. Old Douglas firs support a community based on arboreal lichens, whose mass may be 4 times that of the foliage. The tops of old growth redwoods support another forest 150 feet up. Protruding redwood tops break off in storms and the upper branches bend up, turning the top into a small forest of two foot thick vertical branches, dead stubs, soil (which collects in cavities left by broken-off stubs and in hollows in the branches), with banks of fern, blueberries, seedlings of Douglas fir and Sitka spruce and nesting seabirds. Part of the temperate rainforests that occupy west facing coasts worldwide, most redwoods are found within 10 miles of the sea. They grow where their limbs and needles can comb water from coastal fogs—12 inches during the usual summer dry season—and out of the reach of salt spray. Marbled murrelets nest in their crown forests, ancient murrelets among their roots, the former now in steep decline as their nesting habitat has shrunk from 2 million to a few thousand acres. Southern pines, grown mostly for paper pulp, are cut at much earlier ages, 25-35 years (younger trees make better pulp).

Like browsing, cutting changes the composition of the forest. The cutting rotation in the Northeast is now too short for hemlock, a tree that can live 600-900 years and is probably worth cutting at 150 years. (The crown forests of old hemlocks show some of the characteristics of old red-woods.) Frequent cutting in the Maine woods has favored balsam over red spruce (another slow-growing tree). In 1902 the volume of spruce was 7 times that of fir; today the two are virtually equal in volume. Frequent cutting in the forests of northern Pennsylvania favors trees that sprout (cherry, oak, maple) over softwoods like white pine and hemlock. Early logging probably favored American chestnut, which rose from 4-15% of forest volume in early surveys of New Jersey and Connecticut to 60% (becoming the most abundant tree species) in 1900. While oaks and hickories produced hundreds of pounds of nuts per acre, chestnuts produced thousands, and every year; so the end of the chestnut must have meant an emptier forest.

Cutting can also mimic natural process. In Sweden white spruce usually reproduces itself through small gaps, such as the death of a large tree. Such forests can be reproduced by selective cutting that leaves most of the forest intact and also leaves existing snags, fallen trees and some old trees. One would also leave most deciduous trees, which have important functions in conifer-dominated woodlands. Spruce-fir stands in boreal North America are often fire-adapted communities, reproduced by stand-replacing fires. Such communities can be maintained by a modified clearcutting, leaving wide bands of trees along watercourses, with large, scattered clumps of trees (about old trees) left in the cut as nesting sites, sources of seed and future snags. Tops and slash are left on the ground, for their nutritional value and for shade. Spruce and fir are usually used for pulp in the Northeast and so are cut as soon as they are large enough to make cutting economic (this size goes down as logging becomes more mechanized). Red spruce makes valuable lumber (it is used for the sounding boards of pianos); but it is a slow-growing tree. Leaving clumps of red spruce among the balsam to mature increases the variety and value of the forest. Letting some aspen and birch mature (birch for sawlogs, aspen for pulp) increases the forest's productivity and makes it more friendly to game animals. To minimize blowdown and drying of the forest, clear cuts are usually limited to a width 1.5 times the height of the trees (or 100-150 feet). Such cuts attempt to mimic the effects of catastrophic disturbance in primary forests and leave the landscape conducive to the movement of seeds and animals. In the East, such cuts are not burned (similar, lodgepole pine cuts in the West probably benefit from being burned). Leaving merchantable trees in the woods increases the future health and productivity of the forest. It reduces the volume of wood harvested, and thus current profits to the landowner, but current losses in volume will be more than compensated by the more valuable old trees of the future.

As forests mature, they accumulate dying and fallen trees. Fallen trees cover up to 20% of the forest floor in mature Douglas fir woodlands in California. A recent checklist for old-growth forests in the East included 3-4 logs greater than 16 inches in diameter, per acre, on the forest floor: a much smaller number. Hollow trees become den or nesting trees for various animals and birds; large hollow trees are used by bats and chimney swifts as well as hibernating bears: Audubon counted 9000 chimney swifts leaving one hollow sycamore, an unimaginable number. (Bats now make use of attics and mines, chimney swifts use chimneys, both birds having become in the Northeast—like the nighthawks which nest on the flat roofs of commercial buildings, and whose notes drop from the skies of suburban evenings—animals of settled landscapes.) Downed trees create new habitat on the forest floor for invertebrates, rodents, birds and amphibians. Soil accumulating on the upslope side of a log provides habitat for burrowing insects and small mammals, while the downslope side provides shelter and nesting sites. (An acre of Maine woodland has more biomass of salamanders than of moose: a fact only briefly surprising.) Rotting logs have a greater mass of living tissue (in the plants and decomposers living in and on them) than growing ones. Nitrogen fixing bacteria living in the guts of Pacific dampwood termites fix nitrogen for the termites. The nitrogen ends up in the nutrient-poor ecosystem of the rotting log and fertilizes tree seedlings growing on it. The invertebrates of the logs and the forest floor (earthworms, centipedes, millipedes, spiders, beetles, mites, springtails, psocids, nematodes) are eaten by small mammals, birds and amphibians (shrews, toads, frogs, thrushes, salamanders; also by the less abundant snakes and turtles); the amphibians, reptiles, birds and mammals form part of the forest's food chain. The fallen logs also act as dams to keep forest soil from moving during rains and thaws.

The old trees of a primary forest create a moister, shadier environment. Their large limbs are habitat for lichens that synthesize their nutrients (including nitrogen rich proteins) from the air. These nutrients enter the forest ecosystem through leaching and litterfall, and through the dung of the squirrels and rodents (often different species than the mast eaters) that eat them. It is said that young Douglas firs first grow on the nitrogen banked in the soil by the nitrogen fixing shrubs that follow forest fires, then (as adults) on the nitrogen synthesized by their lichens. A major food of the rodents of old growth woods are the fruiting bodies of the mycorrhizal fungi that are allied with the roots of the trees. The fungal spores pass through the rodents' digestive systems unharmed and are deposited on the forest floor and in their tunnels and burrows, generating more fungi. These rodents are part of a new food chain of the primary forest that ends in predators of the deep woods, such as pine martens and spotted owls. The water that flows out of such woodlands, thanks to the shade and the lack of sediment, is cold and clear (sometimes tinged brown with tannins), nutrient poor, habitat for salmon and trout. Tree roots stabilize the streams; and the trees themselves, falling across the brooks and larger creeks, also stabilize them, creating pools. Salmon or sea-run trout, running upstream, like the seabirds that nest among the trees, contribute nutrients from the sea.

That only a third of the mixed Maine woodland was in primary forest implies a fairly high level of natural disturbance. In much of the Appalachians and Midwest the disturbance regime was several times the lifetime of the trees. The result was old forests with complex mixes of trees of different shade tolerances and ages. In natural forests non-catastrophic replacement is usually in patches. One or a group of trees that rise slightly above the level of the canopy are blown down, or a tree collapses from old age and takes some others with it, and the shade tolerant saplings that have been growing in the understory shoot up. If light is sufficient, the seeds of the shade intolerant species that have been waiting in the soil sprout and grow. Yellow birch and white ash will grow in the gap left by a single tree, while red oak, black cherry, sweet birch and tulip tree need larger gaps. Hurricanes, tornadoes and fires clear larger areas. What develops on a site after a catastrophic disturbance depends on what trees are left alive (the so called seed trees), the presence of trees that sprout from roots (many deciduous trees and shrubs), the seeds in the soil bank, and what is brought to the site by birds, mammals and the wind. So the forest one sees on a site depends on the site's history, its surroundings (another history), and chance, as well as on its soils and climate.

In the conifer and mixed hardwood forest of the Boundary Waters Canoe Area, the original fire rotation time was thought to be 50-100 years. Such short rotations favor fire adapted trees: those that sprout from roots, such as quaking aspen; or that have well dispersed seeds, such as paper birch; those with serotinous cones carried high in the canopy, such as jack pine or black spruce; or trees whose thick bark lets them tolerate fires, such as red pine, which become fire resistant after 50 years. (Short fire rotations must be fairly old in the general area since the Kirkland's warbler is adapted to breeding in fire-succeeded stands of jack pine; it will breed nowhere else and abandons older jack pine stands.) If a new stand escapes fire for 40 years or more, jack pine and aspen slowly succeed to more fire resistant, very mixed stands of black spruce, balsam fir, paper birch and white cedar. Red pines, which are capable of massive recruitment after a fire slowly die out in very old stands (very old: red pines are still present after 300 years). In this case pines will no longer reproduce if disturbance is limited to windthrow and spruce budworm; they need fire. The spread of hemlock into a stand lengthens the fire frequency indefinitely (they draw up groundwater to the surface and moisten the soil), and if a stand escapes fire for a very long time, a forest dominated by hemlock and the shade tolerant sugar maple may develop. By lengthening the fire frequency, the hemlock makes possible the success of the fire susceptible maple. Both hemlock and sugar maple produce a forest floor unfavorable to the establishment of other species, so this forest is somewhat self-maintaining (but in fact no more of a "climax" than the more mixed forests it replaced: felled by a windstorm, it will be replaced by aspen, spruce and birch).

Primary forest in the eastern United States is characterized by fairly large trees (16-20 inches or more in diameter according to a recent checklist; but certainly many first growth trees were larger). Other characteristics include a relatively small percent of intolerant (pioneer) tree species in the canopy; a full compliment of spring ephemerals suitable to the site (flowers that bloom and set seed before the tree leaves come out; these plants are not too much bothered by occasional logging, but frequent logging, clearcutting or grazing will seriously reduce or eliminate them; and since they are pollinated and their seeds distributed by ants, they spread slowly and can take centuries to recolonize a site); by a full compliment of bryophytes (mosses, liverworts, lichens: plants of damp and shaded woods), including those characteristic of old growth; by large old logs on the forest floor (microhabitat for many birds, mammals, reptiles, invertebrates, decomposers, mosses, fungi, tree seedlings; forests in northern regions probably need 12 tons of coarse woody debris per acre to maintain site quality); and by large, standing snags. Many of the large trees of a primary forest are rotten. This was a complaint of woodcutters in early Maine, who would chop a hole 3 feet into an old pine, 50 feet above the ground, to check its condition before felling it. (Such large old pines were "pumpkin pines," their wood valuable for being soft and easily worked.) Many species of birds and mammals in the eastern forests require standing dead trees for perching, foraging, nesting, roosting and denning. The ivory billed woodpecker is said to have specialized in debarking large, dead trees to eat the insects beneath the bark. The loss of such trees as southern bottomland forests were logged and converted to agriculture was said to be a major factor in its extinction. (Audubon however shows a family of ivorybills collecting beetles from under the bark of a small dead stub.) Many writers claim that old growth or primary forest is not absolutely necessary in a functioning natural landscape, even along streams, but no one can be sure of this. Some of the forest's inhabitants (those microbes, fungi, invertebrates) will go extinct without it and the landscape in a larger biological sense will be reduced. Other writers argue that tracts of old growth must be very large to function truly as primary forest. The woodland must contain the large carnivores (grizzly bears, wolves, mountain lions) whose presence indicates an intact food web. Such tracts require 250,000-300,000 acres of contiguous woodland, with connections to other such areas.

Logged forests are not natural forests. Economic considerations will never allow them to move far into the realm of old growth. Their soils may or may not achieve a net accumulation of carbon. Logging disturbs the soil and exposes it to sun and rain, where increased microbial action releases carbon and other nutrients. The decay of logging debris releases carbon. It takes 10-20 years for a logged second growth forest to begin a net accumulation of carbon, 45 years for regrowth to compensate for the carbon released by the decay of debris left from cutting an old-growth forest. And the carbon in the logs removed returns to the atmosphere quite rapidly. A third to a half of the logs end up as waste (sawdust, planer mill shavings, cut ends of boards), which is burned or decays. The rest of the carbon in the logs returns to the atmosphere after a short detour through the man-made world (for much of the material, made into paper or pallets, less than 10 years; that made into furniture or buildings lasts somewhat longer—less than 50 years on average). But it should be possible to mimic somewhat the process of natural forest succession through logging: to provide a variety of native habitats (if not in the same proportions as in the aboriginal forest); to leave some old trees and downed logs; to conserve forest soils; and to protect forest streams from too much sun and from the constant pulses of silt and nutrients that logging produces. All this of course will cost the landowners and loggers money, until, after a century or so, the higher value of the logs compensates for the reduced volume of wood. (This sort of delay, by reducing current income, always results in future losses, under the capitalist doctrine that discounts the future for the present.)

Cutting, like fire, sets the forest on a new trajectory. In the humid forests of the Northeast and Middle West, selective cutting favors the shade tolerant beech and sugar maple; clear cutting the warmth loving pines and oaks, and other pioneer species like aspen, pin cherry and paper birch. Clear cutting in dry western forests can eliminate the mycorrhizal fungi on which the trees depend and make reforestation of west facing slopes difficult. (Adapted to living with certain species of trees, the fungi must find a new host in two years or die.) Clear cuts, and heavy selective cuts, create more or less even aged stands, with more pioneer and early successional species, while light selective cutting creates forests of more shade tolerant species. The early successional forests created by heavy cutting (taking all merchantable trees when entering the woods) are supposed to provide a constant, sustainable flow of wood, at the maximum potential of a site. But, an artifact of cutting, these forests are quite unstable, vulnerable to insects, diseases, climate change.

Logging influences what trees grow in a forest, and speeds up (may double) their rate of growth. Logging also has many negative effects on a forest. The stems and roots of trees are wounded, creating entry points for fungi and disease. Soils are compacted and erode. Compaction lowers soil oxygen levels and reduces water infiltration, slowing growth for some time after a cut. Heavy cuts move water and nutrients into streams. In areas where the natural water temperature in summer is near the upper limit of their requirements, clear cutting may raise stream temperatures above what salmon and trout tolerate. So fish disappear. Cuts along streams remove the trees whose roots hold the bank together. With additional water and soil moving into them, the streams widen, scour, straighten out and take some time (usually more than a decade) to return to a more stable condition; after a few decades, logging starts again. Heavy cutting on steep slopes in the Pacific Northwest makes the slopes liable to landslides following insignificant rains; slides begin 3 years after logging, when the roots that held the slope together have rotted. Cutting near streams also removes the logs that would have become coarse woody debris in the water and on the forest floor. Such logs retard water flow, and help stabilize streams. Logging fills streams with tops and slash. The cumulative effect on streams can be catastrophic for fish, as is the case with many salmon and trout streams in coastal California, Oregon and Washington.

Slope tremendously influences the erosive capacity of water, which increases by the fifth power of its velocity. As the speed of flowing water (down a skid trail, through a culvert) doubles, it is able to carry 32 times the sediment and move particles that are 64 times heavier. Much of the sediment in west coast streams comes from logging roads and skid trails. When these are treated properly after logging is over (which includes closure to vehicles), the sediment loads decrease, the streams clear, and salmon and trout return to them. Hunters and drivers of recreational vehicles object to such closures, which limits back country travel to horse or foot. (Included with the sediment are hydraulic fluids from the logging eqipment, 70-80% of which escapes in leaks and spills, and the chain oil from the saws: arguments for using biodegradable oils.)

A forestry that tried to recreate the forest as an ecosystem would place old growth in the commercial forest along streams. Very lightly logged primary forest (a tree per acre every 5-10 years) would cover the banks of deeply incised streams, and cover the streambank out for 50-100 feet otherwise. (A tree per acre every 5-10 years is 4-8 trees per acre for entries at 40 year intervals; depending on the market, logs from such trees might be worth $1000 or more at the mill, or $300 in stumpage to the landowner.) Probably less than 20% of eastern forests have soils good enough to make the risk of growing trees greater than 100-150 years old worthwhile (coves, valleys, the toe slopes of hills, places where soils are derived from nutrient rich bedrock.) Most of the best forest soils in the East are now in agriculture. So one puts old growth where, whether a good commercial risk or not, it will do the most biological good. Then fingers of old growth penetrate the logged woods. Primary forests are multistory forests in the Northeast and upper Middle West, where most trees top out at 80-100 feet, and large white pines penetrate the canopy to reach 150 feet or more. These fingers of old growth will not function well if they back entirely upon clearcuts (winds will blow the trees over, too much sun dry the forest floor). So the forest that abuts the old growth is a mix. If we follow the Maine example, one third would be in early old growth (cut at 125-175 years, or older, on better sites; at 80-100 years on poorer ones). The value of the water running off the land and the land's value as a carbon store will affect cutting choices. Two-thirds of the forest would be maturing and young forest. If the water that runs off the forest is worth $30 an acre, when close to the presettlement original amount and condition, forest management that keeps the water in that condition should be worth that. Carbon storage in trees and soils depends on the site, the forest and its age; if it amounts to half a ton per acre per year and stored carbon is worth $50 a ton (probably a high figure), then carbon storage is worth $25 an acre. Over a 150 year rotation, $55 an acre a year ($8250 dollars) is likely comparable to the stumpage value of the timber (in the Northeast, about a third the market value of the logs).

Logging is always somewhat destructive to a forest. By opening the canopy, logging increases light, temperature and wind speed; it lowers relative humidity; disturbs and compacts soils (especially the top few inches where most of the tree roots grow; this is why the worst time to log is spring, the best time in winter on frozen ground). Logging damages nearby trees and causes wounds in roots and stems that leave trees susceptible to decay or disease. After logging, erosion increases in the forest as a whole and especially on the 5-10% of it that is in roads, skid trails and landings. But logging also speeds up natural processes, raising the 2% return in growth on unlogged stands to perhaps 4%. Young stands of northern hardwoods will naturally thin themselves from 1000 trees per acre at 4 inches diameter at breast height to 40-60 trees per acre at 20 inches diameter. (Since a site can support only so much basal area of tree per acre, as the individual trees get thicker, there must be fewer of them.) This process takes 200 years in nature, but half that if the stand is thinned periodically. (Put another way, for the trees in a fully stocked stand to gain an inch in diameter, 20% of them must die.) So this is the rationale for periodic logging in ecologically managed forests.

Thinning of long rotation hardwoods in the Northeast usually starts at 30-40 years when they have reached (coincidentally) 30-40% of their mature height. Letting them remain crowded up to then forces the trees to grow taller. Deciduous trees grow upward in response to light and will adopt a rounder, shorter habit if not forced upward by competition; conifers grow from an apical bud that responds to gravity and so continue upward no matter what the competitive situation. Thinning lets the crowns develop (the crown should occupy 35% of the height of the tree) and the boles increase in diameter faster. Trees 8 inches in diameter are usually thinned to 200 to the acre. Thinning then continues at 10-40 year intervals, depending on how the site is being managed. In general the longer intervals, which involve fewer entries of heavy machinery into the woods, are better for the forest. Most commercial forestland in the East is young, and logging to create a mixed age forest with some old trees would have to focus on the trees to be left rather than on the trees removed (the usual focus of loggers). The larger trees in a stand may be the same age as the smaller ones but more vigorous. So one leaves those, whatever their place in the succession. Early and mid successional species that have been overtopped and suppressed are removed. In general, one leaves vigorous, late successional species with good form and no wounds, that are capable of reaching the overstory. The trees that grow earlier in a succession, such as white birch and red maple and the shrubs, put their growth into stems, flowers and fruits rather than roots; their purpose is to reproduce before they are shaded out. A heavy fruiting may kill them. More long-lived species, such as sugar maples and oaks, put their energy into roots, stems, branches and leaves, and fruit later and at longer intervals.

Logging aims at a forest with trees of different ages. Several old trees are left per acre, no matter what their status or form, to become dead snags, hollow trees, and finally coarse woody debris on the forest floor. Creating such forests takes time, in most cases a century before they become capable of producing a steady flow of timber. In an ideal world, returns from the value of the water flowing off the land and from carbon storage would provide additional income over this time. Logging of old growth in a given woodland would not happen all at one time, since one aims to provide roughly the same proportions of each habitat over time: this lets the various inhabitants of the forest (as well as the loggers and lumber mills) survive.

In an ideal eastern forest, one would cut only those trees unlikely to survive until the next cutting cycle (a variation on John Muir's sawmill for dead trees). This will result in a forest of shade tolerant trees: a "climax forest." Cutting mature trees is likely to damage the surrounding ones however. An alternative is cutting the forest in little clear cuts, 0.25 acre to 2 acres in size (2 acres is large), somewhat linearly, along the slope. This allows for reproduction of more warmth loving and intolerant species (oaks, birches, pines). It tends to reproduce the existing forest. About 15% of the forest is left standing in the cut: some green trees for shade and seed, some large trees to become overstory trees, some to become snags and logs on the forest floor. Tops less than 6 inches in diameter and branches are left in the woods. The microhabitat the standing trees and downed logs provide reduces erosion and loss of soil nutrients, the partly shaded habitat is also more favorable for soil microbes and invertebrates, mycorrhizal fungi, small vertebrates and amphibians. The slash helps prevent browsing of saplings by deer. A different mix of trees will grow into the architecture of the remaining forest than into a clearcut. (In former Forest Service guidelines for western forests, stumps were removed and the slash piled and burned, in preparation for replanting fir or spruce. Except where it is very dry, most forests will reseed naturally, but not to the monocultures of plantations.).

As the young trees grow, they are thinned. Timely thinning speeds tree growth but also removes carbonaceous material and nutrients that would add to forest soils (leaving tops and branches from the thinned material helps). The point of thinning in the developing forest should be to create a mixed stand of trees (mixed in age and species), not to harvest the largest and most valuable timber (generally the goal in commercial thinning). The first thinning is done when the trees reach 8-12 inches in diameter. This thinning produces poles, sawlogs and firewood. Trees too small to use are left in the woods. One leaves trees suitable for the site and opens up the forest around the better formed, better adapted (to the site), more vigorous trees. Usually commercial selective cuts involve the removal of 50% or more of basal area and all large trees, but cuts in the ecological forest would remove considerably less than that, leave the larger more vigorous trees for later cuts, and leave a cross section of tree ages. Since selective cutting will tend to favor the more shade tolerant trees (such as hemlock and sugar maple in the East), one would also in good seed years cut some larger gaps to allow regeneration of less shade tolerant species (yellow birch, oaks, pines). Cutting over the long term inevitably reduces the nutrient capacity of a site, but the term is long (probably longer than the climate that supports the forest) and careful forestry can reduce this loss to near zero. Ecological forests are managed for their processes, but (after a time) also produce a steady flow of forest products (poles, sawtimber, veneer logs, fuelwood, chips, pulp). In the modern world the healthiness of a forest may constitute its greatest value (its value as background for other uses of the landscape).

There are many different types of forest in the United States. In the latter half of the nineteenth century, logging and fire converted much of the pinery of the upper Middle West into an aspen forest (aspens sprout vigorously from their roots); or, in places where the topsoil was burned away by the fires, to shrublands of bird planted pin cherry, blueberry, grasses, raspberry and shadbush. (From 1875-1900, logging fires averaged 500,000 acres a year in Michigan and Wisconsin and killed several hundred people.) The white pine forests of the northern United States originally contained several hundred billion board feet of timber. Much of this area had little value for agriculture and ended up as national forest. The aspen forest has been maintained as a monoculture by cutting on a short term rotation (30-50 years) for paper pulp and for chips for composition board. Clearcutting the sites removes the conifers and other hardwoods in the stand, and regenerates the aspen from root sprouts. To recreate a more mixed forest, probably a more stable forest (and one with more commercial potential), loggers would remove the mature aspen but retain the conifers and the other hardwoods. Released from competition, these trees would grow quickly. The aspen would regenerate, but less strongly, in competition with the residual trees. The result would be structural and habitat diversity. Future cuts would treat the forest as a developing old growth woodland, clear cutting the aspen stands as they matured (their greatest value for grouse and deer tends to be when young), and the evergreens and hardwoods as they reached the first stages of old growth. Along watercourses trees would be let mature. Some of the canopy in a cut would be left standing as a legacy of the stand architecture: seed trees, snags, hollow trees, some large old trees, some young and middle-aged trees (these should be left in clumps). The clear-cuts would be small and patchy. The alternating dominance of aspen and conifers would mimic the lifecycles imposed on the forest by tent caterpillar and spruce budworm, the insect pests of each species. It would also exploit the opposite effects of aspen and conifers on the nitrogen cycle. Much of the nitrogen that accumulates in conifer needle litter does not break down and is vaporized in stand clearing fires—especially in boreal woodlands (no stand of black spruce has been found that did not originate in a fire)—while nitrogen in the rich easily decomposed litterfall of aspens is readily recycled.

Some eastern forests burn naturally. Many more were regularly burned by the Native Americans. Burning altered the forest cover along much of the East Coast, its effects extending up the major river valleys where the Native Americans preferred to live. In the Southeast, burning created the savannahs of long leaf pine characteristic of the precontact coastal plain. (Perhaps 3% of this once enormous habitat remains, most having been converted to short lived plantation pines or agriculture.) In the Middle West late summer burning produced open forests of oaks and hickories and pushed the edge of the continental grasslands hundreds of miles east. The sunny dry forests of the mountain west naturally burned, but some were burned by Indians. At European contact, probably 20 million acres of the trans-Mississippi West burned every year, much of it grassland and scrub, but 6 million acres of it forest.

Virtually all western forests arose out of fire: the cool coastal rainforests of Vancouver Island and British Columbia were perhaps an exception. The types of fires varied. Under natural conditions about 40% of western forests experience low intensity fires every 1-30 years. These are so called understory fires. Low intensity understory fires create forests of large, well spaced, sun loving, fire resistant trees, such as Ponderosa and Jeffrey pine and western larch in the Rockies; giant sequoia, redwood and some types of oak forests further west. Ponderosa pine forms a relatively pure climax at middle elevations in the western United States. Ponderosa is moisture dependent and grows more thickly at higher elevations (more mixed with Douglas fir), more spaced out at lower and drier ones. The trees in historic old growth middle elevation forests were 200-400 years old, with 30-40 trees to the acre (one tree per 200 square meters). Grasses and fire resistant herbs and shrubs formed an open understory, which was browsed by elk and deer. Frequent ground fires in such forests scorch the lower branches of the trees (which eventually die and fall off), kill saplings and large shrubs, and thin (often drastically) the Ponderosa pine seedlings and those of shade tolerant trees, such as true fir and Douglas fir. They burn off the litter layer of undecomposed needles, fallen branches and the dead grasses on the forest floor. Thus they recycle nutrients and keep fuel loads low, and by pruning trees and killing seedlings, remove the fuel ladder that lets fires reach the crowns of the trees. They maintain the open forest.

When such forests are logged, most of the old trees are removed. Fires are suppressed. Shade tolerant Douglas fir, true fir and lodgepole pine seed in and grow up into thick stands of young trees, among the remaining older trees. The tree density is 40 times that of historic old growth stands. Competion for moisture and nutrients in the thick forests stresses the older trees. Without fire to keep competion down, the old larch and Pondersosa pine decline in vigor. Diameter growth slows drastically after 30 years and foliage grows sparse after 80 years. The trees loose the ability to manufacture the resins necessary to combat bark beetle infection (the beetles are drowned in their tunnels by sticky resins secreted by the tree) and are killed by beetles or fire. In general, crowded stands are more susceptible to bark beetles, spruce budworm, dwarf mistletoe, root rot and catastophic crown fires. Because of such poor management in western forests, stand replacing fires are twice as common as historically. In forests with a history of fire, where fire is suppressed, insects and fire take control. The resulting catastrophic fires kill both old growth and young trees and endanger human settlements in the forest. Ponderosa pines are adapted to light fires. They have deep roots, thick bark, open crowns, large fleshy buds and long needles spread out to avoid rising heat (not the short densely spaced needles of black spruce that make the tree flare up like a torch). But severe fire in Ponderosa forests makes regeneration difficult. Ponderosas have heavy seeds that fall within 150 feet of the tree, so unless planted by birds or squirrels, the trees take time to spread; the soil after a hot fire forms a water repellant layer that keeps the soil subsurface dry; and the rest of the forest vegetation, also not adapted to severe fires, recovers slowly.

One way to regenerate an overgrown stand of old-growth Ponderosa pines is to log from below. One removes most of the small and medium sized trees and all the shade tolerant trees (such as Douglas fir, some of which may be large: removing these large trees helps pay for the treatment). All the large pines that are vigorous enough to survive are left. After logging, when conditions allow, the forest is burned, to reduce the fuel load and further thin the Ponderosa and fir seedlings. One wants a mix of different age classes of trees amidst an open understory that will allow periodic prescribed burning. This forest over several decades will become an open forest dominated by shade intolerant Ponderosa pines, with some young growth and some trees of other species. Such forests are logged every 25-30 years to remove excess small, medium and large trees (those that wouldn't survive to the next cutting cycle). Some dying old trees are left. The small openings created by each entry allow a new age class of pines to develop. The forests are burned every 10-35 years to keep fuel loads low, recycle nutrients, control Douglas fir and stimulate herb and shrub regeneration. In time, such forests will provide a constant flow of commercial timber: large sawlogs, smaller logs and poles for latilla and viga makers (the traditional southwestern ceiling beams and lattices); chip wood for boiler fuel. The constant flow of wood provides predictable amounts of material for sawmills, pulp mills, wafer board and plywood plants, post and pole plants.

Restoring overgrown forests in 1% of Montana a year (not just Ponderosa forests, but all types of forests) would increase timber productivity by 50%. (In 2000, thanks to a century of poor management, wood production in Montana was 15% of that in 1990.) Such restoration forestry costs about $200 an acre initially. With the second or third cut, the forest breaks even, and after 100 years should turn a profit, which continues indefinitely, and rises with good management. (Fighting crown fires is also expensive, up to $1700 an acre, with no profit in sight.) The United States spends over $2 billion a year fighting fires in the West. This is money that in many cases pays for a history of forest mistakes, a tax levied by short term capitalist management and scientific ignorance. Without putting money into good management (an amount equal to that used to fight forest fires would restore 10 million acres a year), such costs will only rise. One tries to recreate the historical forest because this forest is more sustainable, with a lower risk to its biota and a lower risk of catastrophic fire. It makes economic sense; as several writers have pointed out, a forest's maintenance of ecological process constitutes its greatest value (human or otherwise), its ability to maintain a predictable flow of wood and support a rural population constitutes its greatest social value, while control of its fires raises the value of the land about it as real estate.

The cooler, moister, higher elevation forests of the West are subject to more intense fires (so-called intermediate intensity fires), every 30-100 years. In such forests good fire conditions exist for a short time: a few days or weeks each summer. The fires are intense enough to kill most fire-susceptible trees but the mountainous areas burn unevenly, so much of a site (up to half) remains unburned. Such forests become very mixed in species and in tree ages. Here, logging followed by the suppression of fire leads to dense forests of shade tolerant trees which are susceptible to stand replacing fires. A more sustainable forestry in such overgrown stands, which, as they get older, are favored by spotted owls, martin and fisher, would remove the shade tolerant trees in a patchy pattern, retain the sun-loving trees, and encourage (with a controlled burn) regeneration of fire-dependent herbs, trees, and shrubs. Such treatments would take place every 40-100 years. In the case of spotted owl habitat, a central (fire-unstable) old growth forest (say of firs) would be surrounded by more open forests of large trees (larch and pine), which would serve as firebreaks. Such firebreaks are usually a quarter of a mile wide. When fire destroys the central old growth, or after it is logged, parts of the firebreak are let develop into the thick, shade tolerant, old growth forest favored by owls. Surrounding undisturbed old growth forests with a matrix of semi-natural wooded habitat makes it easier for the animals of those forests to move to other old growth areas.

Moist cold forests on the upper slopes of mountains (the last 20% of western forests) burn occasionally, at 100-400 year intervals, in stand replacing fires. In many ways, they resemble the boreal forests of Canada, altitude here replacing latitude. Stand replacing fires occur when conditions are unusually dry and fuels are available, often during major droughts. Such forests include lodgepole pine; and the white-bark pine, whose plump seeds help support mountain populations of grizzly bears and which is replanted in areas cleared by wind and fire by Clark's nutcrackers storing pine seeds for the winter. Such forests become mosaics of burned and unburned patches, of hundreds to thousands of acres. The trees are adapted to catastrophic fires. Lodgepole pine cones open in the heat of fire. Green cones in fire killed western larch and coastal Douglas fir drop viable seed. After a burn, elderberries, currants, gooseberries, bitterberries, ceanothus, wild geraniums and hollyhocks are planted in clearings by birds; aspen and fireweed seeds blow in; alders replenish the nitrogen in the soil. The forest is full of standing dead trees. Depending on the intensity of the fire, willow, aspen and mountain maple may sprout from rootstocks; and other plants grow from seeds in the soil seedbank. Clear cutting such forests is not a replacement for fire. Clear cutting leaves no standing dead trees to become breeding places for insects that attract birds, or that slowly release nutrients; no whole fallen logs to help slow the movement of soil and ash, and whose shade helps in regeneration of trees, herbs and shrubs. Fires leave irregular patches and strips of surviving trees and an ashy seedbed for the bird and wind planted seeds from outside the burn. Clearcut stands lack the diversity of fire cleared stands. A better way to log such stands is to remove two-thirds of the trees, leaving most of the overstory trees (such as Douglas fir and western larch) and the rest of the stand in irregular clusters. Burning the slash after logging kills most of the remaining trees and leaves snags and a suitable seedbed for regeneration of the conifers, herbs, shrubs and deciduous trees characteristic of recent burns.

The whitebark pine of the high slopes is suffering tremendous mortality from blister rust, an Asian fungus that took almost a century to reach whitebark stands from its initial foothold on the East Coast (where it remains a problem in white pine). Some trees are rust resistant but it may require intensive human intervention to reproduce sufficient rust-resistant whitebark pines. The whitebark's large, oil rich seeds let grizzly bears gain enough weight to go into winter in good condition and keep the bears in autumn at high elevations out of the way of humans. The bears rob caches of cones hidden by squirrels, then crush the cones between their paws to get out the seeds. (Another late fall bear food is cutworms, whose moths are blown into the mountains from Kansas wheat fields, and whose larvae are found under mountain rocks.) A poor competitor with other trees, whitebark pines grow on the thin, stony soils and in the difficult climates of alpine elevations. They shade the snowpack and stabilize the rocky soils, protecting the quality of the snowmelt and rain that flows in brooks and streams down to the lowlands. Clark's nutcrackers collect whitebark seeds and cache them several miles away in open areas for the winter. Those that are not retrieved grow into new stands of trees. (As the number of seed falls as the pines grow more scarce, feeding by nutcrackers become a problem for the pines).

Sustainability is a human concept that may or may not work in nature. It implies permanance—a constant flow of animals, crop plants or timber from a place. Sustainability in agriculture means maintaining agricultural soils and reducing erosion to the levels at which the soil is rebuilt (essentially zero); and making agriculture one part of the larger ecosystem. Sustainable forestry in the long run probably means maintaining all components of the ecosystem. Such components include disturbances (such as fires); large herbivores and their predators (moose and wolf); the small mammals, amphibians, arthropods and fungi of the forest floor; the soil microflora; the arboreal lichens and mosses of old forests. The purpose in retaining large canopy trees during initial logging operations, and lengthening the rotations, is to produce a forest with more large, old trees, a multistoried canopy, a greater variation in tree size (these all lead to a more complex forest architecture), with large woody debris on the forest floor, tighter nutrient cycles, shaded refugia for mycorrhizal fungi and nitrogen fixing bacteria. (The fungi, with which many trees have obligate relations, by connecting trees of different ages and species, are a major factor in controlling what happens in the forest.) Such a forest provides for beneficial predator-prey relationships among forest vertebrates and habitat for plants and animals that require structural complexity and late seral conditions, as well as for the needs of those that inhabit young forest. Some birds in Pacific Northwest forests jump in density as the forest enters a state of old growth, indicating a (non linear) change in the quality of the habitat. (But in general bird variety in old growth forests is low.) In woodlands with a history of fire, maintaining open forests of old sun loving trees, mixed with some shade tolerant trees and with younger stands, minimizes the destructive potential of fires. One can't eliminate fire, only plan around it; the success of Smokey the Bear has caused our present dilemma. Such forests also maintain the place of the forest globally, which a young forest, maintained on a short rotation, doesn't: the forest as a store of carbon, as an inoculae for mycorrhizal fungi, lichens and forest bacteria, as a positive influence on the water that flows through it, and on the beaches and coastal estuaries that lie below.

Some forests should be left undisturbed and a region's forestlands should be connected by corridors and buffer zones that allow plants and animals to move among them. Forests managed for their process as well as for timber will produce less total wood fiber. Leaving large trees standing and lengthening rotations to let the forest become old both reduce the amount of wood harvested per acre. The net return per acre however remains the same or rises, since larger logs from old trees have a much higher value. (About half the wood cut in the U.S. goes to paper, though sawtimber is worth 2-5 times as much, good sawtimber more than that.) So the contribution of the forest to the economy remains the same, or rises; but only after a period of forest recovery that would last for about a century. Such forests will require a new infrastructure to use their logs of different species and sizes (saplings, poles, sawlogs). Some thinning operations would be better done by hand, or with horses or light tractors, than by heavy machinery. Modern industrial forestry, faced with forests of declining value, uses harvesting machinery of immense power and mobility in order to make a profit harvesting wood of low quality and (often) low quantity (per acre). The post harvest machinery tends to homogenize the harvested logs into composite products like finger jointed lumber, which replaces solid boards; or oriented strand board, which has replaced plywood, which requires better logs: in the latter case, resin replaces fiber. Plants to produce composite board are now being constructed around the perimeter of the Cumberland Plateau in the southern Applalachians, in order to use the second growth hardwood forests of the plateau. The forests will be clearcut to feed the plants, at terrible cost to the biota and watercourses of the region. An alternative, regeneratively harvested forest would produce a mix of graded materials for plywood, oriented strand board, flooring, cabinetry, moldings, wall panelling, furniture, construction lumber, tool handles, fencing and so on. Such a forest would require a more far sighted and curious forestry, some public funding and patience.

One faces essentially the same question as in agriculture: whether to focus on the yield per acre per year, or on the forest as habitat and process, a landscape that also produces wood. As with food, one can ask how much wood do we need? Wooden pallets use 1.5 billion board feet of lumber a year in the United States (about 40% of the hardwood cut) and most of them are thrown away after a single use. Remanufactured pallets could provide the lumber for 300,000-600,000 houses a year; or for hardwood flooring, material for composition board, and wood shavings for packing material. A tax on logs (making pallets more expensive) would encourage reuse. A rise in landfill fees would make reuse of lumber in demolished buildings profitable. (Construction debris constitutes 15-40% of landfill contents in developed countries; an increase in Danish landfill fees pushed reuse of construction materials from 12% to 82% over 10 years.) A better recycling effort would reduce the wood needed for paper and cardboard by up to 90%. (Paper can be recycled 9-10 times; 90% recovery probably isn't possible but 75% probably is. Reducing wood use by recycling would also reduce our carbon footprint.) Alternative fiber crops such as kenaf (an African plant), hemp, bamboo, hybrid poplar, native trees that sprout from stumps, grown and harvested as saplings by machinery, produce much more fiber per acre than natural forests (2-5 times more in mechanically coppiced forests), and are easier on the land than cotton or corn. Such artificial shrublands could be managed as part of the larger local habitat, cut when the loss of the their habitat matters least. Hemp and bamboo produce more fiber per acre at a much lower environmental cost than cotton.

In short, we don't need much of the wood that modern forests produce; but under a capitalist regime land must produce value. Wood fiber (not whole ecosystems) is what forests produce, and it is presently economically advantageous to use the fiber and throw it away. (Turning discarded carbonaceous material like pallets into charcoal—biochar—and spreading it on farmland would improve the farmland and remove much carbon dioxide from the atmosphere.) Returning some of the value of regeneratively managed forests to the landowner (their value in slowing soil erosion, protecting streams, controlling water and nutrient runoff, filtering the air, storing carbon, in providing habitat for useful wildlife, in regulating local microclimate and rainfall) would help make managing forests for their processes profitable. Such value, in dollars per acre per year, may be worth more than the wood, but that value only becomes apparent when the forests are degraded or gone. The monies could be raised through a tax on logs. The appearance of healthy forests also increases property values. Costa Rica pays its farmers to maintain and replant their forests for all these reasons. In a capitalist world my notions are forest dreams; enforceable on public land if the government and the people desire.

The usual fate of forestland in the modern world is to become building lots. Inland from the sea along the Gulf of Mexico in the southern United States were so called wet pine savannahs. These were grassy meadows with scattered pines growing on waterlogged clay soils, amidst tupelo-cypress swamps. The wet savannahs formed a narrow band between the drier piney uplands and the brackish marshes of the Gulf. The savannahs and the uplands were partly maintained by fire. The savannah soils were naturally acid and nutrient poor and filtered the water flowing into the coastal marshes. They were the preferred habitat of the Mississippi sandhill crane, now much reduced in numbers. These wetlands, like many southern coastal wetlands, were ditched and drained by timber companies after the Second World War to grow slash pine on short rotations for paper pulp. Fire on the plantations was suppressed. Drainage removed the filtering effect of the savannahs, increasing the nutrient flow to coastal waters. After a few rotations, the timberland was sold for building lots, as air conditioning made life in the Deep South more comfortable.

There are other stories. The Menominee Indian Reservation in northeastern Wisconsin includes 220,000 acres of forestland. The forest has been commercially harvested for 140 years, with approximately 2 billion feet of lumber removed from the forest over that time. The volume of standing timber now is greater than when the reservation was established in 1854. The Menominee management program predates current concepts of ecosystem management. The Menominee must log (to support the tribe) and try to manage the forest so as to maintain a more or less even flow of sawtimber and pulpwood. They try to maximize the quality as well as the quantity of sawtimber, with a sustained yield of material, and a diverse mix of native tree species. Harvests are based on excess stocking of overstocked stands. When the stocking rate becomes too great for the site, trees are cut. Trees are let grow as long as they remain healthy and vigorous. Thus the forest contains many large old trees of great value (habitat for the insects and lichens of old-growth forest, great producers of mast). The Menominee cut trees no faster than the forest can regrow, rather than in response to market conditions (say, a rise in the price of red oak logs). Loggers must attend a class in how to cut and skid timber and contracts are terminated if the cutting methods are not satisfactory. About 65% of the forest is managed as a mixed age forest, with a 15 year cutting cycle. The remainder is managed as even aged forest. To a casual observer the forest looks pristine. Partly because of the old trees, it contains a much more diverse mix of plants and animals than the surrounding commercial timberland. The Menominee forest has healthy streams. Whether such forests are net accumulators of carbon isn't certain: the above ground part of the forest has grown in mass and the soils have probably continued to increase their stored carbon, but 2 billion board feet of wood (about 170 million cubic feet or, counting waste, branches and tops, about double that in woody material) has been removed, and much of its carbon returned to the atmosphere.

In the Acadian Forest Region of the Canadian Maritimes (a division of the Northern Hardwood Region), some writers claim a well stocked 100 acre woodlot will provide full time employment for 4 people, if the logs are sawn into lumber and dried and sold from the lot. The sawdust and slabs that are not burned for heat are used on the wood roads. Four employees may be an overestimate but is comparable to the manpower used in Swiss woodlands. Much depends on the productivity of the site and on the price of lumber. I would think 1-2 people per 100 acres is a more reasonable estimate for much of the Northern Hardwood Region. On the other hand, a modern woodcutter in Canadian industrial forestry needs 800 acres to support his job. (And the dry, communally managed pine forests of Oaxaca in southern Mexico support one worker per 160 acres. Work includes felling trees, growing and planting seedlings, weeding, milling lumber, making furniture.) Such woodlands, like the Menominee forest, would be managed for their long-term productivity. Sawing the lumber on site lets work in the woods be done when conditions (weather, the state of the ground) are appropriate. Logging might occupy 2-3 months a year. The rest of the year would be spent milling lumber, working up firewood, thinning, working on roads, perhaps making maple syrup or maintaining summer rental cabins. The investment per person in such enterprises is much lower than in industrial wood production. Jobs per log or board foot are higher. (Not usually a good thing in the economic world, but it is costs per board foot that matter.) One theory of civilizations' collapse proposes that as the complexity of a society rises, its costs per unit of investment also rise, and the marginal returns on investment decrease. That is, investment produces declining (or negative) returns. Finally, this decrease in returns (from private investment in manufacturing and agriculture, from public investment in education, a military establishment, public safety and public works) bankrupts the society. Improving the yield of woodland, or of the agricultural landscape, so as to provide a constant flow of high value material, provides one way out this impasse.

Chapter 19
Developed Landscapes

Developers are the third shaper of the modern landscape. For most of the twentieth century developed landscapes were a small percent of the total landscape, though their location (in flood plains, near ocean estuaries) intensified their ecological effects. The less dense developments of the latter part of the century (such as suburbs, served by cars, or warehouse districts, served by trucks) have spread the effects of human settlement out. By some estimates land is developing 7 times faster than population is increasing. From 1982-1997 the population of Pennsylvania grew by 2.5% while developed land increased by 47% (from 1990-2000, by a million acres). In the 15 years 1982-1997 developed land in the United States grew by 25 million acres, an amount equal to 25% of all land developed since 1492. Such numbers explain why deer, coyotes and mountain lions have become animals of the suburbs. Developing landscapes near water (most cities are near water) interferes with the ecological function of these landscapes, since their natural variability (flooding; periodically high water tables; river channel migration; erosion of beaches) must be controlled. Such control turns productive rivers into drains.

Building and paving increase a landscape's absorption of heat by changing its reflectivity (or albedo). Cities are generally several degrees warmer than the surrounding countryside (about 10° C. by day, 6° C. at night; the denser the city, the greater the effect). Cities send their thermal effects, in the form of clouds and rain, downwind. It is thought Tokyo's torrential summer downpours have been intensified by the continuing spread of the city (a mini-warming, probably intensified by the larger, global one). Summer rainfall near Tokyo increased 20% from 1979-1995, and the rate at which the rain falls has also increased. The average temperature of Phoenix, Arizona, has risen 5° F. since the 1960s, and as the city grows over the next 30 years may rise 15°-20° more, apart from any rise caused by a changing climate.

Cities, suburbs and superhighways are all sources of small particulates (from burned fuel), sulfur and nitrogen compounds, metal and rubber dusts, benzene, dioxins, furans, PCBs: the products of combustion, electricity generation, drying paints, industrial discharges, automobile use. Their sewage waters are sources of nutrients, hormones, and hormone mimics, as well as of a high and steady water flow into streams. Fish and alligator populations in waters that receive large flows of treated sewage waters decline sharply; in both groups of animals the sexual development of males is compromised by human estrogen in the water. City roofs and pavements shed runoff into streams, eroding and flooding them. City (or suburban) water supplies require dams on rivers or wells to pump out groundwater. Groundwaters (out of sight) are almost always overpumped (including the sandstones about Milwaukee, Wisconsin, and the limestone acquifers of South Florida). Cities also have environmental advantages: public transportation is much more energy and materials efficient than private automobiles; heating and cooling costs in apartments are several times less than in detached houses; per capita water use is less (less car washing and lawn watering; fewer private swimming pools). In general, per capita energy use is several times less in cities than in suburbs, or in isolated rural houses, for the same standard of comfort. While dams interrupt the natural flow of rivers, watersheds protected for urban use are natural oases in a developed landscape.

Modern development usually begins with bare ground, graded to taste: an immediate, radical simplification. Paving or roofing more than 10% of a watershed begins to degrade its streams, but thanks to impervious driveways, roofs, lawns and roads most suburbs are effectively 25-30% paved. (Los Angeles is 70% impervious surfaces and much of northern New Jersey not much less, which explains its problems with flooding.) Movement of soil and nutrients into watercourses is high during a development's construction (this can be ameliorated), then falls as erosion is reduced. After development, roads, roofs and parking lots increase the amount of water runoff and its rate of flow and contribute a mixture of combusted hydrocarbons, motor oil, benzene, anti-freeze, brake fluid, metals, automobile greases, tire dust, aromatic hydrocarbons (from worn asphalt), nitrogen, phosphorus, sodium chloride, sand and silt to the runoff water. Because the land itself absorbs so little water, and because so much of it is carried away in pipes, small rainstorms produce a large flow at any time of year. Piped into watercourses, this flow excavates streambeds and changes the habitat for aquatic invertebrates, amphibians and fish. The rate of runoff amounts to 2-4 times natural background conditions. The chemicals and nutrients in the fluid don't help. The weathering of soils rearranged by construction into a new, more stable equilibrium (re-establishment of the soil profile) can take a thousand years and is slowed down by a lack of deep-rooted plants. (It can be speeded up by appropriate plantings and vigorous organic gardening.) Suburbs lack the energy advantages of cities: energy costs for heating and cooling are higher (they can be reduced by good construction, efficient equipment and trees; and space is available for solar electric panels); public transportation usually doesn't work until houses reach a density of 7 per acre, and then only if the suburb is laid out for it. Most American cities resemble suburbs, with perhaps 4-8 houses per acre. Most are not laid out for public transportation. Residential areas are separate from commercial and light industrial ones, rather than having businesses and other places of employment clustered along the main streets (where public transportation runs), with housing adjacent, above, or behind. The isolation of suburbs from commercial activity is a legacy of the Garden City Movement of the early 1900s, which arose with the suburbs, and segregated areas dominated by coal-fired heavy industry, the domain of males, from the bucolic domestic environment run by females.

The ecological effect of developed lands can be reduced. Drainage water can flow in shallow, vegetated ditches, rather than in pipes. The vegetation slows the flow and cleans the water and returns some of it to ground water. As well as nutrients, cattails and common reed accumulate metals, which can be reclaimed from the harvested plants. Water from parking lots can be led to vegetated areas (shallow wetlands with cattails), or to slightly sunken borders with trees, shrubs and herbs that tolerate some flooding (instead of the raised borders now used); the plants will help deal with the pollutants (their stems and roots act as scaffolds for the microorganisms that degrade them) and take up nutrients. The trees in sufficient numbers will cool the parking area. Parking lots can also have permeable pavements that let the rain sink through, in which case one lets the microorganisms in the soil deal with the pollutants. Roof drainage from individual buildings can be led to aquifer recharge areas near the street or in the yard; on steep slopes, such recharge areas may have to be connected by wide drainage channels to wetlands down the slope. (Rain gardens in front of 600 feet of houses on a Seattle street absorb 99% of the water from storms.) Open drainage structures (small ponds and marshes) that collect sediments from pipes must be periodically cleaned.

Most American cities have enough land available for them to moderate their effect on local watercourses, although some land, right along watercourses, or in very built up neighborhoods, would have to be purchased (for instance, to allow rivers to flood). Such overflow areas become parkland or wild land. City downtowns that are heavily paved can't recharge their ground waters; their streams have long since been put in pipes underground, and the rising ground water from aquifer recharge would flood basements. At some cost, mostly for storage, such cities could capture the water that falls on their roofs, thus reducing their runoff and adding to their water resources. Collected roof water can provide 10-20% of a city's needs, rainfall from the whole area 35-50%; such water is usually used for flushing toilets, which are plumbed with a separate system. (Roof gardens are another possibilty, and besides reducing runoff, lower average summer temperatures in cities by several degrees.) If there is sufficient space nearby, street runoff can be led to constructed wetlands, which are a cheap and effective means of cleaning runoff and regulating its flow (and recharging streambed aquifers), especially in more moderate climates. Sewage water can also be given a final treatment in constructed wetlands. In severe climates, treatment wetlands are housed in greenhouses and used to grow crops like flowers.

The most efficient way to reduce the toxicity of what flows off city streets is to reduce what ends up on them. This means less toxic automobile greases, fuels and fluids. It means less traffic and more public transportation. Getting more miles per gallon of fuel also reduces pollutants on the street, as does the increased use of electric vehicles.

It is fashionable to regard a city as an ecosystem, though one with a long reach, bringing in grain from the Middle West, heating oil from Venezuela, exporting sewage sludge to Texas cotton fields, waste paper to China or Europe, unseparated trash to Virginia landfills (where it will one day be dug up, separated and reused, as today on Nantucket Island). Rearrangement of nutrient flows are a part of this idea. Some wastes can be recycled on site; dying urban trees can be milled into boards, their branches chipped, the mulch used in city parks. Industries can be designed to use each other's waste products. Cleaned sewage water is a potential source of some industrial water: certainly cooling water. (Drinking water in many cities in the Mississippi basin consists partly of treated sewage water from upstream; and more and more in the dry West, urban water supplies are re-cycled sewage water that has made a sidetrip through a local aquifer.) Ship-ping pallets can be turned into flooring, into other lumber, or into shavings for biodegradable packing material. New closed cycle paper mills that use very little water, perhaps very clean sewage water, can recycle a city's waste paper in the city, saving immense amounts of energy in haulage. Plastics and metals can be recycled on site. Returnable glass bottles can be refilled with fruit juices, sodas, spaghetti sauces. In any urban manufacturing proj-ect, the cost of land and the increased volume of truck traffic are problems. (Trucks that ran on natural gas and that were shut off when stopped would make this much less of a problem.)

Yard wastes can be collected and composted with other wastes, such as animal manures (from the 100 million dogs and cats in the United States), and food wastes (from households, restaurants, food processing plants). If the compost is clean and if city soils are not contaminated with metals like lead or cadmium, such composts can be used to grow backyard vegetables. Vegetables can also be grown in rooftop greenhouses. Photo-voltaic solar collectors are a better use of urban roofs; but the two are not mutually exclusive. Photo-voltaic collectors mounted on buildings shade roofs and walls in hot climates, lengthening the lifetime of building mate-rials and reducing cooling costs. Pipes behind them (the collectors function better when cool) collect domestic hot water. Miniature worm farms in city kitchens turn kitchen scraps into worm-generated composts. Shredded gar-bage can also be composted, rather than landfilled, the methane generated during composting collected and used to generate electricity. While the compost will probably not be clean enough to use for growing food (that depends on the efficiency with which things like batteries are removed), it could be used for projects like reclaiming strip-mined land; perhaps for growing fiber or nursery crops. The solution to keeping trash cleaner and making human settlements less toxic to their surroundings is to keep toxic materials out of the stream of our lives. This requires taxes, incentives or regulation. For instance, items that contain mercury (such as mercury batteries and energy saving fluorescent lights) might be purchased with a deposit; this makes them returnable and recyclable. All manufactured goods should be easily recyclable: one ought to be able to separate a burned out (recyclable) compact fluorescent bulb, with its mercury lining, from its still functioning ballast base. (Soon, the much more efficient LEDs may replace compact fluorescents.)

Most of of our current manufacturing technology is replaceable (and is replaced, at 2% a year, so a turnover time of 40-50 years is feasible). While most toxic materials are unnecessary, they part of current industrial chemistry and are supported by considerable investment, which demands its profits. Technical alternatives to current materials are legion, but mean new plants, new manufacturing processes, new investment, some sort of guarantee of a market. Use of biodegradable soaps and cleaners lets the homeowner use household graywater (the waste water from sinks, shower, the washing machine) to irrigate lawns and gardens. In much of the country, photovoltaic panels on roofs, and on walls of high buildings; in corners of yards; over driveways and parking lots; as roofs of garden sheds, would provide all our daytime electricity needs. Solar panels are becoming more efficient, and recent figures indicate that rooftop photovoltaic panels could provide all the household electricity needed by a place like England (not an ideal climate for solar power), plus (from cooling pipes behind them) domestic hot water. The electricity needed for night-time would have to be stored (as chemical energy, or pumped water); or come from other sources. With solar electric power the rain of carbon, metals and sulfur that falls on much of the Northern Hemisphere from power plants, and much of our interference with natural waterways (electric power plants are the largest industrial users of water) would be reduced. (New electic power plants can reduce cooling water use by up to 90%, but cost slightly more and so are not built as long as water is free. Assessing power plants for their effects on waterways would make their water use quite expensive and make conservation economic.) Twenty-four hours after the failure of the power grid in eastern North America in August 2003 the air downwind had 90% less sulfur dioxide, 50% less ozone, and visibility had increased by 25 miles. Automobile traffic continued at normal levels. The decrease in air pollution was a result of power plants (coal-fired power plants in the Middle West) being shut down. Current economic life depends on the sale of huge quantities of unnecessary things; it would be better if they were also harmless things.

In sunny climates rooftop water heaters provide hot water, as they do today in Israel and Austrialia, and in the 1920s did in southern California. In temperate climates hot water typically constitutes 35% of household energy use. If appliances and houses were more efficient (also not a technical problem: the cost of a house would be higher, but since utility bills would be less, the overall cost of owning a house would remain the same or fall), rooftop solar cells would provide considerable excess power; this could be used to generate hydrogen from the splitting of water molecules in the presence of a catalyst, and the hydrogen used in fuel cells to generate electricty (water is the waste product); the electricity would run cars, trains, factories, light cities at night. Or (since the use of hydrogen has many difficulties) the electricity could be used to charge electric cars, whose batteries would provide the backup power (while leaving enough for the morning commute). So cities in sunny regions could be virtually energy independent, eliminating most of that Venezuelan and Saudi Arabian oil, and the considerable costs of keeping the Middle East safe for oil development (before the Iraq war, estimated at $50 billion a year, another cost not factored into the price of gasoline). With a concerted effort to cover parking lots and urban superhighways with (ever improving) solar panels, sunny American cities could become net exporters of energy as electricity, chemical energy or hydrogen gas; they would do this without interfering with the natural environment any more than they already are (which is not true of wind power). This is not to say a new industrial base of solar power and hydrogen would be pollution free; the manufacture of the collectors is polluting (it can undoubtedly be made less so) and they must be replaced on a 20-50 year cycle. The energy advantage of flat-plate solar collectors (the amount of energy they produce compared to what goes into building them) is 4 or more in cloudy climates like England; that is, you get 4 times more energy out of them than went into making them. In comparison, the energy advantage of oil from the Alberta tar sands is also 4; that from the Texas oil wells of the 1950s was about 80, of all U.S. wells at the peak of U.S. production about 50.) Production of hydrogen by electrolysis is energy demanding and unless storage of solar power improves (flow batteries, which store electrical power as chemical energy, are a possibility; as is pumped water storage—say in the reservoirs behind high dams), most night time power would have to be produced by fossil fuels. So would baseline power (power needed to even out fluctuations in the solar supply). But one would need much less of it: perhaps 60% less, perhaps (as energy efficiency improves) 90-95% less. (At that point, capturing carbon dioxide from the smokestacks of power plants is no longer a problem.)

A landscape altered for human convenience does not perform the ecological work of a natural landscape. Many new developments could be avoided by redeveloping old developments; making the human habitation more dense, but with parkland, playing fields, community gardens. Some of the ecological work done by the landscape can be recovered if aquifer recharge areas and vegetated drains are part of this. A drainage pattern shapes a landscape in a way the usual checkerboard of suburban houses cannot: a pleasing case of form (the drainage pattern) determining function. In dry climates, trees and shrubs along drainage areas need less irrigation. Letting drainage water flow in open channels saves the developer money, as do narrower streets that slow traffic and let the tree canopy shade the area more completely. Such shading makes a difference of several degrees in summer temperatures. Vegetation (such as that in parks and playgrounds) also lowers summer temperatures by evapotranspiration. Old developments can also be made more energy-efficient and thus reduce their impact on the natural environment. If zoning allows commercial and industrial development as part of the mix, people can walk to the store, the library, to work, to the cafe. Development friendly to public transportation usually has commercial or light industrial uses along main transportation routes with residential areas behind; so public transportation is a 10-15 minute walk from anywhere. Such communities are more friendly to that considerable part of the population (about 20%) that doesn't drive: the old, the poor, the young, the disabled. If we follow Mr. Odum, wild, undisturbed lands should occupy 40% of a new development (these can include wetlands, headlands and beach fronts, but not agricultural land or parks and playgrounds). Such lands should contain parts of all the ecosystems represented in the development and be connected to nearby wild lands.

Zoning presently is a matter of protecting property values: thus keeping chickens, hanging laundry out on a line, or renting out the apartment over the garage may be forbidden. Biologically-based zoning is a matter of protecting landscapes as ecological wholes. It would be based on geography, hydrology, topography, climate, soils, winds, as well as on economic and cultural matters. Working ecosystems require the protection of their essential features, which are not always easy to identify. Healthy streams and aquifer recharge areas are among the easier things to preserve. And with streams and aquifer recharge areas comes much else. One of the original plans for Los Angeles left several hundred yards on either side of the Los Angeles River undeveloped. This was to become a mix of developed parkland (playing fields, picnic areas) and wild parkland (trails through the shrubby woods). The potential profit from developing the land overcame civic mindedness however and so Los Angeles has no Central Park. What organizes the Los Angeles basin are the freeways. Building went right up to the edge of the Los Angeles River, which, following a serious flood in the 1940s, was turned into a cement lined channel leading directly to the ocean. Now in some places the cement is being removed and the edges of the stream replanted. Here and there water birds use the river. The river is seen as an amenity. Perhaps not yet as an ecological amenity: its water is polluted with viruses from septic systems and parasites from pet feces, as well as chemicals from local industry and road runoff. Ducks swim in a dilute solution of dry cleaning fluid. The material from the river and from storm drains ends up in the ocean off the famous Santa Monica beaches, making a swim there somewhat risky.

———

Habitat fragmentation is another effect of development, again one with unforseen consequences. To keep a full compliment of most species, tracts of Amazon rainforest need to be at least 10 square kilometers, 2400 acres, without roads, powerlines, or much human interference. In general, viable populations of large predatory animals (grizzly bears, jaguars, tigers) require larger areas (250-300,000 acres, or 400-475 square miles). Better habitat is provided by several connected areas of this size.

People get in the way. Railroad lines restrict the migration of Mongollian gazelles in Central Asia, farmers' fields block the movements of wildebeast in the Serengeti, hydroelectric reservoirs hamper the movement of woodland caribou in Canada, highways in North America impede travel by grizzly bears. On the Norwegian plateau, roads, powerlines, summer cabins and hydroelectric developments have reduced the winter range of the native reindeer, which stay several kilometers away from such human improvements, by 50%. In a study of a fenced Connecticut watershed (2400 acres protected for water supply), with use limited to 20 people with permits, populations of box turtles fell steadily over the 10 years of the study. The reason was unknown. Both dogs (let off the leash to run) or crows (attracted to the remains of picnic lunches) may have been the problem (both harass and kill turtles). Box turtles breed at 5-10 years. They lay 6-8 eggs a year, most of which are eaten by raccoons, skunks and opossums. The animals live for 50 years or more, a sign of an animal that reproduces slowly. They have trouble dealing with many of the artifacts of modern civilization: for instance, they are too slow to avoid cars and cannot climb over road curbs. In a recent study of northeastern woodlands, researchers found that woodlands of less than 5 acres (not a small area in the suburbs, where an acre is large and many suburbs are built at 4 houses to the acre, had 3 times as many of the ticks that cause Lyme disease and 7 times as many infected ticks per square meter as larger woodlands. Small woodlots also had more mice. White footed mice and white tailed deer are both alternate hosts of the tick, which the mouse carries at no harm to itself (it is the "reservoir species"). The researchers speculated that mouse densities were high in the smaller woodlands because of the absence of predators. So the continual fragmentation of woodlands by suburban development in southeastern New York and Connecticut, by making life too hard for minks, weasels, foxes, hawks and owls, and too easy for white-footed mice and deer, may have helped cause the rise in Lyme disease there since the 1970s. A more recent study pointed out that the less the mammal diversity (including that of predators) the more white footed mice were infected with Lyme disease. The researchers speculated that with more different mammals around, the more often the ticks are likely to bite animals that don't carry the disease, thus its frequency drops. Both speculations may be right. Of less importance to suburbanites, the same fragmentation has helped reduce populations of migrating songbirds by 50% since the 1950s. In fragmented habitats their nesting success is lower, often below replacement levels, so such habitats become sinks for excess populations rather than sources of new birds. In fragmented habitats populations of the nest parasites (the brown-headed cowbird) and predators of songbirds increase. (Predators on songbirds, especially on nestlings, include jays, skunks, opossums, raccoons, red squirrels and domestic cats; none of these except cats eat many mice.)

Habitat fragmentation is a problem without a solution. It will inexorably rearrange plant and animal populations over large areas. Fragmented habitats are much more friendly to introduced aliens or opportunistic native species. One can make developments more friendly to surface and ground waters, and much more energy efficient, and such developments will have populations of birds and animals (reptiles, amphibians, spiders, beetles, songbirds, small mammals, microbes) that can deal with a fragmented habitat. How such communities will work as biological wholes is another matter; it is largely the habitat shaping plants and the invertebrates and microbes of the soil that make landscapes work, but (as I have tried to show) the larger birds and mammals also play a part. Vegetated corridors along streams provide a degree of connectedness; a sufficient one for some animals (deer, mink, coyotes). Semi-native habitats can be improved. Habitats may have sufficient food (mice, insects, wild fruits) but lack other necessities for breeding (roosts, privacy, hollow logs, dens, tall bushy trees). Some lacks can be corrected. Eastern bluebird populations crashed during the first part of the twentieth century, partly because of competition for nest holes with the introduced starling (starlings are aggressive enough to drive the larger northern flicker away from newly excavated nest holes; competition for food may also have been a problem), partly because of the growing lack of dead trees and decaying fence posts that once held nesting sites, as the landscape became more picked up. Bluebirds have partly recovered thanks to the provision of nesting boxes with entrance holes slightly too small for starlings. Similarly, shrike populations in otherwise good habitat can be doubled by adding hunting perches; at the same time, their nesting success rises (since they must travel less far to find prey).

Weasels in small northeastern woodlands may lack places to hide from domestic cats (which also eat tens of millions of songbirds a year); foxes may lack safe denning sites, as well as sufficient hunting territory (that is, safely connected semi-wild areas). Busy roads are a constant hazard to both species; vegetated overpasses, wide enough and with enough shrubbery so the animals can keep out of sight are one solution; better than culverts under roads. (Such overpasses were built for animals like elk and bears crossing the Trans-Canada Highway.) Hawks and owls may lack nesting trees. Amphibians are killed crossing roads to breeding sites (losses can be considerable; here culverts work), snakes are killed on their way to winter dens (reducing their effect on slugs and mice and their abundance to their predators, which may also eat mice). Loggerhead shrikes are apparently in decline because their young forage for insects along roadsides and are run over by passing cars.

A simple way to control the ticks that cause Lyme disease is to put out cotton balls impregnated with pyrethrum for the mice to build their nests with (the balls are usually stuffed into toilet paper tubes and put at the edge of the lawn). The insecticide kills the ticks on the mice. This will work until the ticks become resistant to the insecticide. One can also get rid of the deer, a necessary link in the tick-deer-mouse-man chain. Because of their effect on forest vegetation, deer should be controlled, and this must be done by people, as people have for the most part excluded their other predators (wolves, mountain lions) from the suburbs. Computer simulations indicate that to lower tick densities much, you must get rid of essentially all the deer. This is difficult, but possible: perhaps it is desirable. Coyotes adapt quite well to suburbs and are not a bad partial solution to the deer problem, but eat small dogs and cats as well as mice, injured adult deer and fawns. People seem to deal with them better than with the resident Canada geese, which take over backyards, golf courses and soccer fields.

The geese, a sign of success, are an illustration of the problem of living with wild animals. Geese can be concentrated in smaller areas by manipulating the landscape (they avoid shrubbery and tall grass, where potential predators lurk: coyotes will eat foolish or young geese and goose eggs). A machine pulled behind a tractor (or pushed by hand) could sweep up their dung from playing fields and golf courses; then it could be composted and used for fertilizer (grass-goose-grass is a natural connection, which the geese have exploited). If they were edible (most are contaminated by lead, from the tailpipes of cars burning leaded gasoline), they could be hunted—I suppose with bird darts, as the Native Americans did. Living with animals as hunting and gathering people did and seeing them as equals (the tribe of mice, the tribe of mountain lions, the tribe of geese, the tribe of insects) means both respecting and killing them. Otherwise deer become tame and eat the shrubbery and mountain lions move in to eat the deer and then us. The lion and the lamb will not lie down with us but will take advantage of the situation, just as we have: that is their biological imperative. Deer carcasses in the backyards of Boulder, Colorado, are part of the natural order.

Some parts of the landscape are more essential to wildlife: breeding areas for amphibians (these are likely also to be aquifer recharge areas); trees in which hawks or owls nest; roosting and wintering areas for bats (often mines and caves); scattered groves of pines among the fields of southern Mexico where migrating raptors rest; the Atlantic capes (Cape Cod, Cape May), where migrating songbirds gather to feed before setting off on their thousand mile flight over the ocean to the Caribbean islands or South America. Because of the geography of the North American coast, staging areas like Cape Cod and Cape May are essential for the survival of viable populations of migrating birds. Much of Cape Cod and Cape May have already been developed. Such developments can be made more bird friendly by planting shrubs and trees the birds use (for their fruits, seeds, or insect pests), and by driving slowly and keeping the family cat indoors during migration. Also needing protection are the pathways the migrants follow in their slow springtime return through Central America, where the flowering or fruiting of some shrubs coincides with their passage; and the forests along the Gulf coast where those that fly across the Caribbean arrive in spring. A coastal forest several hundred meters wide to greet the birds and to absorb the force of storm surges would be a good thing from many perspectives. Perhaps a mile or two deep: Hurricane Katrina destroyed 90% of the structures within a half mile of the Mississippi coast in 2005, the same structures that were destroyed by Hurricane Camille 37 years before, making rebuilding more than pointless. The same landscape, left in its natural state, could do much useful work as habitat protecting the coasts, filtering water before it reaches the sea. Offshore islands, coral reefs, tidal wetlands, coastal mangrove forests all dissipate the energy in storm waves and limit the destructiveness of storm surges. They are less effective in blocking tsunamis, which are very long waves (the tsunami in southeast Asia in December 2007 was 8 miles long and rolled in for an hour). Coasts exposed to storm surges or hurricanes (where forests are useful in breaking the force of the wind) would be better left without permanent human settlements, and with their forests and wetlands intact. This would benefit not only homeowners and insurance companies, but birds, sea turtles, manatees, satwater crocodiles, and major coastal fisheries.

Another reason for the collapse of songbird populations is that the so called wintering grounds (actually their native grounds) of many neotropical migrants in Central America and the Caribbean have been developed, logged and increasingly fragmented. Birds are adaptable; many neotropical migrants winter well in shade grown coffee or cacao plantations. The trees in shade grown coffee plantations include banana, guava, citrus, trees used for firewood, and many native trees and herbs (up to 300 species of plants have been documented.) Their ground is covered with leaf litter and the invertebrates that live in the litter, the prey of birds that forage on the forest floor. Shade grown coffee plantations have 60-70% of the diversity of wild tropical forest. Sun grown coffee produces 5 times as much coffee per acre but the cost of fertilizer and pesticide means it costs 6.5 times as much to grow. Their higher production per acre however (higher gross returns) makes the land in such plantations more valuable, so 40% of the coffee grown in the western hemisphere is now sun grown.

Along the shore, native plantings, with stretches of mangrove forest, as well as beaches (the mangroves are often removed to make the beach), would make Caribbean resorts much more warbler friendly. Would it be good for business? Golf courses could be made much more ecologically friendly, with less water use (only the greens irrigated), fertilisation with composts, mowed along a narrow part of the central fairway, most of the course in native vegetation (sand, thorns, palms, bunchgrass: the rough more adventurous). As bird numbers are reduced we lose the value of the work they do in North American fields and forests. This amounts to 10-20% of the yearly growth of the forests, probably something similar for field crops. Such calculations help put a value on the lands needed during their migrations. As time goes on, more and more such lands will be discovered for more and more species. For instance, some effort is now made to preserve wetlands in the United States, but wetlands are intimately connected to nearby uplands, not only for their water supply but because animals like pollinating bees breed in nearby uplands. Wetlands with more intimate connections to uplands, rivers and the sea have a more diverse mix of species. (Finally, I note that 'swamp' is still a perjorative term in English.)

Air blows across the fields and picks up nitrous oxide from bacterially manipulated fertilizers, carbon dioxide from respiring soil, methane from manure. Growing plants scavenge some of the carbon dioxide from the soil and from the air. Cornfields, like forests, deplete the air above them of carbon dioxide in July when they are strongly growing and also transpire soil moisture into the air. (The total moisture transpired during the summer would cover the cornfield with water five feet deep.) Water condenses out of the rising air and forms clouds on mountains. In the temperate zone the average height of the bottoms of such clouds marks the level at which deciduous trees give way to evergreens. Cloud ceilings rise as forests on the slopes below are cleared: perhaps 30 feet per decade lately in the Appalachians; until there were no clouds left in some cloud forests in Costa Rica (and then the amphibians of the forest, which depended on the moisture from the clouds, died). Water drains off fields and forests into rivers and spirals downstream through river channels and riverside wetlands, the river water in intimate connection with huge pools of water that lie under the river valleys themselves. Long sections of these rivers have had their banks reconstructed with riprap or concrete, their vegetation removed, their beds altered to increase the water's depth, all of which changes the temperature of the river, its rate of flow and its horizontal connections with its valley landscape. Development changes nutrient relationships among river water, riverside soils and ground water. River water flows downhill over dams, with side trips into fields, through power plants and through municipal water systems, until it reaches an inland basin (the Great Salt Lake, the Dead Sea, the Caspian Sea), or the ocean. From such watery places it evaporates and is carried by the atmosphere until it falls over the mountains as rain. This tale leads us to our next chapter.

Chapter 20

The Ocean

So we come to the ocean, into which everything flows. The sea is enormous and delicate. Ocean waters vary in temperature, density, salinity, the presence of faint electrical fields, dissolved gases and minerals, scents: subtle chemical variations over horizontal and vertical scales. Sea turtles may follow the scent of specific ocean basins, as well as the earth's magnetic lines, in their thousands mile migrations along ocean currents. The turtles that migrate between the west coast of North America and Japan remain mostly in the top 20-120 feet, which makes it possible to set up a protected zone. The Gulf Stream travels up to 3.5 miles per hour and is 2000 times the flow of the Misissippi. Young fish and larvae from Caribbean waters are caught up in it and deposited by its turbulent swirls along the south coast of Long Island, where they mature but don't survive the winter. (They may in the future. So new environments are colonized.) Deep waters upwell as currents pass over seamounts, or collide with coasts, fertilizing the sunlit surface, whose organisms are the basis of their productivity. Where nutrient rich coldwater currents meet oppositely flowing warm ones, plankton grows, and fish (along with seabirds, turtles and marine mammals) thrive. Off the coasts of Maine and Atlantic Canada, where the southward flowing Labrador Current meets the warm waters of the Gulf Stream, the sea blooms with plankton most of the year. Forage fish and invertebrates like squid feed on the plankton and become food for cod, whales, seals, swordfish, bluefin tuna, porpoises and Atlantic salmon. Strong swimmers like tuna cross huge expanses of open ocean to reach such places. Flows of fresh water also deliver nutrients to surface waters—the Amazon to the ocean off South America, the Mississippi to the Gulf of Mexico, the Niger and the Congo to the Gulf of Guinea; and before they were dammed, the Yellow River to the Yellow sea, the Nile to the eastern Mediterranean, the Colorado to the Gulf of California. Such places are also productive, their productivity rising and falling with the flow of nutrient rich fresh water. Biologists divide the sea into biogeographical provinces that depend on local processes delivering nutrients to sunlit surface waters; but the boundaries of such provinces are dynamic. If the meeting places of currents shift, marine reserves that protected them (or the migrations of turtles along currents) would have to move with them. The sea is also noisy. Whales sing low songs across great distances (the bass rumbles of the blue whale cross the Atlantic), the noise of sea urchins grazing on rocky reefs peaks just before dawn and just after dusk, a chorus of shouting fish greets the dawn off the California coast.

The edge of the sea also constantly changes. A large wave breaking over a mud flat removes the chemical signature that attracts the larvae of the sedentary animals of the flat (polychaete worms, clams) to settle, dig burrows, build feeding tubes, oxygenate the sediments, grow. Without the presence of that chemical scent, the larvae drift off elsewhere. Larvae must settle within a given period of time (their development is time limited). Thus chance is involved in the colonization of a flat; but the chemical signature is restored in a day. Rivers bring down silt, sand, nitrogen and other nutrients into estuaries; the flow of fresh water, against the tides, determines the level of salinity in a given spot. Salinity moves up and downstream with the tides and seasons and is used as a developmental cue by many plants and animals. How deep the mud flats lie below mean high tide is determined by silt loads and tidal flushing. The water depth over the flats determines what animals live there. Rapid siltation raises the flats, suffocates the existing benthos and means different suites of animals (or for a time, none), a slow re-colonization, a gap in the food chain. The silt sent down the Sacramento and San Joaquin Rivers by the gold miners of 1849-1852 raised the bed of the Sacramento River by 10-30 feet and reshaped much of San Francisco Bay. Hundreds of square miles of farmland along both rivers was flooded, as was downtown Sacramento. As a result hydraulic mining was curtailed, but not before most of the gold was gone. (To date about 35% of San Francisco Bay has been lost to sedimentation and land reclamation.)

People have purposely affected the sea's edge for at least 1000 years, diking and draining for harbors and fields. Salt marshes have low species diversity but high productivity. They accumulate nutrients with each tide, trap and supply nutrients to surrounding waters, contribute to both aquatic and terrestrial food webs and are habitat for commercially important species of fish. Their slow accumulation of peat lets them grow with sea level rise; or slowly invade the ocean. Higher ones were once mowed for salt hay. Ditching and flood-gating for mosquito control on east coast marshes in the 1930s greatly changed the marsh habitat. A major problem was the pile of spoil left at the side of the ditch by the dredges, which prevented complete drainage of the marsh at low tide. (New dredges that send the spoil flying out over the marsh avoid this.) In the stagnant pools, phragmites (reeds) began to replace the stands of smooth cordgrass. Cordgrass was eaten by wintering ducks and geese. Detritus from the grass, whose beds delimited the reach of high tide, fed microorganisms, plankton, fish and crabs; the reed beds were much less useful to the birds and other organisms. About 50% of east coast marshes were lost to invasive phragmites through draining and diking. Similarly, in San Francisco Bay many organisms have been introduced, either intentionally, such as striped bass and shad, or unintentionally: the latter often arrive in the ballast water of ships, which ought to be pumped out and exchanged in midocean, but which, because of the time involved (a cost to shipping companies), is not. Half the fish in San Francisco Bay are alien species along with the majority of the plants and animals on the bay's floor. In parts of San Francisco Bay 99% of the plants and animals are non-native. In healthy ecosystems, most introduced aliens find a small niche among the natives; some however, released from their usual predators and parasites, find a competitive advantage, expand geometrically, and take over the habitat. This happens more easily in stressed ecosystems.

In the late 1940s and 1950s, DDT, sprayed to control mosquitoes, tremendously reduced the crustaceans (crabs, shellfish) and other invertebrates of the marshes, whose populations never really recovered. The oceans and their top predators, the marine mammals, are sinks for DDT and other persistant organic pollutants. DDT was banned in most industrial countries in the 1970s but is still manufactured in the U.S. and is used in many underdeveloped countries. The DDT burden in marine mammals fell through the 1980s with the ban, but then leveled off, as the chemical was recycled in the environment and as new supplies eroded off uplands, seeped out of dumps, condensed out of the air (from current use), were released from melting glacial ice, or rose on currents from the depths of the sea. Levels of persistant organic chemicals in male marine mammals tend to rise throughout their lives but those in female mammals level off when they reach reproductive age and begin transferring the chemicals to their offspring during pregnancy and nursing. Persistant organic pollutants such as DDT and the PCBs have been linked to reproductive failure, lowered immune function, and skeletal abnormalities in marine mammals such as seals, walrus, whales and dolphins. Lowered immune function probably explains the periodic collapse of dolphin populations along the East Coast of the United States, and of seals in the Mediterranean and North Sea, from infective viruses.

The land meets the sea in coral reefs, salt marshes, sea grass meadows, mangrove forests, beds of kelp. These aquatic forests and grasslands, fertilized by the tides and by nutrients washing off the land, are the nurseries of the sea. Along the Gulf of Maine a brown kelp forest grew near the coast, while further out the rocky bottom was covered with shaggy red algae. Kelp forests are maintained by a balance between the invertebrates that graze on them, such as sea urchins, and the animals that eat the grazers (sea otters, cod, lobsters). If predation pressure on the grazers is reduced (by trapping out sea otters or by overfishing cod), the grazers multiply and the kelp forests disappear. Kelp forests provide nursery habitat and shelter for many fish and marine invertebrates. Beds of giant kelp still circle the northeastern Pacific from Japan to southern California. The beds off California and the Aleutians, that once supported grazing manatees, and may have been one sea route from Asia to the Americas by people in boats, are now maintained by the balance between grazing sea urchins and urchin eating sea otters. (Abalones, fished to bioeconomic extinction, were once part of the pattern.)

Mangrove forests and coral reefs, like the salt marshes of northern latitudes, and the meadows and islands of the Mississippi Delta in the Gulf of Mexico, protect the coast from high tides and storm surges. Mangroves are as effective as sea walls costing $300,000 or more per kilometer; are an important nursery for fish; habitat for many terrestrial insects and animals, including tigers and jaguars; and provide approximately 10% of the organic carbon that enters the ocean from the land. This carbon is resistant to breakdown and so constitutes carbon removed from the atmosphere and withheld from the oceans; carbon put in medium term storage (making mangrove forests candidates for a carbon storage payment). Such coastal areas are often used for aquaculture. Mangrove forests are cut to dig shrimp ponds, replacing the natural environment of coastal wetlands, deltas, lagoons, tidal flats, with manmade brackish ponds one third of whose water (half of that fresh) must be replaced every day. So levels of fresh ground water in coastal regions with shrimp ponds drop rapidly. As fresh water is pumped out, salt water intrudes into the aquifers. Waste water from the ponds, full of nutrients and chemicals, contaminates surface and ground water, overfertilizes offshore coral reefs and ruins coastal fisheries. While the existing habitat of coral reefs and mangrove forests are capable of self-sustaining production of timber, fish, sand and rock if not overexploited, shrimp ponds, because of the build-up of shrimp diseases, are abandoned after five years, leaving an unproductive landscape behind, that mangroves have difficulty recolonizing. Mangrove forests are also cleared for shoreline development, the stands of trees (which grow in coastal shallows) replaced with beaches. About half of coastal mangrove forests worldwide have been cut (about 70% in the Phillipines). The benefits of mangrove forests are not linear; that is, they can be exploited. Perhaps 20-30% of a stand can be cut without losing its benefits (as a fish nursery, source of carbon, protection from storms).

Coral reefs lie between land and sea in nutrient poor tropical and subtropical waters, drawing in nutrients from land and sea in a very efficient recycling system. Coral reefs are adapted to a low nutrient environment. Marked by a telltale line of breaking waves, the stony skeletons of coral reefs shelter the coastline from waves and storms. The waves bring nutrients to the reef, wash away sediment, oxygenate the water. A symbiotic algae that lives in the coral polyp secretes sugars for the polyp, whose nitrogen rich wastes are used by the algae. The reefs' mineral skeletons are built of calcium secreted by the polyps so reefs do well in waters supersaturated with that mineral. Coral reefs need clear water for the algae (or zooanthellae, a dinoflagellate) to photosynthesize. Sedimentation from construction, agriculture or sewage outflow pipes will damage or kill them. Reefs are commonly mined for building blocks or sand, damaged by fishermen, run into by boats, dredged for shipping channels. When overwhelmed by nutrient runoff from the land, they turn into algae covered rocks. The algae grow over the corals, smothering them. Sugars from the algae fertilize pathogenic bacteria that infect the corals and kill them. (The bacteria are always present but multiply more rapidly in the presence of algal sugars.) Algae are always present on coral reefs (they are one of the bases of the food web of the reefs) but reefs were formerly grazed of algae much more heavily by turtles and fish. For instance, between 35-100 million green sea turtles once grazed Caribbean coal reefs, accompanied by 33-39 million hawksbill turtles. Adult green sea turtles turtles weigh 220-500 pounds. They eat crustaceans, seaweeds, starfish and mollusks. They move slowly in shallow water and like to bask in the sun near the high tide line, waiting to be refloated by the tide. Their grazing, along with that of herbivorous fish, especially the larger fish, kept the corals clean of algae. The grazers (especially the fish) recycle the productivity of the reef. As nutrient runoff over the reefs increased during the twentieth century, there weren't enough grazers to keep the algae in check. By the 1750s most of the turtles in the Caribbean had been fished for food for slaves on the islands' sugar plantations. (Many islands' upland habitats were totally transformed long ago by the cultivation of sugar. Forest cover was gone from Barbados by 1665: the runoff from such changes must also have stressed the reefs, which however recovered.) Manatees, the other large, edible vertebrate of Caribbean reef systems, were more or less gone as organisms of ecological importance by 1800. Both animals also ate seagrass and so helped renew those underwater meadows that serve as fish nurseries. But the reefs maintained themselves until modern times, when increased nutrients running off the land, along with increased fishing pressure, began to overwhelm them.

An example of such reef deterioration comes from Jamaica. As prosperity increased in Jamaica after the Second World War, live coral near the north coast of the island fell from 60% of the reef in the 1960s to less than 5% in 2000. Kelp and other seaweeds grew over the coral. The grazing turtles were mostly gone, algae eating sea urchins died of a disease in the 1980s, grazing reef fish were fished more and more intensively for restaurants. As a market for sharks developed in Asia, fishing for sharks (which reproduce slowly) let more midsized predatory fish like groupers and barracuda increase; the groupers ate the smaller grazers (the smaller parrot fish), the barracuda small and large ones; at the same time the larger grazing fish (the larger species of parrotfish, which might have helped control the algae) and the groupers themselves were still fished down to supply restaurants. The seaweeds were nourished by silt and nutrients washing off golf courses, from the lawns of resorts, from sewage. So the reefs died. A final effect in this cascade is a continuing loss of beach sand. Sand in the Caribbean is renewed by several species of coralline algae that assemble carbonate grains out of seawater; when the algae die, the grains become sand that help renew the white beaches. Such algae are adapted to low nutrient situations and are overgrown by bacterial slimes in the presence of too many nutrients. So Jamaican beaches also lose sand.

The effects of overfishing on corals can be reversed through the use of ocean reserves, where fishing is limited or forbidden. In protected Bahamian reefs, groupers increased by 7 times and ate the smaller parrot fish, those less than 6 inches, reducing their numbers considerably. Large parrot fish, too big for the groupers to swallow, increased in numbers and (being larger) caused a net doubling of reef grazing. The result was a four fold reduction in seaweed on the reef. Jamaica is of course not alone in its loss of marine habitat. Deforestation in the highlands of the Dominican Republic sends silt into Samara Bay, one of the most important fish nurseries in the Caribbean. The estuary, the Caribbean's largest, produces 40% of the fish catch of the Dominican Republic and is a sanctuary for humpback whales. Silt also runs off the coast of southern Florida; water saturated with lawn fertilizer and septic tank effluent seeps into Florida bays. Sewage effluent is emptied into the sea. The result is that many, if not most, Florida reefs are dying or dead.

The pastures of the sea are microbial. Despite their vertical extent, compared to terrestrial pastures they are sparse: standing biomass of terrestrial plant life is 200 times that of marine plants. (But the mass of microbial life on the seafloor may dwarf that of terrestrial plants.) Cyanobacteria (formerly blue green algae, a member of the pico-plankton) are among the chief photosynthesizers and cycle both carbon and nitrogen. They are eaten by single celled animals (the protists) or by shellfish (which filter them out of the water). Among the known bacteria and viruses floating on the surface of the sea are thousands or tens of thousands of unknown species, most in very small numbers, that represent vast stores of genetic diversity, with the ability to take over planetary functions after massive global changes (increases or decreases in temperature, changes in the relative abundance of atmospheric gases), that increase their competitive advantage. Photosynthetic activity occurs throughout the sunlit 600 feet of the upper ocean but much of it occurs on the water's thin surface skin. Dense blooms of plankton absorb sunlight that warms the oceans surface, increasing (in tropical latitudes) the frequency of cyclones. Fish also concentrate in the top 600 feet of the sea, especially in the area over the continental shelves, fertilized by water from rivers and upwelling. About 90% of fish are caught here and catches are two orders of magnitude (a hundred times) over those in the open ocean. Winds drive currents, and upwelling of nutrient rich deep waters along coasts or seamounts or in the wake of hurricanes and typhoons (as well as the daily action of the tides) returns the nutrients depleted by the photosynthesizers to surface waters. Winter storms stir up nutrients from the bottoms of shallow seas, such as Europe's North Sea, setting the stage for the plankton growth of spring. Much of the upper deep ocean, constantly depleted by microbial photosynthesis, is nutrient poor. Populations of microbes respond rapidly to changes in the nutrient situation, expanding and contracting with changes in their food supply; or their chemical and physical environment. Domoic acid is a neurotoxin produced by a photosynthetic diatom that accumulates harmlessly in fish and shellfish but kills seabirds and mammals (sea lions, whales, dolphins, people). Increasing amounts of urea (from fertilizer and sewage) and copper (from boat paint) in seawater, along with rising water temperatures, seem to stimulate the algae to produce more of the chemical. Domoic acid first was found off the California coast in 1991 and has appeared regularly since 2001. Thousands of sea mammals have died and some fish and shellfish are unsafe to eat. Similarly, increasing nutrients and warming temperatures increase the relative numbers of the dinoflagelates that cause red tides, toxic to people and fish. Increasing nutrients increase the incidence of the fish eating bacteria *Pfiesteria piscida* that causes fish kills in east coast rivers polluted by runoff from hog farms. (*Pfiesteria* toxins also affect the nervous system of people.) The microorganisms are responding to environmental changes that favor their competitive position.

The land is the ultimate source of nutrients for the sea. Estuaries, which receive relatively large volumes of riverborne nutrients, are the only marine biome that competes with the land in productivity. Two-thirds of marine life begins in shallow coastal waters. (Perhaps 98% of commercially important fish in the Gulf of Mexico begin their lives in the gulf's estuaries.) Estuarine plants and animals are adapted to seasonal changes in the levels of silt, nutrients, salinity, pulses of fresh water. Developing the terrestrial landscape upstream changes the levels and timing of the these pulses and the old systems break down.

Florida Bay, a major nursery for fish of the Gulf of Mexico, is fed by the Everglades, the great swampland, a shallow river six inches deep and 40-60 miles wide, with a drop of 2-3 inches a mile (comparable to the lower Mississippi), that once flowed the 80 miles from Lake Okeechobee (15 feet above sealevel) across southern Florida to Florida Bay. Wet prairies bordered the sawgrass flats of the swamp. On the west coast the land slid down into coastal mangrove forests fed by the water, while on the slightly higher and rocky east coast the land met the sea in barrier islands. The sawgrass was rooted in marl precipitated out of the calcium rich water (limestone lies under the Everglades) by the periphyton (a mix of algae and zooplankton) that lived on the roots of the grass. The dead leaves of the sawgrass formed peat, in which bay trees and willows might root. These plants produced more peat, whose acid decay dissolved the limestone and deepened the pools in which they lived, giving them more space and forming bayheads and willow heads. The pools varied the habitat and stored water for the dry season. Deeper pools were occupied by cypress trees. Mounds of peat that accumulated above water level, sometimes thrown up by alligators weeding their home pools, became dry hummocks with pines, hardwood trees and palms. Waterlevels in the Everglades deepened in autumn as the summer rains from central Florida filtered into them, letting their fish populations grow, and delivering a slow pulse of fresh water to Florida Bay, and then fell in winter and spring, concentrating the fish in small pools, where they were easy prey for wading birds. Most of the pools were dug by alligators as refuges for the dry season. (The modern Everglades would probably not have worked without the tens of thousands of pools dug by alligators.) Water from the chain of marshes and lakes in central Florida flowed into Lake Okeechobee down the wide floodplain of the Kissimmee River, then spilled out of the lake into the Everglades. The Kissimmee Basin, Lake Okeechobee and the Everglades were a 9000 square mile hydrological system. The slow autumn pulse of fresh water maintained Florida Bay's mangroves and seagrasses, and their associated bacteria, plankton and fish (all organisms adapted to a mix of salt and fresh water). The seagrass meadows were the nurseries for the growing fish. The

Everglades are a low nutrient system, whose characteristic phytoplankton and sawgrasses are overtaken by other plants (especially cattails) when water levels are stabilized and nutrient levels in the water rise. The Everglades sawgrass is a plant of nutrient-poor, hydrologically unstable regimes. Its roots support the mix of algae and zooplankton—the paraphyton—that is the basis of the Everglades' food chain. Shrimp graze on the paraphyton, fish eat the shrimp, and wading and diving birds and alligators eat the crustaceans and fish.

Everglades National Park occupies 20% of the historic area of the swamp. Half the Everglades has been drained for agriculture, much of the rest diked off to use for housing for the expanding populations of southeast Florida. Phosphorus from fertilizer used on sugarcane grown on the several million acres of drained swampland near Lake Okeechobee, along with phosphorus and nitrogen from agriculture and from urban areas carried down the Kissimmee from central Florida, now enter the Everglades. After floods in 1928 Lake Okeechobee was leveed off and its excess water drained away into the Atlantic, or used for irrigation and urban water supply. Five million people depend on water from Lake Ocheechobee. In the 1960s, the Kissimmee River, once a meandering 103 mile stream was straightened and turned into a series of five pools with locks, making the river usable for recreational motorboats and its formerly swampy floodplain for cattle pasture. Channelization destroyed the filtering capacity of the riverside marshes and the Kissimmee began delivering nutrients from upstream to Lake Okeechobee. The numbers of wintering waterfowl on the floodplain fell by 90%. (Channelization of the Kissimmee cost $35 million. Restoration, which began almost immediately, as the extent of the disaster to the Everglades became clear, cost $20,000 an acre or $512 million for restoring two-thirds of the former Kissimmee wetlands. But the restoration seems to have been successful.)

Further diking and draining and the construction of a highway across the Everglades, turned the Everglades into a series of managed pools, rather than a continuous shallow flow. Water levels were more stable, which, along with the raised nutrient levels, favored cattails. In years when water is scarce, water is delivered to farms and cities rather than let flow into the Everglades. Water bird populations in south Florida have fallen 95% from the 1930s, and the fisheries that depend on Florida Bay have also fallen. The effects of prolonged water shortages and nutrient pollution are cumulative and in the early 1990s Florida Bay turned from a clear water ecosystem dominated by sea grasses and manatees to an ecosystem of turbid water dominated by algal blooms. The underwater meadows suffered extensive mortality from diseases. Excess nutrients and increasing salinity because of the lack of fresh water let algae outcompete the grasses, and their supporting suite of organisms (juvenile marine fish, zooplankton) that eat the algae. Some writers speculate that grazing (by sea turtles, waterfowl or manatees), which renews the meadows, may—especially in the present situation—be necessary for their long term health. Green sea turtles favor the tender tips of young grasses and clip away and discard the tough older tips as part of their feeding behavior. This stimulates the sprouting of buds lower down on the stem and keeps the grasses fresh and productive. Such trimming may be especially useful under high nutrient regimes. But green sea turtles and manatees are no longer present in numbers that matter ecologically: they are functionally absent.

Another tale of the land meeting the sea can be read in the history of the Mississippi Delta. About 75% of marine fish in the Gulf of Mexico begin life in the wetlands of the Mississippi Delta. Until the 1950s these wetlands grew, against the constant erosive action of the sea, and against subsidence under their own weight, from silt brought down by the river. Dams on the Mississippi's tributaries, especially the Missouri, reduced silt loads at the Mississippi Delta by half. (The Ohio—*la Belle Rivière* of the French—once a wide clear shallow stream, with mussels plating its riffles, is now a narrow deep muddy one thanks to channelization, agricultural runoff and deforestation. It carries 10 times the silt of formerly, but its contribution doesn't make up for that lost from the prairie streams.) There are 8000 dams, large and small, in the Mississippi basin. A major shipping channel (rarely used) that led directly to New Orleans from the gulf, and 10,000 miles of oil exploration canals opened up the delta marshlands to wave erosion. As the tides penetrated inland, the freshwater marshes turned salty, and the land washed away. The freshwater marshes were feeding grounds for overwintering waterfowl. Some supported cypress and oak forests, which are more efficient than marshlands at reducing the height of storm surges and at protecting the coastline from storms. Sending silty Mississippi water, with its fertilizers, pesticides and industrial chemicals directly out into the gulf, instead of letting it flow over the natural river levee and into the Louisiana marshes, turned the silt and fertilizer into something toxic, rather than something that would build new land and grow useful biomass (trees, fish, crayfish, ducks); and whose newly built marshes would further protect the shoreline and release cleaner water downstream. Instead the nutrients created a dead zone at the mouth of the river.

Cores from the seafloor off the Mississippi Delta show a long period of stability in algal biomass before the Europeans arrived (for several hundred years of this time large sections of the Mississippi Valley near the river were being farmed by Native Americans), an increase in algal growth from the 1850s as the Middle West was settled by European Americans, then algal blooms and seafloor hypoxia on a regular basis from the mid-1950s on. Nitrogen use increased 6 times in the United States from 1955-1980 and the concentration of dissolved nitrogen and phosphorus in the Mississippi doubled. New levees after the Second World War narrowed the river's floodplain and eliminated much riverside swampland that once filtered the water. (The Mississippi is separated by levees from 90% of its floodplain.) The dead zone of summer hypoxia in the Gulf of Mexico doubled in size after the 1993 floods, which brought down huge amounts of nutrients, to 7000 square miles (an area about 80 by 85 miles). Excess nutrients accumulate in marine or lakebottom sediments and are recycled again and again back into the water column, especially under low oxygen conditions, making such conditions somewhat intractable, but usually much more intractable in lakes than in the ocean, whose waters are churned by tides and storms and mixed by currents. At present, perhaps afraid of damage from a major hurricane on a coastline less and less protected by marshlands, the Corps of Engineers is considering siphoning some silty river water over the levee into the marshes. Some may not help; and in order to protect the delta as a whole, some newly settled lands may have to be abandoned to seasonal floods.

Nutrient related seasonal hypoxias (areas of the ocean in which there is too little oxygen for oxygen breathing organisms like fish and shellfish to survive) are now found at the mouth of the Mississippi, in large portions of Chesapeake Bay, about New York City (including the western end of Long Island Sound), in the Adriatic, the North and Baltic Seas, the Inland Sea of Japan, the Yellow Sea off the Chinese coast, and the Persian Gulf.

There are about 200 such areas worldwide, most in places important for spawning marine fish. Future archeologists will call nitrogen fertilization, like disturbed soils or pioneer ecosystems (or rising levels of carbon dioxide), is a sign of human occupation. Dams on rivers that feed these seas reduce the silicaeous sand needed by diatoms (a base of the food chain), and shift plankton populations to the potentially more toxic dinoflagellates, which thrive on the increased nitrogen. Dams also block fish spawning runs, greatly reducing or eliminating many species and shortening the food chains of estuaries and bays. Shorter food chains, with their fewer prey species, make populations of predatory fish more vulnerable to yearly variations in the supply of plankton. Plankton production is dependent on the weather, which is always variable. The catch of alewives (river herring) in Long Island Sound in the early 2000s was 3% that of the 1960s. The causes of the decline are probably overfishing and the dams that block nearly every alewife nursery stream entering the sound. Alewives are forage fish for larger fish and an important food fish for ospreys, whose colonial nesting sites—some held several thousand birds—on Long Island have never really recovered from the birds' poisoning with DDT.

Dams on the rivers that feed estuaries change the amount and timing of the pulses of fresh water to which the life cycles of fish are adjusted. Algal blooms from excess nutrients carried by the rivers smother seagrass beds and shut off their light, making them vulnerable to diseases. Shellfish populations collapse from overfishing, lack of oxygen and introduced or stress related diseases. The frequency of red tides from dinoflagellate blooms increases. Red tides occur naturally but in modern times are also a sign of excess nutrients (from land, from aquaculture, from shoreline reclamation) or of unusual warmth. A red tide off the west coast of Florida from January to October of 2005 covered 2000 square miles (an area 200 miles by 10 miles). The toxins in the dinoflagellates kill fish and the rotting fish deplete the oxygen in the water and supply more nutrients for the bacteria. In Florida, hundreds of tons of dead fish washed up on the beach. The toxins from the dinoflagellates come ashore on the wind, so people living near the coast experience respiratory problems (coughing, sneezing, itchy eyes, difficulty in breathing).

Dams and excess nitrogen are two major ways in which people affect the marine environment. But increased human activity degrades the marine habitat in many ways. Traffic from pleasure boats in coastal waters increases shoreline erosion and damages shellfish reefs. Turbidity from boat traffic is a major problem in busy estuaries. Large, fast ships increase underwater noise, a problem for some species, especially marine mammals. (It may be involved with beaching whales.) Underwater noise from ships has increased 10 times off southern California since the 1960s. Antifouling paints used on boats and docks to kill the marine organisms that colonize ships' hulls or bore into the supports of docks, kill many nontarget marine organisms and contaminate harbor sediments with tributyl tin. (Tributyl tin is also used as a slimicide in the cooling systems of power plants and so contaminantes the rivers from which power plants draw their water.) Industrial chemicals, oil, metals, and organochlorines wash off the land and also pollute harbors. Much of this ends up in the sediment, which in extreme cases (such as New Haven Harbor on Long Island Sound) becomes toxic to the burrowing organisms of the benthos, such as polychaete worms, which in healthy environments oxygenate the sediments and recycle their nutrients. The larvae of the worms may settle but in penetrating the sediment they ingest chemicals that kill them. Most oil pollution in North American coastal waters comes from land and not from the occasional (but spectacular) spills at sea, or from the 180 ships a year that sink at sea. (This oil, if not lost immediately, will be lost over time. Continual low level releases of oil from vessel operations, including oil tankers, may have the greatest effect on seabirds, and small spills by the oil industry worldwide average more than one per day.) Of the 29 million gallons of oil that enters North American waters yearly, 85% comes from end-users, such as the owners of trucks and cars, and service stations, not from the oil industry. The oil is put down drains, washes off roads and is carried by rivers to the sea. The presence of oil reduces the survival of eggs and larvae of many marine species, including fish. Effects may be long term. Herring populations in Prince William Sound collapsed four years after the Exxon Valdez spill there and fish living near old oil spills (several decades old, the oil now several inches down in the revegetated mud, little degraded in the anoxic environment) show elevated levels of liver enzymes associated with chronic oil pollution. Minute amounts of polycyclic aromatic hydrocarbons (PAHs) in the oil cause deformities in developing fish embryos that kill many of them before they reach adulthood.

Tiny plastic pellets, feedstock for the plastics industry, dumped or lost at sea, are dispersed throughout the oceans, at a density of 1-4 per square meter. The pellets are mistaken for fish eggs or zooplankton by seabirds, sea turtles, and fish. The pellets cause ulceration in the stomachs of seabirds and reduce the functional capacity of their gizzards. PCBs and other persistant organochlorines are adsorbed on the surfaces of the pellets at concentrations up to a million times those in open water and ingested with them by fish and seabirds: the chemicals are then absorbed into the animals' bodies (and into ours, if we eat them). Smaller pellets, so called microlitter, used in products like exfoliant creams, wash through sewers into the sea and are carried back onto the beach, where they clog the digestive systems of organisms like sand fleas and lugworms. Larger pieces of plastic, blown or washed off the land, or dumped with other refuse at sea, cover the ocean's tropical gyres (approximately a quarter of the planet's surface) with a mass several times that of the zooplankton on the ocean's surface—in 2008, in the North Pacific Subtropical Gyre, an area of water twice the size of the continental United States, the plastic weighed 46 times the zooplanckton. Such bits—of toothbrushes, cigarette lighters, plastic containers, plastic bags, all of which also adsorb toxic hydrocarbons—are eaten by seabirds and turtles, who starve to death with their stomachs full of plastic. As a writer has remarked, plastic, like salt or calcium, is now a component of ocean water.

Increasing ultraviolet radiation (from the Arctic and Antarctic ozone holes) causes mortality among the juveniles of some fish, such as the northern anchovy, an important forage fish. Human disturbance limits the reproductive success of animals that breed on the beach: sea turtles, terns, piping plovers. Harvesting horseshoe crabs for fishing bait has decimated their population in Delaware Bay, once the center of their distribution on the North American East Coast. The crabs were trawled from the bottom of the bay and also collected by the truckload from the beach. The female crabs were sold as bait to commercial fisherman. Horseshoe crab eggs, laid on the beach at spring high tides, in densities of 100,000 or so to the square meter, are a major food for shorebirds migrating north. Flocks of red knots (a shorebird) arrive on the beach after a 7000 mile flight over the ocean from the bulge of Brazil; the nonstop flight takes a week and the birds land in a state of exhaustion and hyperphagia, or extreme hunger:

they need immediate, easily available food. Their feeding barely affects the abundant supply of eggs but the eggs must be abundant for the birds to survive. (They also need to put on weight for the next leg of their trip north.) Despite all this, the nutrients that flow into estuaries from growing human settlements is one of their main problems; and overfishing the main problem for ocean fisheries.

"If you lose the hills, you lose the sea," a scientist has remarked. The sea also has problems specific to itself. Early forms of trawl netting were opposed in England 600 years ago because of fear of damage to the "flower of the sea": the plants and sessile animals that grow on the ocean bottom, waving blooms of pink, green and brown, in many nearshore locations. Living in these tangles were healthy populations of juvenile fish and of invertebrates, food for adult fish. As in aboriginal North America, fish in Carolingian Europe were caught by hook and line, in seines, in dip nets or traps. River fish, especially anadromous fish, which spend a major part of their lives in the sea, were first exploited, but by the 1050s mill dams and siltation from expanding agricultural settlement had reduced their habitat and, along with overfishing, their populations. Weirs and nets across rivers let people catch most of the migrating fish. From 1000-1300 the human population was growing and the demand for fish increasing. Diking the Rhine Delta to stabilize its watercourses eliminated much of the breeding habitat of North Sea sturgeon, once a mainstay of the northern European diet (studies of middens indicate sturgeon made up to 70% of the fish eaten in the Baltic States in the 700s). Christians were obliged to avoid meat 130-150 days a year, so fish was a major part of the diet of Christian Europe. As anadromous fish decreased, fish production turned to fishponds (often established in the still water behind dams), and the sea, which was also fished with hook and line, traps (such as those for migrating tuna along the shores of the Mediterranean, which date from Phoenician times) and nets.

The trawl was first mentioned in a complaint to the English king in 1376. The petitioners argued the trawl nets destroyed the plantations on the sea bottom and thus the little fish and other animals the big fish ate. While the trawl was spectacularly successful at catching fish, dragging nets behind ships destroyed the bottom habitat of rock outcroppings, boulders and cobbles; the shell-like structures of algae, worms, brachiopods and bryozoans; the beds of mollusks (now harvested by specialized trawl); the vertical structures of anemones, sea pens, coldwater corals, rooted algae. (Deep coldwater corals extend for tens of kilometers along oceanic gravel ridges, reach 180 feet in height and shelter 1000 or more species of organisms.) Trawling also eliminates the cycling of seafloor sediments by a wide variety of worms (sediments smother the worms) and releases nutrients to the upper waters. The North Sea originally had many oyster beds and extensive reefs of tube building worms (shallow waters along the German and Dutch coasts still have these) and was much more clear than now. Despite 800 years of fishing and the collapse of several fisheries, until the 1870s European seas still were full of life; then powerful, steam-driven trawls began to transform their bottoms and catch too many fish. Much of the North Sea was dry land during the last ice age and is shallow, productive and easy to trawl. Fish catches (measured by effort put into fishing) soon fell. Then the absolute size of the catch began to fall. By the 1920s the effects of motorized trawlers were felt worldwide. The average size of cod landed in the Gulf of Maine decreased by 66% (from one meter to one third of a meter) following the advent of mechanized fishing in the 1920s.

Every year shallow banks are reworked by storms and must be recolonized by worms and other benthic organisms but below about 80 feet (less near sheltered coasts) most animals and plants survive the storms. Coastal banks are also scoured by tides. Tides increase the flow of oxygen, organic matter and plankton over the banks, making them good habitat for filter feeders, and their sediments and gravels can be stabilized by animals like corals, sea fans and crinoids. Stabilization of the sediments lets other plants and animals establish themselves: starfish, snails, sponges, sea squits, crabs, lobsters, sea anemones, prawns. In shallow northern seas such as the North Sea or the Gulf of Maine, whose cold, nutrient rich, well oxygenated waters make them excellent places for fish, elaborate reefs of cold water corals covered the rubble left by melting glaciers. Trawling dug up the bottom, broke up the oyster beds and the crusts of shells scallops live among, scooped up young fish, crayfish, and other invertebrates the cod and haddock eat. In the North Sea, 16 pounds of marine invertebrates are killed for every pound of marketable sole. Currents also sweep over seamounts, bringing nutrients and oxygen, creating another favorable environment for fish and for filter feeders of the bottom. In the Tasman Sea, corals and crinoids cover 90% of pristine seamounts. After trawling, the figure drops to 5% and the seamount loses half its biomass, and much of its potential as a fishery. Recovery takes 50 years or more. (So trawling here resembles clearcutting. With 30,000 seamounts in the Pacific and 6,000 in the Atlantic, such fisheries can go on for some decades.) Modern bottom trawls trap and kill almost all fish, mollusks, and invertebrates they encounter. Much of this is unwanted or too small to keep legally and is thrown back dead or dying into the sea. Thus trawlers sorting their catch leave a characteristic trail of dying fish and feeding gulls.

Large bottom trawls, pulled across the seabed at 4 miles per hour, leave trails of mud visible from space. The fertilized water column above the trawl is good for algae, and the muddy bottom good for breeding shrimp. (In an early argument trawls were said to "plow the sea" and so increase fish production; they may increase production of shrimp.) Trawls are set to run along the bottom. Two boats fishing a rough bottom (rough, say, with boulders or underwater corals) drag a heavy chain between them over the bottom to level it and then fish. Scallop trawls are set to excavate the bottom for scallops, flounder trawls to dig out the fish, which lie half buried in the mud. So the habitat left by trawling is not good for fish: their prey is gone, along with the shrubland of corals and crinoids of the ocean floor; and the mud clogs their gills, interferes with their vision and causes algal blooms in the waters above. Global positioning systems and competing fishing fleets mean likely places for fish are trawled much more often. In 1900 (a century ago!) it was estimated every trawlable part of the North Sea (an area of 100,000 square miles) was trawled twice a year. Near the end of the Grand Banks fishery for cod 80 years later each spot on the banks was being trawled every four months. Small hills were levelled and the seabed turned into a vast mudflat. When fishing stopped, the cod population was perhaps 0.3% of the original population: in Canadian waters the original cod population was something like 7 million tonnes, while the population when trawling stopped, perhaps 22,000 tons—for a fisher who encountered a school, still a lot of fish. With the cod gone, invertebrate populations exploded and a lucrative fishery for snow crab, northern prawns, lobsters, rock crab and sea urchins—all once prey of the cod—appeared. The legally saleable bycatch of cod from fisheries now allowed in Canadian cod waters amounts to probably 90% of the cod population, which shows no sign of recovery.

Biological production and fish catches change by a factor of 10 by trophic level. (From phytoplankton to zooplankton to small fish to still bigger fish to the fish we eat; so the fish we eat are 100-10,000 times less abundant than their zooplanckton, invertebrate or fishy prey.) Most large fish eat at several trophic levels. Marine food webs are dynamic and the fish taken every year (by other fish, whales, seals, weather, people) are replaced at varying rates. If one animal (fish, crab, squid) becomes scarce, others are available as prey; and a change in its favored prey by a predator allows another population of animals time to recover. (Human fishers are just one predator in the sea; fish and sea mammals still eat several times more fish than people, though the proportion is falling). Heavy human fishing changes the structure of fish populations and simplifies food webs in the oceans, interfering with the complex food webs that provide ecosystems with a greater degree of resilience. Under the selective pressure of heavy fishing, fish become smaller and breed earlier. Fishermen in heavily fished waters now catch fewer large fish, fewer predatory species of fish (such as cod and tuna), more fish at the middle of food chains (so-called forage fish, such as pollock). Industrial fishing has reduced populations of large predatory fish (blue marlin, tuna, swordfish, sailfish, cod) in the oceans by 90%. Populations of predatory fish tend to stabilize at about 10% of their pre-harvest populations under industrial fishing. Along heavily fished coasts, which are all coasts in the developed world, large predatory fish such as cod, jewfish, swordfish, sharks and rays are functionally or entirely absent, with implications that run down the trophic levels to the plankton that feed the sea; such fish join the large reptiles and mammals—whales, sea turtles, manatees, dugongs, sea cows, monk seals, salt water crocodiles—as ecological ghosts.

Fished fish also become smaller. The overfishing of cod under industrial trawling in the 1920s, reduced the proportion of large fish, decreased the length of fish of a given age, and decreased the size at which fish spawned (thus reducing the total production of eggs and the fertilisation rate). Tuna weigh half of what they did 20 years ago, marlin one quarter. Fishing for so-called forage fish, some of which remain abundant, reduces the food available to large predatory fish, the fish people in general prefer to eat. The forage fish are processed into animal feed and oil, some of which is fed to farmed fish, like Atlantic salmon, which convert the processed fish pellets to salmon flesh at a theoretical efficiency of about 33%; that is, at a loss of about 67% of the fish protein, which ends up as salmon excreta (a pollutant) concentrated below the densely stocked salmon cages. (For many reasons, actual losses from forage fish to salmon flesh are much larger; all intensively farmed carnivorous finfish and shrimp are net consumers of protein and require 2-5 times the protein they produce; if our desire was food, and not the taste of shrimp or salmon, we would be better off eating the forage fish: as a nutritous broth, for instance.)

Large fish have become rare partly because industrial fishing catches them all, and partly because catching all the large fish creates a selection pressure for smaller, earlier maturing fish; such fish, because they are smaller, have several times fewer eggs and sperm, making recovery of the population from fishing more difficult. (Large old fish have larger and healthier eggs, and more eggs, which hatch into faster-growing young.) Size selection in a population of Atlantic silversides (a minnow), in which the larger 90% of the fish were removed ("fished") before breeding, led over 4 generations to a population of fish that weighed half that of a population that was anti-fished (the smaller 90% of the population removed before breeding). The biomass of the fished population also fell, to about half that of the anti-fished population. The same process seems to happen in the wild. So fish in an over-fished population become smaller and the biomass of the population declines. The results of such selection pressures are not limited to fish. In an isolated population of bighorn sheep in Alberta, 25 years of trophy hunting, involving the taking of 57 animals (2 per year), reduced the mean body weight of 4 year old rams from 200 to 160 pounds, and their horn lengths from 28 to 20 inches. Most of the rams taken were 8 years old or younger; they had horns with four-fifths of a curl or more; most of them probably had not yet bred. Presumably smaller animals with less perfect horns did breed and so the population got smaller. The speed of the effect is startling. But populations of seed eating finches in the Galapogos show small yearly variations in bill sizes depending on what seeds are available; two years of large seeds is reflected in a change in the mean size of their bills. So populations of animals respond rapidly to selection pressures.

Weather and the size of the parental population determine reproductive success in most marine species, more than the availability of resources. Environmental factors in any one year can overwhelm the effect of population size on recruitment of young fish—one reason the theory of maximum sustainable yield, which used the size of the parental population as the major determining factor in managing populations, didn't work. Summer storms bring up nutrients from deep water, fertilizing the surface waters and improving the survival of algae eating young fish, making for good catches of (say) Bering Sea pollock a few years later. Salmon populations in the North Pacific are in general favored by stormy winters, and extensive ice cover improves the survival of larval snow crabs in the Labrador Sea. Fish like cod and herring prefer cool water, so do better in the southern parts of their range (off the Massachusetts coast, off the coasts of Holland or Sweden) when winters are windy and bitter, while the opposite is true of cod at the northern limits of their range in the Lofoten Islands of Norway or the Labrador Sea. Herring eat plankton, the size of whose bloom varies with weather and sea conditions. And cod eat herring, perhaps as many as 29 billion of them in the North Sea in the mid-1800s, or 10 times more than people. Most fish and marine invertebrates release their eggs and sperm into the sea. Fertilisation is by chance. Since it helps if eggs and sperm are abundant, many species gather in groups to spawn. Spawning locations are traditional and partly determined by currents that carry the fertilized eggs and larvae back toward good juvenile habitat near shore. After spawning, the eggs and larvae float to the surface and drift back to inshore waters, where the juveniles mature. (So gravid female lobsters congregate where currents will wash their larvae back to the shallow, cobbled bays juvenile lobsters favor.) As they grow larger, the young fish begin moving offshore. Fish like cod and herring generally remain over the continental shelf, in waters less than 600 feet deep.

When fish congregate to breed, the abundance and health of the animals and of their spawn matter. In terms of energy allottment (the use to which food is put), small fish are growing, large fish are reproducing. Big old fish play an important role in the ecosystem. A 25 inch female red snapper produces 200 times as many eggs as a female 16 inches long, or two-thirds her length. Among Pacific rockfish, older fish produce 10 times the eggs, and the survival rates of their larvae are nearly 3 times higher and their growth rates 3.5 times faster. (The larvae are larger when they hatch from the eggs, which probably contain more nutrients.) Fish eggs and larvae lead very uncertain lives. Birds and fish eat fish eggs, including fish that are normally prey of the adults. (People do too, but kill the adults to get them.) So abundant populations of fish must produce abundant spawn. Herring spawn along the English coast up through the 1800s covered gravelly bottoms in drifts 3-6 feet deep. Haddock eating them acquired a distinctive flavor.

The currents that deliver the larvae of marine animals to inshore waters depend on oceanographic conditions; currents that move lobster larvae in the Gulf of Maine are influenced by ice melting in the Arctic, by cloud cover and by winds. (Ice, by damping waves, changes the movement of currents.) Some populations of Caribbean reef fish that live near each other as adults show genetic differences indicating they are from separate breeding populations. The different populations spawn in different places and different currents carry their larvae inshore to nearby, but separate, rearing grounds. Tides and winds are somewhat predictable, but dynamic, while local currents are dependent on the weather, and storms and sudden shifts in currents wash many fish larvae out to sea. The very great majority don't survive. Red sea urchins along the Pacific coast, fished by hand by divers, went into a decline under controlled fishing. Red sea urchins need a sufficient density of adults for efficient fertilisation. Young that settle near adults also survive better (the so-called nursery or canopy effect). To maintain the fishery it was suggested that each area be fished once every 3 years; and that the largest 20% of the population be left (they are the most fecund, and provide the most canopy habitat), along with the smallest 20% (for the next harvest). In such animals, small amounts of random variation in recruitment lead to highly variable future populations. Low levels of exploitation can cause a continual decline in the population, with no stabilization. In many echinoderm species (and marine species in general) periods of relatively low recruitment are followed by years of abundance; the population must be able to take advantage of the abundant years to maintain itself. (That is, when conditions are favorable, there must be enough large, old individuals to breed abundantly.)

The collapse of cod populations across the North Atlantic was a signature event of twentieth century fisheries; cod from the North Sea and the

North American banks had been a cheap source of protein in Europe for over a thousand years. The original cod populations had many large fish. Cod caught off New England in 1602 were larger than those caught off the Grand Banks of Newfoundland, which had been fished by Europeans for 100 years. These large fish produced enormous amounts of eggs and sperm. Their reproductive potential allowed for recovery of the population from declines caused by fishing or bad weather. Many species of animals that gather in groups to breed (such as lekking birds, like sage grouse) will not breed unless enough individuals come together. Original spawning aggregations of cod at the edge of the North American continental shelf probably consisted of hundreds of millions of fish (the Grand Banks population is thought to have been several billion fish). Old cod knew the way to traditional spawning sites, following channels of warm oceanic water through canyons in the continental shelf. The underground forest of an undisturbed bottom probably aided in the survival of the fertilized eggs and larvae of cod, which are large zooplankton (and food for many of the fish that will eventually become their prey). After spawning, the adult cod followed capelin, a major prey species, inshore to feed over the summer. Most bays along the New England coast had their local populations of cod, which were connected to specific spawning sites. These populations were fished out from the 1930s to the 1950s.

Mechanized fishing for cod (steam trawling) began in the 1920s. Trawling was aimed at large old fish and thus selected for faster-growing, earlier maturing fish. Reducing the numbers of old fish reduced in geometric proportion the abundance of eggs and sperm and made maintaining the population more difficult. By the 1960s trawling had reduced the proportion of large fish in the population, decreased the length of fish of a given age and decreased the age at which fish bred. These are all ominous signs, but were not recognized as such by fisheries biologists, who believed half the population of a fish like cod could be removed every year indefinitely: the high rate of human predation would make the cod population respond and grow faster. This theory of maximum sustainable yield ignored the effect of environmental factors on fish recruitment and the effects of fishing for large fish on the size of individual fish. Some inshore populations of fish (those fished more heavily and whose spawning sites were known) began to disappear. As cod declined in the southern Grand Banks in the 1950s and 1960s they were replaced by flatfish (flounder). Those were fished down with specialized trawls in the 1980s and 1990s. The collapse of the cod, the apex predator of these northern seas, may have improved the lobster fishery (one of the few modern sustainable fisheries) in the Gulf of Maine (cod eat young lobsters). The decline in cod ended 5000 years of stability between cod and kelp forests. Cod ate the sea urchins and other invertebrates that grazed on the kelp. When cod were abundant the invertebrates fed only at night. When the cod disappeared, the sea urchins and other invertebrates along the Maine coast increased and grazed down the kelp forests. After the sea urchins were fished out for the Japanese market, the vegetation grew back, but it was now dominated by introduced species (thus the bright yellow green of the modern intertidal). These had been there all along, and took advantage of the reduction in the native plants to make their move. The new seaweeds changed the camoflage background for young fish and for invertebrates and probably the nutritional status of the shore.

The primary reason cod off eastern North America have not recovered is probably that they are still overfished. Their legal take as bycatch in other fisheries is too high. Once fish stocks fall below 10% of their unfished populations (some would say 15%-20%) recovery becomes difficult. Predation and competition from other fish may prevent recovery. Meanwhile other problems appear. A warming climate creates various mismatches among connected species—hummingbirds migrating across the Mexican desert arrive in the high altitude gardens of the Rockies before the flowers have opened; the late autumn freezing of sea ice keeps polar bears marooned on shore in Hudson's Bay, losing weight, out of reach of their prey, the seals that live on the ice. The time of algal blooms in the ocean is determined by daylength, while the development of fish and invertebrate larvae is determined by temperature, so many mismatches are possible. In the North Sea the quantity and quality (calculated by size) of zooplankton available to larval cod has declined since the 1980s. The sea has warmed by almost 1° C. While the spring bloom of diatoms and dinoflagellates (the photosynthesizers) is determined by daylength, and occurs more or less at the same time, the large zooplankton (such as larval cod and the larvae of copepods, a food of larval cod) emerge in response to temperature and now hatch up to 2 months earlier. Young cod are faced with food too big or too small for rapid growth. Until evolution corrects this mismatch between producers and consumers the local cod population may disappear. In the North Atlantic, the collapse of the cod population has let its prey species (such as shrimp, crabs and herring) increase. These species feed on large zooplankton, including juvenile and larval cod, and their predation may be helping prevent recovery. Also, forage fish, such as capelin, on which cod feed, are still overfished, among other things, for food for farmed salmon. Finally, most of the adult cod that are left are small. The females produce many fewer eggs and less well nourished eggs than large old fish.

There are many explanations for the lack of recovery in collapsed fisheries but in almost all cases fisheries reduced below an economically exploitable level (so breeding stock was, say, 1%-5% of unfished levels) have not recovered during the period of observation (until now, 15-25 years). In general, there is little evidence of recovery in fish stocks fished down to 10% of their reproductive biomass, after a period of 15 years. All major herring fisheries in the North Atlantic and Pacific collapsed in the 1960s and 1970s; after 25 years there has been some recovery in Norwegian stocks, but little or none in the others. Catches of Peruvian anchovies fell from 11 million tons in the late 1960s to 100,000 tons in 2000 and remains there. (Herring and anchovies, like salmon, flounder and cod, are opportunistic species capable of rapid reproduction, that can better stand fishing pressure than slower growing species like marlin, shark, or grouper.) Healthy populations of fish can survive climate cycles, such as periods of cooling or warming, if they don't last too long. Fish are adaptable. Salmon are colonizing rivers on the Bering coast of Alaska north of their former range. Cod are also extending their range north and if trawling is stopped, may do well there. With a warming climate, the fish that stayed in the old range would survive on other prey (perhaps favored by an earlier bloom) and if not, follow the changing water temperatures north or south. Populations of any large animal (cod, ospreys, passenger pigeons, moose) probably build up through a series of lucky events; several good years for recruitment; then good years for survival and growth of the young animals; that is, through a historical process, which over enough time converges in many abundant populations. The population reaches a size that can buffer itself against normal environmental perturbance. But once reduced from that size (by fishing, hunting or natural catastrophe) it may no longer be able to buffer itself; its breeding stock may be too small, its individuals too scattered (the case with Pacific abalone), its individuals too poor at reproducing, their numbers too low for cultural behaviors such as breeding aggregations to work well. Its food supply may be limited or its habitat less favorable; perhaps polluted. Predation that the population could once withstand (and that had evolutionary benefits) may reduce the population further, perhaps to zero (the effect of wolves on a failing population of elk; perhaps of herring on larval cod). All this is true of cod populations (whose habitat for instance, has been diminished by trawling, and whose larvae are being eaten by their former prey), but their primary problem is that they are still overfished.

Fishing quotas in the North Sea are a political matter on which governments, scientists and fisherman cannot agree, so limits are set too high. This is true in most commercial fisheries. While the Canadian Grand Banks have been closed to cod fishing for more than a decade, cod caught during legal fishing for shrimp, flounder and skate (such fish are legally salable) are thought to amount to 90% of the breeding cod population. Industrial fishing is very efficient. In general, in new fisheries, it reduces a population of large predatory fish by 90% in 10-15 years. So this is the life of new fisheries, such as that for orange roughy, or Chilean sea bass, a deepwater fish of the continental shelf that matures at 20-30 years and lives to 150; or for round nosed grenadier, another deepwater fish that matures at 8-10 years and lives to 75. Such fish, taken in their spawning aggregations, could probably be fished at 1-2% of their populations sustainably, an economic impossibility. Long life and low fecundity are typical of deepwater fish, which, living in the dark waters of the continental slope, depend on the rain of nutrients from the sunlit ocean above. Modern deepwater fisheries (there are several) are both extremely profitable and completely unsustainable. Cod, a fish of the sunlit waters above 600 feet (like most fished fish), matures at 5-6 years and lives to 20, reaching up to 90 pounds in weight. Cod, along with anchovies, sardines, herring, salmon and flounder, are so-called opportunistic fish; these are species with short maturation times, that produce a large number of young. Since under favorable conditions they have a capacity for rapid recovery, they can be more heavily exploited. During the industrial fishery after World War II, when fisheries were managed for so-called maximum sustainable yield, it was thought half the population could be taken in any year. This would allow the remaining individuals, with more food available, to grow and reproduce more quickly and increase the yield of the stock. Unfortunately such fishing, with its minimum size limits, selectively removed the large old fish, the best breeding stock. Assessments of the size of stocks were also often too high, influenced by the ability of fishermen to find fish in an emptying sea, pressure for profit, and wishful thinking. The effects of weather and currents, which can greatly reduce reproductive success in fish, and also that of bycatch were not taken into account. (Bycatch constitutes 30-40% of most fisheries; most of the fish—too undesirable or too small to keep—thrown back die. Thus the line of gulls trailing trawlers.) Probably too many variables affect the size of fish stocks for such simple calculations to work; or catch limits would have to be set much, much lower. (It is now thought 20-30% of the stock is a more reasonable catch limit.) So-called competitive species of shallow water fish, such as marlin, shark, grouper, sturgeon and halibut, are slower to mature. Their populations recover less quickly from exploitation and they withstand fishing pressure less well. Most stocks of such fish are seriously depleted or becoming so. Deepwater species can barely withstand any fishing pressure; their continuing presence on restaurant menus comes from locating new stocks. A recent study of data on fish and invertebrate catches from 1950 to 2003 in 64 large marine ecosystems (which together comprised 83% of global fisheries yield over the last 50 years) points to total collapse (stocks fished down to 10% of previous levels) by 2048. The results surprised the researchers, who had not expected such a grim result. Close to 30% of fished species have already collapsed.

Abundant populations of animals may be more vulnerable than their abundance indicates. A rule of thumb is that any animal population that declines by 20% over 10 years is at risk of further depletion. A high standing biomass of plants or animals may be associated with a low capacity for renewal, that is, a low reproductive rate, which makes the population vulnerable to a rapid reduction in size under stress. Some populations of land animals (moose, for instance, which have a relatively high capacity for renewal) seem to flip between populations of high and low abundance; once low, the population remains low. At European contact, passenger pigeons comprised 25-40% of the biomass of terrestrial birds in the United States: perhaps 3-5 billion birds. Pigeons ate mast (oak, walnut, elm, chestnut); they preferred beech, an oily nut of a size easy for them to handle. Beech is a common species of eastern and middle western primary forest. Like many nut trees, beech produces crops every other year, huge crops (allowing good survival and recruitment for pigeons) every 3-7 years. (The birds also ate berries, grasshoppers, insects and fruits.) The pigeons bred from April to June, on overwintered mast. Nut trees produce crops at periodic intervals, as a defense against animals that eat mast, and often those in a large area— thousands of square miles—produce in the same year, so the previous year's breeding location will be unusable. The pigeons came north in huge flocks soon after the snow melted. The flocks spread out over a wide front. This was perhaps a strategy to find the quantities of mast a breeding aggregation needed. Breeding among the pigeons was synchronous: 3 days of courtship, 3 days of nest building, 1 egg, 13 days of incubation, 14 days of nestling care, abandonment of the site as a group. After 3-4 days the squabs also left in one flock. Losses during breeding were high. Branches collapsed under the weight of roosting and nesting birds; eggs and squabs fell out of the flimsy nests. The large breeding flocks attracted predators from far away. (This probably explains why the breeding process was so choreographed and so fast.) Most tribes of Native Americans would not hunt the birds when they were breeding, and pigeon bones in Indian middens are scarce. (So were the pigeons that abundant in 1491? The Senecas performed a dance to celebrate the pigeons' return.) Losses in the juvenile flocks were probably also high. Passenger pigeons laid 1 egg and took a relatively long time (perhaps 10 years) to reproduce themselves, that is, for each adult to produce another breeding adult. Adding an additional predation rate from American hunters of 10-20% might have sent them into a decline, which would have accelerated as the population fell and demand for pigeons re- mained high. (Canned pigeon in the 1870s was something like canned tuna fish today: a staple. A railroad siding was built to the last major nest- ing to ship out the birds.) But it was habitat loss that finally made their lives impossible. The middle western hardwood forests on which passenger pigeons depended were being turned into cornfields in the nineteenth cen- tury at a rate of about 5% a year. Approximately 95% of Ohio was forested in 1800, 10% in 1910. Beech trees begin to yield mast when 40 years old. Clearing began in earnest in the Middle West in the 1820s and peaked in the 1880s. The last big nesting by pigeons was in the 1880s in Michigan. There was less food in the second growth forests for the birds and their colonial habits made them vulnerable to continued hunting. When their numbers were reduced below a certain point, they likely ceased breeding.

Men with shotguns and axes did in the passenger pigeons. Flightless seabirds and island tortoises were eliminated by men with clubs, populations of whales by men with harpoons in wooden boats. Whales in the Bay of Biscay were eliminated as an economic population between 1000 and 1400 by Basque whalers in rowboats, right and bowhead whales in the Strait of Belle Isle off Newfoundland between 1500 and 1600, bowhead whales off Spitzbergen between 1607 to 1670 (at least a million animals total from the eastern Arctic by 1800, with sail and hand tools). In the 1760s, ships sailed in summer from New England to the islands off the Labrador coast where colonies of nesting seaducks took advantage of the lack of predators and abundance of food in the sea to breed. The hunters arrived during the molt of the adult bird's flight feathers, which occurs when the ducklings are still in the nest. Unable to fly, the adults were herded into stone pens and clubbed to death, their body feathers plucked to supply the demand for featherbeds in the American market. The ducklings starved. Within a decade the seaduck populations had collapsed (one species, the Labrador duck, went extinct), and the voyages were no longer worthwhile. Great auks, another flightless seabird of islands, originally ranged along both Atlantic coasts from Florida and the Mediterranean to Greenland and Norway. Great Auks were eliminated from the Mediterranean and from much of the European coastline early, but in 1500 they were still found on offshore islands along the North American coast from the Carolinas to Labrador. The feathers of the great auk were also used for bedding and its fat for lamp oil. The plucked bodies were boiled to extract the oil, what was left used to fuel the fire. Its flesh was salted away in barrels and sold to the poor in place of pork. Like many island birds, it also went extinct.

The stories scientists tell change. The collapse of the California sardine fishery in the 1950s was thought at the time to be from overfishing. In 1936, 726,000 tons of sardines were landed from Monterey Bay, and the sardine fishery there was the largest fishery on the North American west coast. (The anchovy fishery off Peru, now also mostly gone, was 15 times as large). The sardines didn't disappear; they became too few to be worth catching. Their place in the food web was taken over by a larger fish and fisheries biologists thought recovery was prevented by competition with that fish, since predation and competition commonly prevent recovery of drastically reduced populations of animals. More recently, it was suggested that the sardines of Monterey Bay didn't recover because of the draining of the wetlands that encircled the bay, several hundred thousand acres of which were drained in the 1940s-1950s for farmland. The wetlands had supplied iron to the bay; iron is a limiting nutrient for phytoplankton, the algae at the bottom of all sunpowered oceanic food chains. Reducing the iron in the bay would lower its productivity, reduce all its fish populations overall and perhaps preclude the recovery of the sardines. However by 2004 the sardine population had somewhat recovered and 50,000 tons of sardines were landed. Monterey Bay has warmed by 3° C. over its long-term average in the last few decades. The warming of the bay may or may not be connected with global warming (which would make the warming greater and faster). It is however associated with a 60-year cycle of cold and warm periods in the bay that was unrecognized in the 1930s. Sardines do better during warm periods; anchovies during cold ones. (Similarly, a 20-30 year temperature cycle between the north coast of California and the Gulf of Alaska seems to be associated with a variation in salmon abundance between northern California and the Gulf of Alaska. A period of strong, persistent low pressure systems in the Aleutians diverts more of the cold North Pacific current into the Alaskan gyre, less down the coast of California. This means warmer ocean temperatures and less upwelling of nutrient-rich water off the California coast, fewer krill and small forage fish for the salmon, more competition and predation by warm-water fish, and fewer salmon off California, more in the Gulf of Alaska.) All the explanations for the crash in sardines may be correct, and declining amounts of iron in the water together with competition from other fish and commercial fishing may keep the sardines from recovering to anywhere near their former abundance.

The ocean is the planet's final receptacle. Its saltiness and proportion of minerals is determined by what has washed off the land during the last 4 billion years, as well as by what has been stored away in salt and mineral deposits. In modern times oil also washes off the land, carried down to the sea by rivers and spilled directly from ships, more oil than comes naturally from seeps, and more dispersed. The oil weathers into tar balls, that (when small) resemble fish eggs and are eaten by juvenile sea turtles as they enter the Gulf Stream off the Florida coast. The tar balls are concentrated by currents used by turtles; young sea turtles make the same mistake with the plastic pellets used as feedstock by the plastics industry, while their parents mistake floating plastic bags for jellyfish, with disastrous gastrointestinal consequences. The Sargasso Sea, a mid Atlantic gyre, the breeding place of eels from Europe and North America, is full of tar balls. Ocean gyres off the west coast of the United States and Mexico are also full of bits and pieces of weathered plastic, probably carried on currents from Japanese waters. Most seabirds found dead on the beach have bits of plastic filling their stomachs. (Some floating junk is useful. A load of sneakers lost off Korea was used by oceanographers to track currents. The shoes float well and were still being found on North American beaches years later.)

A fluid like the air, the sea also absorbs carbon dioxide, which is slowly increasing its acidity. The sea is thought to have absorbed half the anthropogenic carbon dioxide released since 1800. It now absorbs about 2 billion tons a year, 20 times more than its net absorbtion of 100 million tons under natural (current, postglacial) conditions. This carbon dioxide will eventually be neutralized by calcium carbonate on the ocean floor but because ocean circulation is slow, this will take thousands of years. In the meantime the ocean's acidity rises, a tenth of a pH unit so far in its surface waters (a 30% rise in acidity). Tall seamounts in the ocean are covered with a calcium carbonate ooze, like snow, from the constant rain of shells from phytoplankton. Below 4000 meters, where the ocean becomes more acidic, the ooze dissolves. As the acidity of the upper ocean rises, these deposits will also dissolve, and organisms that make shells of calcium carbonate will have more trouble making them; the process will be energetically more difficult. The rain will cease. Creatures with calcium carbonate shells or exoskeltons are common in the ocean and include algae, shellfish and other crustaceans, and corals. As acidity rises further, the shells of shellfish and skeletons of corals will tend to dissolve.

The ocean also absorbs heat from the atmosphere; this makes it grow in size, increases the speed of its chemical reactions, melts its ice, and changes the flow of its currents. The freezing of sea water in the Arctic and Antarctic leaves behind a layer of very cold, dense, salty water that sinks.

Along with the circumferential Antarctic winds (the so called "roaring forties"), and variations in saltiness and temperature caused by other factors (such as winds blowing moisture out of ocean basins and input of fresh water from rivers), the freezing and thawing of sea ice is thought to drive deep ocean circulation. Oceanic circulation drives the upwelling of nutrients from deep water and oxygenates the deep sea. Antarctic krill, which hatch in the abyssal depths, rise to the surface as larvae and feed on the algae that grows under the ice; the abundant krill (now overfished for salmon food) are what make the seas off Antartica so rich in fish, seabirds and sea mammals. In summer the soft, slushy underside of Arctic ice has a thick coat of algae that supports an ecosystem of nematodes, bacteria, ciliates, rotifers, copepods, a much richer ecosystem than that on the sea surface: ice supports the abundance of Arctic and Antarctic waters. (Algae live on the top, sides, bottom and within ice floes, dormant during the winter, growing and dividing as the light returns.) Upwelling, caused partly by ice, and also by winds, replenishes the nutrients in surface waters. The daily upward movement of krill in Norwegian fjords at dusk to feed on plankton causes some mixing of the nutrient-rich deeper waters with the surface ones. (So one could argue the krill create their own dinner.) The freezing and thawing of sea ice about Antarctica once involved 7 million square miles of ice and was the greatest global climate cycle. Shrinking of the ice sheet began in the mid 1950s, but in an era without satellites, wasn't recognized until scientists began to study the logbooks of whalers. The ice had shrunk by 25% by 1970. Perennial Arctic sea ice has shrunk about 8% per decade since the 1970s. There was a 14% drop from 2004-2005. The Arctic Ocean is expected to be icefree in summer by about 2050 (some say 2030), except for areas about northern Greenland and the Canadian Arctic Archipelago. Ocean productivity overall, measured by chlorophyll content, has fallen by about 6% since the 1980s, perhaps from reduced mixing of seawater, perhaps from higher temperatures in the polar oceans.

Open ocean in the Arctic reached its greatest extent in modern times in the summer of 2007. This was the result of a confluence of events, none of them predictable: the sort of historical accident that can also result in the rise or fall of animal populations. The Arctic ice pack normally rotates around the Pole, where it thickens as it ages. Thick ice resists summer melting, especially at high latitudes. In 1989 a periodic flip in the Arctic

Oscillation changed the normal pattern of winds and air pressures over the Arctic. The weather settled into a phase that carried sea ice out into the Atlantic rather than circulating in a gyre around the Pole. The proportion of 10 year old ice fell from 80% in the spring of 1987 to 2% in the spring of 2007. New sea ice, being thinner, melts more easily. The same shifts in air pressure increased the flow of warm water from the Bering Sea into the Arctic Ocean, with a corresponding flow of cold water out, and increased the flow of the deep warm currents that run north from the Atlantic near Scandinavia. In the summer of 2007 a high pressure system settled over the Arctic and caused unusually sunny skies in June and July. Warm winds from Siberia pushed the melting ice floes offshore, where currents and winds carried them out of the Arctic Ocean. These events worked together with the increasing rain of soot (from burning coal, oil, cow dung, forests, firewood) and dust (from construction, disturbed drylands, agriculture) that now falls on Arctic ice, and which, by absorbing solar radiation, increases its rate of melting; with the warmer air temperatures from ongoing warming of the Arctic; and with the feedback loop of more open ocean absorbing more solar radiation, and melting more ice. Once ice starts melting, the meltwater on its surface absorbs heat and makes it melt faster. For all these reasons the loss of sea ice in the summer of 2007 was the greatest ever measured. Such losses set the stage for more positive feedback (less ice in the fall, thinner ice, more open ocean absorbing more heat) and a rapidly increasing loss of sea ice in the coming years.

———◆———

A healthy ocean needs all its parts. Diverse ecosystems are more resilient, recover more quickly from environmental stresses and can better withstand nutrient pollution (a generalization with exceptions). It is thought one reason the North Sea withstands such heavy fishing pressure (and such heavy industrial pollution) while the North American cod fisheries have all collapsed is that the North Sea is a more diverse habitat. Without its full compliment of sea mammals, sea birds, large predatory fish, forage fish, zooplankton, benthic organisms and invertebrates, the ocean becomes a different place. Some think ecosystems, once formed, evolve as wholes, to states of greater abundance and resiliency, with benefits for all the ecosystem's inhabitants. At any rate, the health of an ecosystem should be a primary consideration in any management decisions affecting that ecosystem. In the 1950s marine mammals caught 7 times more fish than the human fishery, this dropped to 3 times in the 1990s, a result of whaling and increased fishing by people. The majority of the species eaten by marine mammals are not fished by humans, though there is some overlap in human fisheries with those of marine mammals and baleen whales off the northeast coast of the Americas and in the North Sea. While the effects of abundant whale populations on fisheries isn't really known, when great whales were common in the oceans, they not only affected the plants and animals of the water column but provided a huge amount of organic matter to the ocean bottom. Their carcasses supported a variety of specialized invertebrates, such as polychaete worms that live on the fat in whalebone. Besides being a general source of nutrients for the deep sea, dead whales may have supported entire novel ecosystems, in the same fashion as oil seeps and hydrothermal vents. Before whaling, 850,000 whale carcasses may have been on the sea bottom at any one time.

One can see some of the effects of eliminating keystone species like whales in the marine ecosystems of the northeast Pacific. When whaling resumed there after World War Two, the biomass of fin and sperm whales in the North Pacific was reduced from 30 million to 3 million tons. In the waters off the Aleutians 500,000 whales were killed from 1949 to 1969. Killer whales, or orcas, once preyed on all the great whales. During the period of the hunt, killer whales may have increased in number by feeding on injured or dead and abandoned whales. They would have been directed to the slaughter by the sounds of ships and exploding harpoons, the way seabirds follow trawlers. When whaling ceased, this food resource ended. Live whales were also now much less abundant. Orcas first turned to eating harbor seals, which began declining in the 1970s, then to fur seals and Stellar sea lions, both of which began declining in the late 1970s. Stellar sea lions are an historic prey of orcas. It is thought once such animals reach 1% of the orcas' diet they decline. (Estimates are that 26-40 orcas eating sea lions would have caused their observed decline.) Hunted almost to extinction early in the twentieth century, then managed successfully for their pelts by the native Aleuts of the Pribolof Islands, fur seals still continue to decline. The decline in seals and sea lions was probably also helped along by a decline in their preferred prey of oily fish, such as perch and herring. These species were overfished and also faced a warming in the North Pacific that began in the late 1970s. With the warming seas, perch and herring met competition from an increasing number of pollock, a less oily (and therefore less nutritious) fish. The pollock may have been responding to the decline in whales as well as to the increasing water temperature, as both whales and pollock feed on plankton. Anyway, as the fishery for pollock developed (the "white fish" in a fast food fish sandwich), they also began to decline. In the 1990s, as other prey became scarce, some killer whales turned to eating sea otters, a rather undesirable food for them. By 1998 they had eliminated 90% of the sea otters in a 1000 kilometer stretch of the central Aleutians: tens of thousands of otters. (Killer whales have enormous metabolic demands and three or four whales specializing in sea otters could have done the job, though more likely several pods started hunting otters as the chance arose.) Sea otters eat sea urchins and sea snails, which graze on coastal kelp forests. Unlike the plants of warmer waters, kelp have no chemical defenses against grazing. Without sea otters to eat the herbivores, kelp forests decline, and with them the fish and invertebrates that live in the forests (sponges, corals, mussels, abalone) and the predators of these animals (bald eagles, harbor seals, seagulls, seaducks).

This story began with whales and ended with sea otters and kelp. The more traditional story about sea otters and kelp begins with us: that is, with the trapping out of sea otters along the northwest coast of the Americas by Russian and American trappers in the 1700s and 1800s for the Chinese fur trade. Sea otters were once extremely abundant in the Aleutians and along the northwest coast of the Americas down through much of California. They were not wary of people. The Europeans who wintered on Bering Island in 1741-42 killed tame otters for their fur with clubs. Their fur was extremely valuable. In most of their habitat, relentless hunting removed them as an organism of ecological importance. In the twentieth century, as the significance of sea otters in the nearshore environment was understood, otters were reintroduced and had reestablished themselves successfully in many environments until lately, when killer whales in the north, and a disease spread from the feces of domestic cats in the south, began to lower their populations.

The Black Sea is a lesson in the future of ocean ecosystems. The Black Sea is an enclosed basin, fed by huge rivers from the wet Eurasian north, and draining out through the Mediterranean. It is naturally anoxic below a few hundred meters. (Hydrogen sulfide increases with depth; the sea's color comes from a layer of sulfur reducing bacteria on its bottom.) It occupies an area of about 160,000 square miles and is up to 7000 feet deep. Modern vertebrate life in the Black Sea (its fish, birds and mammals) depends on the kelp forests and seagrass meadows of its shallow northeastern shelves. Red kelp (*Phyllophorra*) once covered 5800 square miles of the shelves (about 3% of the sea's total area). The kelp forests were a source of food and a nursery for many fish species; they were also the principal source of oxygen on the shelf, producing about two million cubic meters of oxygen a day. The tops of the kelp were harvested for agar, which is used as a thickener in processed foods, such as ice cream. Beds of sea grass (*Zostera*) and oysters ringed the sea, with beds of mussels below, to the level of the anoxia. Sea grasses also oxygenate the bottom muds and the water. Schools of anchovies circumnavigated the sea, wintering in the warmer southern waters, and fattening on the northeastern shelves. The anchovies were food for bonito, mackerel, tuna and dolphins, fish of open water. Twenty-six species of fish were caught commercially in the 1970s. The Black Sea was famous for its sturgeon.

The ecosystems of the Black Sea were weakened by overfishing and by nutrient driven algal blooms that prevented light from reaching the sea grasses and kelp forests, sending them into decline, and thus reducing nursery and feeding areas for fish and the production of oxygen. The nutrients were brought to the sea by its rivers, which flowed down through industrialized Central Europe and western Russia. Half the nutrients entering the sea come from the Danube. These are largely derived from agricultural runoff and sewage. Riverworks that bypassed large swamplands in Hungary and in the Danube's delta eliminated much of the function of nutrient removal that these wetlands had formerly performed. Besides nitrogen and phosphorus, the rivers brought down oil, mercury (present in high levels in marine mammals of the Black Sea), and organochlorines (also found in the sediments, fish and marine mammals). Dams for hydropower and river transport removed sand from the river water, so the algal blooms in the sea shifted from diatoms to dinoflagellates (diatoms are larger and lead to a more direct transfer of energy to large zooplankton and fish). Road construction and hotel development degraded the coastline, adding nutrients and silt to the nearshore water. Bottom trawling for fish caused siltation and destruction of the bottom fauna. Microbes in untreated sewage infected marine mammals.

Pelagic fish populations fell in the late 1970s. The final blow was the accidental introduction of a comb jelly, *Mnemiopsis*, a carnivorous ctenophore, from the east coast of North America in ships' ballast water, probably in 1982 (some say in the late 1960s). *Mnemiopsis* feeds on the eggs and larvae of fish. Freed from its usual predators and competitors, and helped by increased fishing for anchovies in the late 1980s (which removed a competitor for large zooplankton), *Mnemiopsis* reached densities of one kilogram per square meter in the open ocean, five kilograms per square meter over the shelves: huge numbers for a jellyfish. Swimming in the sea became impossible or unpleasant. *Mnemiopsis* has since declined (another introduced jellyfish feeds on it), but only six fish species were still commercially fishable in the 1990s. The sturgeon populations (fish famous for their flesh and eggs) have collapsed, partly from overfishing, partly from loss of spawning habitat in the Danube Delta. Without further rehabilitation, the future of the Black Sea is of a sea of jellyfish feeding on phytoplankton. Because of the input of nutrients, overall productivity in the sea is probably up. (Of course, both jellyfish and plankton can be harvested for food.)

———

In an ideal world low trophic animals would be used for aquaculture: edible seaweeds, bivalves such as oysters and mussels, sea urchins, sea cucumbers, sea snails, fish like mullet that graze on bacterial films (and were cultivated in tide pools by precontact Hawaiians). Growing the more valuable shrimp and carnivorous fin fish (such as salmon) in pools and pens causes many problems. Aquaculture works by keeping costs down: by crowding the fish (saving on cage space), treating the salmon with chemicals (to control diseases that spread through the crowded and stressed fish and the sea lice that thrive on them), by letting pollutants sink to the seafloor. A typical Maine salmon farm has 250,000 salmon in 20 pens. Each pen produces 2 metric tonnes of waste a day, or 40 tonnes total, the waste of a small city. This sinks to the bottom suffocates the plants and animals there. Its decomposition produces hydrogen sulfide. Carnivorous fin fish require 70% of their diet in fish oil and meal; as a rule, they consume 2-5 times the protein they produce. So large numbers of so called forage fish (such as sand eels in the North Sea, capelin in the northwest Atlantic, menhadin further south) are caught to feed them, 10-20 times the weight of the salmon (much more than wild salmon would have eaten on their own). Such forage fish are an important link in the food chain of wild fish, seabirds and sea mammals. Catching forage fish puts us in direct competition with wild fish like salmon and cod, reduces the food available to them and makes their populations more vulnerable to yearly changes in the abundance of plankton. Concentrating the forage fish in fish meal and oil raises the level of persistant organic pollutants in farmed salmon to 10 times the level in wild fish. A writer has suggested raising salmon on insect larvae cultivated in grain wastes, instead of on fishmeal (this would eliminate much of the problem with persistent organic chemicals in salmon, and reduce the problems associated with fishing so far down the trophic levels). Another has suggested that rapeseed oil, which contains the omega-3 fats salmon need, could be substituted for much of the fish oil. Raising them in better cages would minimize escaping fish, which swamp the declining wild stocks with attractive large males whose genes do not adapt them for survival out of the cage. (But salmon are survivors: fish escaping from Chilean farms have dispersed around the tip of South America and colonized rivers on the eastern side of the continent, which formerly had no salmon.) Raising salmon in conjunction with oysters and mussels would use some of the escaping nutrients (ideally salmon numbers would be adjusted so the bivalves took up most of the nutrients: this may or may not be economically possible). The marine pollution problems caused by salmon farms might be solved by raising them in saltwater ponds on land; and recirculating the purified seawater; a notion with its own problems—where to put the ponds, how to supply the seawater, the costs. Raising fish in concrete pens can work: Calvisius caviar comes from white sturgeon raised in warm water from the steel mills in Brescia. (All of which brings us back to the integration of aquaculture and horticulture in the small farms of the Pearl River Delta, where nutrients flow through several cycles and the wastes from the fish are used to grow vegetables.)

Artificial reefs are a way of recycling biological production in high-nutrient (nutrient polluted) areas (which include all the coasts of developed nations). They are an alternative to raising fish in cages. The reefs concentrate fish by creating new feeding grounds and nursery areas. Thus they enlarge fish biomass. The fish feed partly on the plankton blooms of the nutrient-rich waters, partly on the algae and other organisms growing on the reefs, partly on each other. Artificial reefs also support sessile communites of marketable filter feeders (oysters, mussels, sponges); these filter the water column of algae and bacteria directly. Up to a point, such structures (like delta wetlands, or the oyster reefs of shallow estuaries) can turn nutrients from pollutants into useful biomass. In Japan concrete reef structures have been used to create entirely new fishing grounds. (Netted plastic bags suspended in the water also work; filter feeders colonize the inside of the bags and their protuding parts are nipped off by feeding fish.) Paid for by the state, or by organizations of fishermen, in the interest of making estuaries more productive, but populated by wild fish, such structures require a different look at who owns wild stocks of fish and shellfish.

Biological production, and fish catches, change by a factor of 10 by trophic level. So phytoplankton (green plant plankton: the base of all sunlight driven marine food chains) have 10 times the mass of zooplankton (the small animals of the sunlit surface); zooplankton 10 times the mass of the small forage fish that eat them, small fish 10 times the mass of larger predatory fish that eat them, and so on. Trophic levels are more an intellectual device for understanding differing biomasses of predator and prey than an exact delineation of the natural habitat. Fishing at lower trophic levels (typically for fish of low value to be used for oil or meal: that is, animal or fish food) reduces the forage fish available to larger fish and makes them more dependent on the smaller fish that feed directly on plankton (or in some cases on the plankton itself). This exposes larger fish more directly to seasonal and yearly changes in plankton abundance. The ecosystem as a whole loses resilience and the population of large fish fluctuates more strongly.

Large predatory fish (tuna, cod, sharks, rays) and marine mammals (dolphins, whales, seals, sea otters, walruses) play keystone roles in marine ecosystems. Predators promote a variety of species (species richness) by holding down competition among prey species (so more different species survive); and help maintain a balance between herbivorous animals and the plants they eat. Removing too many top predators causes a cascade of changes through the ecosystem (the trophic cascade). So overfishing of cod, a large, predatory fish, let sea urchin populations off the New England coast explode and graze down the kelp forests which were one of the bases of the ecosystem (many other fish and invertebrates and their associated predators such as ducks and ospreys depended on kelp). Over-fishing of sharks lets mid-size predatory fish increase—such as groupers, a fish of Caribbean coral reefs. Groupers prey on the parrot fish that keep the reefs clean of algae. (And it turns out the size of the parrot fish, especially in the absence of large reef grazers like turtles, is important: bigger is better.) Similarly, killing wolves lets coyotes increase. The coyotes (like the wolves) reduce the numbers of midlevel predators, such as foxes, oppossums and skunks (a good thing for ground nesting birds).

Fishing, like forestry or farming, should try to fit into the ecosystems it uses: that is, to catch the right kinds and sizes of fish, while avoiding other kinds and sizes, and avoiding dolphins, seabirds, turtles, whales, and damage to the seabed. The Maine lobster fishery is one of the few healthy industrial fisheries. (The Alaskan salmon fishery may be another.) Fishermen are limited to 800 traps each. Small lobsters and lobsters large enough to breed must be thrown back. Any female with eggs must be thrown back and her tail notched; any female with a notched tale must be thrown back. So the focus is on catching lobsters of the right size and sex to let the population thrive. Recent videos of traps on the seafloor seem to indicate that lobsters go in and out of baited traps at will and are caught only if the trap is pulled up when they are inside; so lobsters are caught by chance, and the fishery may be feeding lobsters. (Eight hundred traps is a lot. Lobstermen in northeastern Canada use half that number of traps, close the season for several months a year and catch virtually the same numbers of lobster per man per year.) Such small scale fisheries (2 men in a boat) catch more fish for human consumption, rather than for meal or oil, catch higher value fish and expend more labor per fish. So the sea supports more fishermen. The catch of forage fish for processing is about equal to that of fish caught for human food, but the table catch produces 94% of fishery revenues. With small scale fishing, more of the value of the fish stays with the fisher rather than going to shipbuilders, equipment manufacturers and fuel companies. With more fishermen, more money goes into the community as mortgage payments, car payments, payments for food, clothes, schooling, medical care. Because small scale fisheries are coastal and local, they are more amenable to effective management and regulation. For one thing, everyone knows everyone else. (The North American lobster fishery is largely self-regulating. The lobstermen know each other's traps and the area where a lobsterman sets his traps is considered his.) Such fisheries also use passive gear (traps, seines, dip nets, hook and lines of limited length) which, while effective (traps at river mouths will catch virtually all migrating salmon), tend to catch the right fish and do not harm the ocean bottom. (Harvesting shellfish involves digging up the bottom. This may or may not be harmful depending on where and how it is done.)

Fish must be protected during their spawning and nursery stages, in places where they gather before migration, and on the seamounts and along the thermoclines (meeting places of warm and cold water, sought out by schools of anchovies and sardines), where fish of the open ocean, such as tuna, sharks and rays, gather to feed on the forage fish. Trawling, except on historically muddy, gravelly bottoms where recovery of the bottom animals is not expected, should be banned; or perhaps eliminated entirely. Trawling brings up the bottom itself: stones and mud with the scallops, shoveled off the deck; a ton of deepwater coral for every 2.5 tons of orange roughy. Purse seining for species like tuna involves tremendous bycatches of both juvenile tuna and other species. The other species often include large slow growing fish which are slow to reproduce (such as sharks and rays), as well as turtles, dolphins and whales. Schooling fish like tuna, when drifting along a convergence zone, tend to gather under floating objects. Once these would have been drift logs, many hundreds of thousands or millions of which sailed through the preindustrial oceans. Now they tend to be manmade structures equipped with fish finding sonar and satellite beacons, which notify the fishing boat of the presence of fish. Tuna seines are 6000 feet in circumference and 850 feet deep. Each haul of the net brings in fish worth $250,000 to $750,000, and many fish and sea mammals besides tuna. Long lines in the Pacific, up to 60 miles in length and holding 30,000 hooks, catch endangered albatrosses, turtles and sharks. Longline fisheries catch about half the loggerhead and leatherback turtles in the Pacific each year. They have significantly reduced populations of albatross. Adult leatherback turtles weigh 1500 pounds and eat jellyfish. They are useful animals in the modern ocean. The numbers of females returning to nesting beaches fell from 90,000 in 1980 to fewer than 5000 in 2006, a 95% reduction in 25 years. The bycatch of industrial fishing (turtles, dolphins, underage fish, endangered fish, unwanted fish) is generally about a third of the catch. Most of it is abandoned dying. In some fisheries, such as those for tuna and cod, much of the bycatch is juvenile fish of the target species. These fish are thrown back dead, or in the case of bigeye tuna, a threatened species, end up in the can with the legal species. Making fishermen keep all their catch, and stop fishing when they reach their quota of rare or underage fish, is one way to eliminate bycatch.

Closing areas permanently to fishing in marine reserves is probably the best way to restore marine ecosystems. Unlike fishing equipment or quotas, reserves are relatively easy to police. To restore fisheries, large areas may have to be in reserves: from 20-40% of coral reefs, for instance; and 20-40% of continental shelves. Marine reserves outperform fished areas in egg production by 10-100 times (the spawning fish are larger and produce more eggs) and fish in reserves increase in biomass several times over their biomass in fished areas. Without fishing, the bottom habitat can recover. The selective pressure for small, early maturing fish is eliminated. Fish from reserves will populate the areas about them; these areas can be sustainably fished. In birds it is thought that only 10% of the population produces an excess of surviving young; these birds breed in specific places; so the places to protect are important (though places that produce an excess of young may shift). Something similar may be true for fish, with weather on the spawning grounds or during larval development the determining factor. It is thought that fishing effort in the North Atlantic must be reduced by a factor of 3-4 to allow fisheries to recover; and that protecting 40% of the North Sea from fishing would produce the most fish for the least fishing effort, and thus the most profit for the fishers. The fishing fleet worldwide has something like 2.5 times the fishing capacity the seas can withstand. Most fleets are supported by government subsidies. To reduce fishing effort much of the fleet would have to be bought out and sunk (not re-sold). Scuttled boats make good reefs. Buyout payments would go into retirement accounts, not new ships. Trawling would end (so much of the fleet would be obsolete). The number of fishermen would not necessarily decrease, as more labor went into catching bigger fish. Some fisherman might be hired to restore fisheries, such as oyster reefs in Chesapeake Bay, or local populations of cod in the bays off the New England coast. The cost of restoring fisheries would not be more than the current subsidy to the fishery, which amounts to 25-35% the value of the fish caught. The money would still fund jobs. The story is the same as for forests: the largest long term profit is made by catching fewer, larger fish.

Overfishing by sport fishermen is also a serious problem. In some fisheries, the sport fishery may equal or exceed the commercial catch, but catches by sport fishermen are not included in the fishery statistics used to set catch limits. (From 1960-1980 commercial fishermen took 1.5 million pounds of fish per year from Long Island Sound, recreational fishermen 23 million pounds, including 15 million pounds of bluefish. The recreational fishery was worth several billion dollars.) From 1972-1988 the average weight of 5 sport fish in the Southeastern United States—red snapper, gag, snowy grouper, scamp, speckled hind—fell by 75%. The only place to catch trophy fish in Florida at the present time (2008) is near the defacto reserve about the space station at Cape Canaveral, closed to fishermen for reasons of safety.

———————

In the United States, the heraldic fish connecting the land with the ocean is the Pacific salmon: the totem fish. Salmon are born and mature in rivers but gain most of their size and weight in the ocean. I think salmon have become so symbolically important partly because they were extinguished so recently, and so deliberately. Every person sees as normal the state of the world—the trees, fish, birds—of his childhood; and salmon were still abundant half a lifetime ago. Changing perspectives let us adapt to a less and less abundant wild world; but the change cannot be too fast. In the Pacific Northwest the sea and the land mingle; along unsettled coasts, forests come down to the shore; after storms, stranded salmon sometimes dangled from the branches of firs. In California, Oregon and Washington, salmon are now at 6-7% of their former abundance. But salmon are an opportunistic species. Under good conditions 4-6 fish return to a river for each spawning adult. They are thus capable of doubling or tripling their population with each spawning run. Their needs are simple: access to healthy rivers, a congenial ocean and protection from overfishing.

In 1600 the northeastern United States and Canada also had abundant populations of Atlantic salmon. These faded away in the nineteenth century and, like the clouds of migrating shorebirds, are no longer part of folk memory. Salmon were said to be commercially extinct in New England by 1850 but by the 1770s New England fisherman were sailing to Labrador for their (more abundant) salmon. Atlantic salmon along the east coast of North America were never as abundant as Pacific salmon. The riverine habitat was less favorable (though some spawning fish reached six feet in length); and salmon also shared east coast streams, and the nearby ocean, with huge runs of other fish like herring and shad. Aboriginal fishing pressure on salmon was light. In precontact times in eastern North America one writer claims salmon outnumbered people by a thousand to one (for 2.5-5 million fish, a hundred to one is more likely). As Euro-American settlement expanded, sedimentation from logging and agriculture buried spawning gravels and smothered eggs and larval fish. Dams blocked rivers. Small mill dams numbered in the tens of thousands by 1850, when it is thought half the original salmon habitat in eastern North America was cut off by dams. Nutrient pollution from sawdust, silt, sewage and manure reduced the oxygen in river water. Rivers flowing through cleared land warmed. High water flows coming off cleared land scoured out salmon nests, as did water released from temporary dams (splash dams) used in log drives. Travelling downriver with the spring high water, the logs scoured out the gravels and killed juvenile and adult fish. In Oregon, log-driving, which ended in 1954, scoured some small riverbeds to bedrock and made them useless to fish. (Without gravel, there is no place to spawn. So gravel mining in riverbeds also reduces salmon habitat. But spawning gravels can be replaced.) Europe's populations of Atlantic salmon, which were also enormous, were reduced by the eleventh century (though in the 1700s German apprentices still complained of being fed salmon several days a week in season). In the late twentieth century a few salmon were still making it up the Rhine to Switzerland. The long reduction in the English and Scottish fisheries was accompanied by laments and much good advice; when the advice was followed, the fishery would temporarily recover. Traps set at the entrance to English salmon rivers were very effective at catching salmon.

Atlantic salmon fell before overfishing, dams and development. Salmon are born in rivers but gain most of their weight in the sea. The final blow to them was the discovery of their main foraging area in the Labrador Sea, north of the Grand Banks, between Labrador and Greenland. Salmon from both North America and Europe converged here, where the mixing of cold and warm currents produced a continual growth of plankton and forage fish that also supported the North American cod. The salmon fed on capelin, sand eels, herring, squid and amphipods (shrimp-like crustaceans) for 1-2 years before returning to their natal rivers to spawn. Netting salmon there from the late 1950s on reduced returning populations of fish to very low levels. This ocean fishery was regulated in 1984 but the catch was set too high to do much good. Anyway, the requirements of salmon were well known when the Pacific species were discovered by Euro-Americans.

———————

Pacific salmon die after spawning and their bodies contribute nutrients to streams and streamside forests. Scavengers remove half the dead coho salmon from small streams on the Olympic Peninsula and carry them up to 200 yards from the stream; scavengers like black bears leave about half the carcass uneaten (they prefer the innards and brains); the marine nutrients in the remains of the salmon and the urine and feces of the scavengers feed the forest. Trees grow 3 times faster along healthy salmon streams. Nutrients from the sea once formed the bones of California grizzly bears. Up to 30% of the nitrogen in bottomland forests with salmon streams is marine in origin. In lakes that hold sockeye salmon up to 90% of the nitrogen in the algae on the lake bottom and up to 70% of the nitrogen in the lake's plankton and in the juvenile sockeye come from decaying adult salmon. (Juvenile sockeye have been seen nibbling on the carcasses of their elders.) Salmon and their river valley habitat constitute a pool of shared nutrients brought by the fish from the sea. Spawning salmon rearrange river beds, carve away gravel bars, even out the stream bed, turn over the gravels; their yearly efforts change the width and shape of a stream. One could say the shapers of a river with salmon are streams, trees, weather and salmon.

Abundant runs of salmon make small streams overflow. They rearrange the streams by digging nests, filling them with fertilized eggs, then covering the eggs over with gravel, so the eggs and larval fish develop within the protection of the stones. Nest building by female sockeye salmon move virtually the same amount of sediment as that moved by currents. Nests are dug below the usual level of winter high water scour. Such knowledge of how deep to put the eggs is set by evolution in the different races of salmon that use a watershed and evolves as conditions change. (Eggs set at the correct depth survive.) The size of the fish also determines the depth of the nest, and the size of the gravel it can deal with. Small fish can only spawn in small gravel streams, with relatively low winter flows; large fish can spawn anywhere there is room for them. The five species of Pacific salmon reach different sizes and return to spawn at different times, from late summer to early winter; thus they fully exploit the varied riverine habitat. Chinook and coho spawn in rivers, sockeye in lakes or in streams draining lakes, chum salmon in small channels near estuaries, pink salmon in estuaries. Juvenile salmon live in fresh water for several years, then spend from 1-4 years in the ocean, where they gain 90% of their weight. Atlantic salmon are less differentiated than Pacific salmon, but vary somewhat in 'body style' from stream to stream.

Settlement changes rivers. Undeveloped rivers in the Pacific Northwest are bordered by huge old trees. When the trees fall, some fall in the channel. Drift logs accumulate against them and the deflected river excavates a deep pool under them, a resting place for migrating salmon. The river may also excavate a channel around the jam on the shore side; the slower water there is good habitat for young salmon. Gravel accumulates at the tail of a pool, creating spawning habitat. As a river meanders, pools and riffles are excavated at the outside of bends, and cutoffs and side channels provide still water habitat, but snags greatly increase the number and variety of pools and side channels, especially in steeper river valleys. Pools dug around snags are 2-4 times deeper than those dug by the current alone. Sometimes soil accumulates on the log jams and they become islands. Over time, such natural rivers develop complex patterns of channels in their lower reaches. The main channel is used mainly by adults for migration and spawning, the side channels by young salmon. The more pools and the deeper they are, the more room there is in the river for large fish, and the more downpool spawning habitat. Chinook salmon (the largest of the five subspecies of Pacific salmon) spawn in the main channels of large rivers, such as the Columbia. The abundance of chinook and coho salmon in a river (coho spawn in smaller, steeper streams) is related to the number of pools.

Over the last 150 years log jams have been removed and streamside trees cut for fuel or timber, to make rivers navigable, or their banks settleable, and 65% of the deep pools in rivers draining into Puget Sound have been lost. Thus much habitat for fish and for large fish in particular is gone. A log raft with trees 2-3 feet in diameter growing on it blocked the mouth of the Skagit river when settlers arrived. The river flowed under it. (A similar jam was used as a bridge in Louisiana where the Mississippi joined the Atchafalaya.) In winter the log jam flooded 150 square miles of the valley: this expanded rearing habitat for juvenile salmon but prevented human settlement. (So the jam was removed.) Before development, most rivers in forested landscapes had stretches where they flowed as a complex network of channels. Those in parts of the Rhine valley were famous; infamous to boatmen. Interaction among the flowing water, the topography of the floodplain and riverside trees created the riverine habitat. The Nisqually River in Puget Sound (an undisturbed river) has 2000 logs per mile of channel (one every 2-3 feet), most of them in jams, and a complex network of channels in its lower reaches. The Army Corps of Engineers removed 65,000 logs from the Willamette River in Oregon (880 logs per mile) from 1870-1950 to improve navigation. A million snags were removed from the lower Mississippi and 180,000 streamside trees were cut from 1864-1884 to prevent them from becoming snags. (Many, many more trees were cut for fuel for steamboats.) Steadily flowing, single channel rivers, of a certain depth, without sidechannels, backwaters or swampy riverside wetlands, are a modern creation. (Sometimes an inadvertent one, as with former mill streams in the northeastern U.S.) Fishing reduced the number of salmon in the Pacific Northwest, but land development reduced the ability of the landscape to support salmon.

Heavy fishing for the largest fish reduces over time the number of large fish, which are the more successful breeders. In the Pacific Northwest in 1700, 50,000 Native Americans along the Columbia lived on salmon, catching 20-40 million pounds of salmon a year, or 1-2 million fish. This amounted to 5-20% of the precontact run of 11-16 million fish. Industrial fishing after the 1880s took about 90% of the runs. The salmon maintained their population under industrial fishing for 40 years. Their numbers started to fall in the 1920s, shortly before dam construction began on the Columbia, perhaps because of a change in ocean conditions to those less favorable to salmon, perhaps a result of cumulative changes in the river habitat, perhaps from growing competition from the introduced shad, another anadromous fish. Most likely, all three influenced salmon numbers. Forty years of heavy, size selective fishing would have an inevitable evolutionary effect on a population of fish, making them both smaller and less abundant. With salmon, this reduces the fitness of the fish for larger streams, higher water flows and larger gravel sizes. Removing log jams from rivers reduces the size of the salmon habitat. The runoff associated with logging and land development increases siltation in gravel beds, making them less suitable for fish, raises river beds overall (as silt accumulates), reduces the number of pools, and increases flooding and the size of winter flows. The increased flows scour out spawning gravels more deeply, destroying salmon nests (dug less deep by smaller fish). Silty runoff from logging buries spawning gravels and smothers salmon eggs and larvae. Unscreened irrigation diversions lead juvenile salmon (migrating downstream to the sea) into cornfields, where they die at so many to the acre. Logging as little as 5% of a watershed increases streamflows by 10-55%. (Steep slopes and road ditches emptying directly into streams account for the larger numbers.) Five years after a clearcut, stream flows are typically up 50%; streamflows remain 25-40% higher for 25 years; and summer water temperatures remain high for several decades. Logging, together with selecting for smaller breeding age salmon, would have had a long term effect on salmon populations. In rivers about Puget Sound affected by urbanization and agriculture, and thus with high winter flows, fall spawning salmon tend to be replaced with spring spawning cutthroat trout—there is less rain and runoff and thus less scouring in spring and summer, when cutthroats develop.

Salmon are associated with cold, clear, forested streams. They became abundant in the Pacific Northwest 2000-3000 years ago. By then, the warm, muddy, postglacial rivers had cleared the glacial debris out of their channels, the climate was cooling, and forests were expanding. The expanding forests retarded the spring snowmelt, extending the spring rise in many rivers and lessening its force, and also shaded the streams, cooling them in summer, and releasing cooler ground water into them. (The effect of forests on water temperature can be dramatic, reaching 10° F. in summer in a shaded reach.) The forests reduced runoff and sedimentation. Streamside trees fell in the rivers and, anchored by their roots, forced the water to excavate pools around them. This was true even in major rivers like the Columbia. (And during all this time, people lived along the rivers of the Pacific Northwest.) All the reasons for the salmon's association with forests are not clear, but salmon do better in forested suburban streams than in unforested rural ones. (Since salmon productivity depends on water temperature, this is likely because, especially in the more southerly ranges of the Pacific salmon, unshaded streams commonly reach 70° F., too warm for the fish. Other candidates for the difference are the increased siltation of spawning gravels in agricultural rural areas; and the effect of farm chemicals on the invertebrate life of streams.)

Salmon also vary in abundance with ocean conditions. The Pacific Northwest and Alaskan stocks of salmon tend to alternate in abundance in cycles of 20-30 years. Good years for salmon in the Pacific Northwest occur when a strong North Pacific current runs south along the California coast, causing an upwelling of cold, nutrient rich water, and creating an abundance of krill and small forage fish for the salmon to eat. Such conditions keep fish of warmer waters, competitors of the salmon (such as the Pacific mackeral, which eat young salmon) south of the main feeding grounds of the salmon. Alaskan stocks of salmon are favored when strong Aleutian lows, associated with strong El Nino events, divert more of the North Pacific Current into the Alaskan gyre, and less down the coast of North America. (Strong El Nino events are becoming more and more common with the warming of the atmosphere.) Then the Pacific Northwest stocks have less food, more competition and suffer more predation, while the Alaskan stocks prosper. A warming climate may make life more difficult for the southern stocks of Pacific salmon.

Dams for electric power, flood control and navigation on salmon rivers usually mean the end of salmon. Bonneville Dam, the first on the Columbia's main stem, was completed in the 1930s, part of the Works Progress Administration's effort to revive the American economy during the Depression. While salmon on the Columbia were already in strong decline from overfishing, the fact is that dams block salmon runs. Mill dams across English salmon streams were required to leave gaps as big as a well-fed three year old pig, put sidewise to the current, to let the fish pass. Laws requiring fish passage in New England in the early 1700s exempted existing mill dams; they also weren't much enforced. Fish ladders ('salmon stairs') were invented in the 1820s in Scotland. If dams are not too high, fish ladders let the salmon surmount them. (Ladders also work for some other species of anadromous fish, such as shad and alewives.) Some dams are too high for ladders. Grand Coulee on the upper Columbia blocked the runs above it. (The dam is 550 feet high and shut off 1500 miles of spawning streams, but the fish that used that water amounted to only 5% of the Columbia's salmon run.) Dams also change the river habitat into a series of still pools, whose water temperatures and chemistries are more like those of lakes. Summer temperatures rise in the pools to levels near the tolerance of salmon and oxygen levels fall. Warm water from the upper levels of the reservoirs flows into the fishways, with the result that the temperature in many fishways is near the upper limit salmon can stand. (This may be one reason salmon have trouble locating the ladders: they are avoiding the warm water.) Dams flood the stretches of spawning streams above them, the former gravelly deltas of creeks in the river's main stem. In the now slow water, spawning gravels in those creeks silt over, as do those in the slow sections of the main channel, where the largest salmon of the largest run spawned. Gravels downstream of dams erode away. Thus much spawning habitat disappears. About 55% of the area and 31% of the stream miles of salmon and steelhead spawning habitat in the Columbia Basin has been eliminated by dams.

Young salmon are adapted to traveling downstream on the spring highwater (facing upstream, steadying themselves with their caudal fins). During their downstream ride they begin the physiological transformations necessary for them to survive in salt water. These changes are timed and finish in the brackish water of the estuary. Young salmon must reach salt water at the right (evolutionarily determined) time. In a dammed river, young salmon move downstream through waters that are warm and unshaded. The more favorable cooler waters deeper in the pools are low in oxygen. Since the spring flow is controlled, they move more slowly, under their own power, not the river's, and must stop to feed. A journey that once took two weeks now takes two months. Their travels expose them to predation by birds, mammals and other fish. Juvenile salmon moving downstream are also killed by their passage through the turbines in dams. From 10-15% are killed at each dam; so passage through 10 dams means a cumulative loss of 66-80% of the fish. While losses of juvenile salmon are always high, these are not small numbers. Losses through dams can be mitigated by new construction; problems in the pools can be reduced by making the river flow closer to normal conditions (so called run of the river power generation, which means loss of some summer generation capacity).

Counter intuitively, dams also make it more difficult for salmon to move upstream. Coho swim up the Columbia at 1-2 miles an hour for their 1000 mile journey to Redfish Lake in Idaho. Fish can move relatively effortlessly upstream through turbulent water by using slight movements of their heads to let opposing currents push them forward. In the long pools between dams they must swim, using their main swimming muscle. Salmon usually don't feed during their upstream migration and this additional effort uses up more of their stored fat. As they congregate below the fishways, they are exposed to predators like gulls and seals.

So the physical conditions of modern rivers are less favorable to salmon, partly because of changes in the rivers themselves, such as dams (less spawning habitat, fewer pools, warmer water, lower flows), partly because of changes in the watershed that affects rivers (fewer streamside trees, less forest cover, higher flows of water entering feeder streams, levees blocking off valley habitat, more silt entering the stream from agriculture, logging, mining, commercial and urban development). The waters of modern rivers also contain many industrial chemicals, some of which affect salmon. Nonylphenol is a surfactant used in detergents, insect sprays and various industrial products and processes. It is common in sewage effluent and the effluent from paper mills and textile factories. Its use in spruce budworm sprays strongly correlates with a decrease in returning Atlantic salmon in Atlantic Canada. Nonylphenol seems to interfere with the changes young salmon undergo to adapt to salt water (so-called smoltification). It also acts as a hormone on larval fish, turning males into females. Other harmful chemicals include the copper-based algaecides and fungicides that leach from the treated lumber in docks and other river structures. These compounds interfere with the young salmons' vasonomeral system (a primitive sense related to smell) and makes them more susceptible to predation. Chemicals diffusing from broken salmon skin cells (released by the strike of a kingfisher's bill or the bite of a mink) cause nearby young salmon to freeze (and thus be less likely seen). In the presence of the copper-based compounds used in treated lumber this warning system doesn't work.

Then there is fishing. Native Americans and early modern Europeans fished for salmon (Atlantic and Pacific) with traps at the mouths of rivers. Fences of stakes and brush led the fish into enclosures in which they could be caught. Both peoples also used nets, dip nets, hook and line, and spears. They built weirs of stone and brush in the rivers to herd migrating fish. The different races of Pacific salmon are specific to their streams and the different strains adapted to different parts of their rivers. The fish suffer tremendous natural mortality when young: this is part of the evolutionary lability that lets them adapt to new conditions. Fishing at the entrance to a stream or in the stream itself can be managed to ensure a constant supply of fish in that stream, while fishing in the ocean, where all stocks mix, is another matter. Salmon vary in abundance with the weather during their lifetimes. Weather helps determine good spawning conditions in rivers and good growing conditions in the ocean. The size of a given year class of a salmon stock depends on the reproductive success of the salmon in their natal stream that year. Conditions vary from year to year and from river to river. Conditions in the river for the several years the juveniles are growing in the river, and ocean conditions for their years in the ocean determine how many salmon of a given stock return to a given stream. In the ocean the different stocks and year classes of salmon mix. The result is that limits on ocean fishing are impossible to set: there is no way to determine what strain of fish or what year class of that strain one is catching. One can't know how successful a particular group of fish has been until they return to spawn. Ocean fish, not yet full grown, are also smaller (so there is pressure to take more of them). Most native peoples along the Pacific coast waited several days or weeks after catching the first, celebratory salmon, before beginning to fish. This let many large fish escape upriver. They also caught less than half the fish.

Finally, salmon must compete with introduced species of fish, like shad and walleye pike. In 1990 the Columbia River produced about 16 million pounds of shad and 20 million pounds of salmonids (salmon and trout). A historical estimate of salmon production on the Columbia is 50 million pounds. So perhaps 36 million pounds of fish is what the river can now produce. This however supposes salmon and shad (another anadromous fish) compete directly and are not merely complimentary inhabitants (like prairie dogs and buffalo) of the same habitat. They compete for space: salmon and shad, migrating upstream together, crowd each other at the fishways below the dams, where both are eaten by seals and sea lions, and delay each other's passage upstream. Perhaps a less degraded Columbia (or one with manmade spawning channels) could produce 16 million pounds of shad and 40-50 million pounds of salmonids. The introduced walleye pike, a large freshwater predatory fish, eats young salmon as they migrate downstream through the dammed pools. Smolts that make it to the ocean normally spend some time in the Columbia's delta, where many—estimates are 30-40%—are eaten by a growing population of gulls and terns, who nest on new islands of dredge spoil in the delta. Diking and draining the delta wetlands for agriculture, as well as improvements to the delta for river navigation, have made the habitat of the delta less favorable to salmon and more favorable to shad. Many of the water plants and the invertebrates that fed on them—shelter and food for salmon—have disappeared and been replaced by plankton, which shad eat.

Bristol Bay in the State of Alaska has a healthy industrial salmon fishery. Part of this is by accident: most of Alaska's rivers are undeveloped, little polluted and in a more or less natural state. Fishing is allowed only at the mouths of rivers (thus catching fish specific to the river) and only for set times. The goal is to let 50% of that year's fish escape upriver. The large escapement and the restricted timing of fishing let salmon escape the evolutionary pressure associated with catching very large percentages of large fish. (So called split regulations, which require the release of fish smaller and larger than a certain size, are supposed to do this for recreational fisheries.) The State of Alaska also does not allow pens for farmed salmon in its coastal waters, since the tame fish escape and breed with wild fish, to the stocks' genetic detriment. The penned fish also become infested with sea lice, which attach to, and cause severe mortality in, young native salmon that pass by on their way to the sea. Under somewhat similar management, Icelandic runs of Atlantic salmon have increased during the late twentieth century. The Icelanders limit the times of netting and rod fishing and forbid ocean fishing in their territorial waters. Lately an organization of the owners of fishing rights on Icelandic, European and North American rivers has been buying out commercial ocean fisheries for Atlantic salmon. It was thought drift netting for salmon in their foraging areas off Greenland and the Faeroes and in ocean waters near their natal rivers has helped reduce Atlantic salmon to their current very low levels. Three years after netting near the Faeroes was stopped, the number of salmon returning to Icelandic and European rivers doubled. A serious plan for salmon recovery in the Pacific Northwest would buy out the ocean fleet and sink it. It would also take account of the salmon caught in the recreational fishery, which is worth several times the commercial one. The current decline of chinook salmon in the Columbia River is thought to be a result of ocean fishing as much as of dams. (Chinook spawn in the main channel and large parts of the main channel of the Columbia are still suitable for spawning.)

In the United States the idea of the new world still lurks behind the Puritan origin myths, but the actual aboriginal New England landscape of Abenaki villages, great flocks of migrating shorebirds, cornfields, clam beds, whales, forests managed by fire, rivers of shad and Atlantic salmon, has been largely forgotten. That once settled landscape has been replaced by the Puritan villages and fields of a later myth, a landscape at first not so different. In this myth the welcoming Abenaki walk out of the empty forests bearing gifts: in fact the few coastal natives left in a horticultural landscape emptied by European diseases were trying to negotiate a political arrangement with the newcomers, to prevent their land being taken over by tribes to the west. But the idea of the wild land remains. On the west coast the Pacific salmon still hang on as wild fish, with surprising political power. In large part their political power has translated into hatcheries. Hatcheries were once viewed as compensating for the loss of river habitat (a new dam, a new hatchery). They were supposed to let people develop the rivers for economic use and also have salmon. Unfortunately, hatcheries help only under very specific conditions. If used sparingly, they can help augment wild runs in the early stages of recovery. Hatcheries eliminate the evolutionary pressure salmon face in developing from egg to fry to smolt (a young salmon of a size capable of returning to the sea). They select for fish adapted to hatcheries: fish that can deal with a mob of other fish, that have a rapid response to being fed, and that have no awareness of predators. Most hatchery fish put in rivers die. They are larger than wild fish of the same age and compete with them for food and eat them, but they lack the genetic and learned adaptations to life in a stream. The story is the same for trout: a river with wild trout into which stocked trout are placed will end up a year later with half the number of trout it started with. The stocked trout harass and compete with the wild ones, but don't know how to feed, how to conserve energy in the current of the stream or to avoid predators. The result, after a year, is fewer trout. While several states gave up stocking viable trout streams in the 1950s, and Canada in the 1930s, modern management of American trout streams generally consists of stocking hatchery trout. Most of the fish are caught by fisherman in a few months. More hatchery fish are stocked the next spring. The cost of an adult hatchery Pacific salmon, returning from the sea, now varies from $10-$100, depending on the run. Salmon hatcheries were abandoned on rivers in British Columbia a long time ago, but in the continental United States they have remained, as a way of avoiding hard political decisions.

But salmon runs can be restored. In the 1990s the winter run of chinook salmon on the Sacramento River shrank to less than 200 fish and the run was classified as threatened under the Endangered Species Act. (The original salmon runs on the Sacramento were second only to those on the Columbia.) As a result, two large irrigation districts and pumping stations at state water projects in the Sacramento Delta installed modern fish screens on their intakes. This cost a relatively small amount of money the districts had seen no reason to spend on protecting salmon before. Pumping schedules at state water projects in the delta were adjusted to better accomodate downstream migrating smolts (the pumps are powerful enough to reverse the current in the river). The operators of a diversion dam also changed their practices to accomodate migrating smolts. Obsolete dams on two former spawning tributaries were removed. Hatchery practices at Shasta Dam (which blocks any further upstream migration and closes off much upriver spawning habitat) were altered to use eggs and milt from captured wild fish. After 10 years 3000-7000 winter run chinook were returning to the river. The population was still increasing. This is an increase of 10-20 times in a decade. If suitable spawning habitat were available, the salmon population could reach 300,000-2,800,000 fish in 30 years. A run of 200,000-500,000 fish would provide an excellent sport fishery. Such an increase is unlikely without further habitat restoration and control over ocean fishing.

Salmon are a test of our environmental seriousness. Salmon need cold, unpolluted water, without excessive runoff from logged or developed areas. They need forested floodplains so rivers can interact with trees and salmon to create salmon spawning and rearing habitat. If levees are moved back and rivers allowed to occupy their former flood plains (the "migration zone" over which the river moves as sedimentation and downcutting alter its bed and flow), the rivers and forest will recreate themselves. Such floodplains are places for both salmon and people. The mere presence of people doesn't bother salmon, as it does grizzly bears, wood turtles and North American wolves. Urban runoff can be controlled with many small constructions (one for each parking lot, each roof), the same way it was increased; the cost of each construction is not great (a few hundred dollars per roof). Dam operations can be changed to let salmon pass and to release water when salmon need it. Making water available for salmon probably means growing varieties of irrigated crops that need less water; and generating less electricity. (Run-of-the-river dam operations on eastern trout rivers increase the populations of fish, of the invertebrates and insects the fish eat, of freshwater mussels. The absolute number of animals, their population density and their growth rates all increase. Power generation under such regimes is dictated by the natural flow of the river, rather than by demand for electricity.) Obsolete dams, or dams whose benefits are worth less than a restored fishery, can be removed. Removal of a dam is usually cheaper than repairs. (The higher the dam, and the more the accumulated sediment behind it, the greater the expense.) Protecting salmon pays off in the long run, even in a capitalist world. Property values rise near natural rivers. Salmon runs on the Snake River, a tributary of the Columbia, were obliterated by four dams built to facilitate river transportation between the State of Idaho and the sea. The barge channel, built and maintained by the government (barge fees don't even pay for maintenance), is a cheaper means of moving Idaho logs and irrigated corn from the Great Plains to the Pacific coast than railroad, and thus made logging and irrigated agriculture more profitable. Removal of the four navigation dams on the Snake would amount to the cost of 2-3 years maintenance (such costs include the ongoing costs of salmon conservation associated with the dams). A small transportation rebate would make up the difference between river and railroad transportation. A recreational salmon fishery on the upper Snake, maintained for free by the river and the salmon, would be worth more than the value of the barge industry and the electricity produced by the dams. Unlike modern logging and farming, a sustainable salmon fishery is not an extractive activity, and will remain productive as long as the habitat remains.

Tule Lake is a natural lake in northern California that was turned into a reservoir by a dam on the Klamath River. Klamath River and (Tule) Lake are managed for electricity, irrigation water, waterfowl habitat and salmon. The water is over-subscribed and in dry years there is not enough water for the irrigation district and the salmon. In their natural state the seasonal marshes about the lakeshore were flooded through early summer with the spring runoff. As the water receded in summer, food plants important to migratory waterfowl grew in the wet mud. The stabilized water levels of the reservoir eliminated this cycle. Falling lake levels were no longer seasonal but corresponded with the need for irrigation water and electricity. Water management also eliminated the natural disturbances of drying, flood and fire that wetlands need to renew themselves. The reservoir stored silt carried into it from upstream. From 1958-1986 sedimentation in Tule Lake amounted to about 14 inches. This reduced floodwater storage and deepwater fish habitat. It reduced the water depth in the emergent marsh areas, eliminating most of the nesting habitat for diving ducks and colonial nesting waterbirds. The quality of the agricultural lands about the lake also declined. Soil nutrient levels fell, soil tilth worsened, while root knot nematodes and fungal diseases increased. Looking for a way to deal with these problems, the refuge managers began introducing rotations between cropland and wetland about the lake. The rotations recreated the juvenile marsh habitat, broke agricultural pest cycles, and improved soil fertility and condition. The rotations are of 2 lengths. One rotation consists of 3-4 years in marsh followed by 3-4 years in crops. The second consists of 15-30 years in each. The short rotation creates early marsh, the longer one late succession marsh. The short rotation eliminates pests and adds nitrogen and phosphorus to the soil. Little additional fertilizer is needed to grow satisfactory crops. The long rotation adds nutrients and organic matter to the soil. Combining such agricultural rotations with water management that favors salmon lets one have crops, fish, migratory waterfowl and water supply, all under some semblance of human control. (Ideally yearly water withdrawals should be set at 20-25% of longterm average river flow, with a provision to reduce use in dry years. This is precisely the sort of limitation capitalist management abhors.)

Part IV
Energy, Population, Hope

Chapter 21

Thoughts on Energy and the End of our World

If cars got 90 miles per gallon, the straw burned in the fields of Denmark and France, converted to hydrocarbons, would fuel the car fleets of those countries: this is one of the more startling claims of *Natural Capitalism*. Today's cars can probably reach 100 miles per gallon, 200 if the steel in them were replaced by carbon fiber resin and their engines by an electric generator. In Sunshine Farm, an imaginary Iowa spread, 25% of the cropland is put aside to raise food for draft animals, or for material to be converted into fuel for tractors. One could not fuel the current American transportation fleet on crops grown on 25% of the country's land but a writer claims that algae grown on 20 million acres of ponds, a small percentage of U.S cultivable land, would do the job. If the Danish straw were converted to fuels, its fertilizing elements would be lost to the fields, unless the residue of the conversion process were spread on them. Some mashes left from fuel conversion (such as that of corn converted to ethanol) can be used as animal feeds, and the manure spread.

The transportation sector of a modern economy produces about 14% of its carbon emissions (nearly the same as agriculture), almost all of it from fossil fuels. Increasing a car's mileage lowers its production of carbon dioxide, and thus lowers its effect on global warming; so efficient cars would warm the world, but less. Increasing the mileage sufficiently, but still well within current technological limits, lets a nation's car fleet run on waste biological materials; that is on wheat straw, peach pits, apple pomace, walnut shells, cotton waste, tree trimmings, sawdust, waste cooking oils, spoiled grain: materials that are transformable by bacterial action into fuels. (Waste cooking oils can be burned directly.) So the car fleet can be powered by renewable fuels. Since the carbon in these fuels comes from modern (not fossil) plants, which grow again the next year, taking up the released carbon, no net carbon dioxide is added to the atmosphere. (Burning in an internal combustion engine simply speeds up the conversion of the plant based carbon into carbon dioxide, which goes up in flames, rather than being transformed in a slower, bacterial combustion.) Some may be lost from the soil by the agricultural and forestry practices that produce the feedstock. But increasing car mileage sufficiently means that vehicles need so little fuel that its source becomes almost irrelevant.

Transforming wastes into fuels involves costs in energy and materials. Trucks and trains must be used to move the wastes, which are bulky, and one must build the manufactories to do the fermenting, the distilling, the purifying. Processing wastes locally eliminates some of these costs and lets the material left from fermentation be returned to the soil (rather than, say, landfilling it). Energy gains in converting plant wastes to hydrocarbon fuels are often neutral or negative. With oil in its heyday, fuel oil contained 60-80 times the energy that went into mining and manufacturing it, but with corn based ethanol the ratio is 1:1. In other words, it takes 10 gallons of ethanol equivalent fuel to make 10 gallons of ethanol. If the fuel one uses is ethanol, the process isn't worth it economically. (That one can use the leftover mash as animal feed, and the animal manure as fertilizer, helps.) If one uses fossil fuels to make the ethanol, the process is pointless from the point of view of carbon emissions: one might as well burn gasoline in one's very efficient car, and subsidize the planting of a tree each year to soak up one's carbon emissions. The whole chain of conversion must use renewable energy for the process to be worthwhile in terms of reducing carbon emissions. The greatest fuel use in ethanol conversion comes from the distilling process; and this might be cut considerably by the development of appropriate semi-permeable membranes, such as are now available for making maple syrup or desalinating water.

Making fuel from grain (a food and a row crop) does not seem like a very good idea. However some food crops perform better than corn: making ethanol from sugar cane yields an energy surplus and sugar cane ethanol fuels most of the motor vehicles of Brazil. Making fuel from agricultural or forestry wastes, or from a sod crop like switchgrass, perhaps from a native plant like cattails (harvested in winter, from the ice)—so called cellulosic ethanol—is a better one, but makes more biological sense if the manufacturing sites are decentralized (Georgia cars running on peach pits and peanut shells, Oregon ones on fruit waste, Minnesota ones on cattails) and if cars are much more efficient (so little material is required). In all industrial processes, reducing energy use (which also reduces the energy embodied in manufactured materials) is the key to a low carbon economy. The energy costs of transportation make one wonder whether subsidizing those railroad cars of New York City sewage sludge heading to Texas cotton fields is worth it; or those trucks of recyclables one sees lumbering along the highways. The alternatives are even less desirable—ocean disposal of sewage sludges, or landfilling of sludges, bottles and cans. The production and transport of fossil fuels also involves energy costs, as does the manufacture of virgin metals and plastics, which are from 2-10 times more energy consumptive than recycled materials.

It was a dream of the Sixties to run the world on natural products: wood, sand, sunlight, straw, organic food. Trees cut on short-term rotations would produce fuel for electric power plants. Some plants were built but the landscape that resulted in the forest was neither esthetically pleasing nor ecologically functional. This was true even where chips were produced as part of a logging operation that also produced sawlogs; that is when just "junk" trees and tops went into chips. Landowners appreciated the fact that less mess was left behind in the woods, but the woods are naturally chaotic. The mess had included the nutrients in the tops and limbs (structures where above ground nutrients are concentrated); and the shade the cut branches provided, that kept the soil of the logged forest cooler, reduced nutrient losses, helped in the survival of invertebrates, fungi and amphibians and in the regeneration of the plants of the forest floor. The mess also included wolf trees, whose shape makes them unmarketable as sawlogs, and other large old trees, partly rotten and unmarketable as sawlogs, both which produce crops of mast, are habitat for birds, bats, fungi and insects, and are essential parts of the structure of the forest. When so much of the forest is removed, its place in the local hydrology is substantially altered. Erosion increases because of the loss of forest cover and the destruction of the soil surface by heavy equipment. Soil and nutrients flow into forest streams. Besides all this, agriculture or forestry that produces hydrocarbons for fuel is subject to the same economic pressures as one that produces sawlogs and food. Such landscapes are ruled by economics, not biology. Only so much land in a watershed should be put into producing food and fiber. If such landscapes can produce biofuels, without compromising the ecosystems of which they are a part, that would be wonderful. But for the most part biofuels, unless derived from waste (peach pits, cooking oils), are a disaster. The the main advantage for the larger landscape comes from reducing fuel use, not from using biofuels. In a sense, we are already growing our fuel, in that our enormous corn exports help pay for our imported petroleum (6 barrels per acre with oil at $100 a barrel, and corn at $4 a bushel and 150 bushels to the acre): that is, our eroding landscape pays for our oil.

During the 1970s I followed a debate in the margins of the scientific literature about the relative energetics of styrofoam and paper coffee cups (lab workers get their coffee from machines). I think the fundamental basis of the debate was between the old and the new; between good old, seminatural paper and new, forward looking, petroleum based styrofoam. (A better insulator, styrofoam kept the liquid warmer longer; the downside was burning your tongue on the first sip, or your fingers picking up the cup.) At the time, thanks partly to DDT, petrochemicals were in poor repute (this hasn't improved), but papermaking, because of its use of mercury as a biocide in the pulp mix and of chlorine as a bleach has been a major polluter of waterways and of their food chains; of waterbirds, fish and people. The fish in large parts of major Canadian river systems have been made inedible through papermaking. After many calculations of the energy involved in the manufacturing chain, styrofoam came out with a slight advantage in energy and water expenditure. However, it turned out that pottery cups (the old old thing) were the best, if wash water use were kept low and if they lasted long enough. Rinsed cups that lasted several years involved by far the least expenditure of energy and materials. On a similar theme, glass bottles, if reused 8 times or more, are superior in terms of energy consumption to plastic or aluminum, even if these materials are recycled. Glass bottles are however (like pottery) heavy; and for the numbers to work refills cannot be transported more than 150 miles. So optimists like me picture local bottling plants (a current feature of the American landscape) refilling local bottles. Ideally these are standard bottles, in several sizes, with glued-on labels (nontoxic inks, water soluble glues), usable for any fluids (soda, fruit juice, spaghetti sauce). At the end of their useful lives the bottles are ground up and remelted into new bottles, at a considerable energy savings over manufacturing new glass. The wash water for the bottling plant is cleaned by a manmade marsh next door (with a green-house for winter use in cold climates, that grows flowers, bedding plants, fish and edible greens). The greenhouse and marsh also handle the runoff from the roof and parking lot. The cleaned water sinks into the ground or flows through a vegetated channel into a local waterway. Power comes from photovoltaic panels on the flat roof of the super insulated building.

The main way to make cars more efficient is to make them lighter. Better aerodynamics also help. (One reason trains are so much more efficient than cars is that they are long and thin, with little frontal area per person.) Currently, with fleet mileage in the United States a little over 21 miles per gallon, about 1% of the gasoline burned in the engine goes into moving the passengers. But internal combustion engines can only be 25% efficient; most of the energy in the fuel ends up as heat (therefore car radiators). So cars that run on electricity (some sort of battery) immediately become 75-80% more efficient. Powering cars with electric motors also saves weight by eliminating the engine (the electric motor is much lighter), the drivetrain, clutch, driveshaft and transmission. Much of this weight would be replaced by that of the batteries. The lithium-ion battery pack in a Tesla roadster weighs about 1000 pounds, about a third the weight of the car. The Tesla has a range of 220 miles and uses about one fifth the energy of a fossil fueled car, though the actual carbon emissions and efficiency of the car depend on how the electricity is produced: coal, wind, photovoltaic panels.

Very efficient cars (90-200 miles per gallon) can't be made of steel; it's too heavy. Heavy cars require a large engine for acceleration. Plastic composites are stronger than steel and can absorb 5 times more energy per pound, and thus are as safe in crashes. Composite cars weigh 2-3 times less than metal cars, perhaps 1500 pounds, 10 times more than a person. If car bodies are manufactured of carbon fiber and resin, something like 90% of the steel and 30% of the aluminum in a car can be eliminated. U.S. resin production would rise by 3%. All that metal that doesn't have to be made involves tremendous savings in fossil fuels (less carbon dioxide, metals and hydrocarbons in the atmosphere), in mining (less atmospheric pollution and earth-moving), and in transportation (ditto). Thus making cars lighter also (in general) helps with the carbon footprint involved in building them, which for a midsize car in 2007 was equivalent to burning 900 gallons of gasoline. According to the authors of *Natural Capitalism*, the savings for car manufacturers in going to composite bodies are also tremendous: the capital costs of manufacture (tooling costs, fabricating plants, the time needed to bring a new model out) are reduced by 50-90%. Our need for oil would drop precipitously. If we saved the oil we now import from the Persian Gulf (around a billion barrels a year—making cars that get 40 miles to the gallon would save this), we could also save the $50 billion a year we now spend, on average, to keep the Persian Gulf safe for oil production. We would save the costs of the steel that goes into the military ships and oil tankers, the coal and oil needed to make that steel, the diesel or uranium that powers them, the costs involved with weapons manufacture. Money that went for oil could now be invested at home, into infrastructure projects that saved energy (home insulation, energy-efficient lighting, more efficient heating and cooling equipment); into better public transportation; and into supporting renewable energy (solar thermal power plants, solar cells on roofs). This would further reduce our need for oil. Our balance of payments situation would improve by $80-$100 billion (now, in 2008, $125-$135 billion) per billion barrels of oil. The materials used in car bodies have changed once before; from 85% wood in 1920, car bodies shifted to 70% steel by 1927.

The point of making cars that get 100 miles a gallon is to reduce fuel use and carbon dioxide put into the atmosphere, as well as to reduce air pollution generally (nitrogen oxides, hydrocarbons, ground level ozone); that is to reduce the environmental impact of cars. In the United States it also reduces our international indebtedness and our dependence on foreign oil. But the effect is to make room for more cars. Since many of the environmental effects of cars are worldwide, in a more just world this room would be in under-developed countries, but in fact it will be in the developed world as well. In our current capitalist world, cars will expand to fill the available space; or to use up the available fuel. Oil prices rose in the

1970s and the average energy intensity of the United States' economy (the energy required per dollar of economic output) fell by 34% from 1980-2000. (Energy costs are typically 1-2% of industrial costs and so aren't worth worrying about, but high enough costs make them so, and so energy efficiency doubled from 1975-1985.) At the same time the population increased by 22%, mostly from immigration. The economy also grew. If the average percapita Gross Domestic Product had remained constant, the United State's energy use in 2000 would have been 20% below the 1980 level, but the percapita GDP rose 55% and so total energy use was up 26% in 2000. The point is that growth in the economy or in population will soon swallow up any energy gains. This is the Malthusian dilemma, which human ingenuity sometimes lets us evade. (Thus Malthus didn't foresee how the exploitation of fossil fuels after 1800 would let human populations grow uninterruptedly for two centuries.)

In a typical herbivore irruption (meadow voles introduced to a temperate island, sheep in the late 1500s to the highlands about Mexico City) the animals, now without predators, parasites or diseases, and with an unlimited food supply, increase in population exponentially until they overshoot the landscape's carrying capacity. This phenomenon has been documented mostly in hooved animals, the ungulates, but beaver and muskrats, unchecked, will eat themselves out of house and home, then move out or starve, and the vegetation recover—quickly in the case of a muskrat "eat out" of a cattail swamp, more slowly with the poplars and willows about an abandoned beaver pond. With ungulates and voles, the population crashes as the animals run out of food, then reaches an accommodation with the (now degraded) food supply. The plant population stabilizes under grazing pressure at a lower density, height and species diversity. The population of animals (sheep, voles) is also much reduced and begins to oscillate about the food supply. In some cases the landscape is so degraded it will no longer support any grazers; or hold much water, which runs off the bare ground. Efficient predation (by microbes, predatory starfish, sea otters or large fierce animals) is thought to prevent such eruptions and keep the earth green.

The human situation is something like this. When people were hunters and gatherers, they slowly filled the world. Some niches were difficult and took a long time to exploit, such as the Arctic lands of the Inuit. Some landscapes, such as Antarctica, or altitudes above 16,000 feet, were unexploitable. The constant expansion shows population pressure remained a factor in human life. Polynesians living on small islands kept their populations low, partly by exposing infants, partly by forced emigration of young people (which is how the distant islands of Polynesia were settled). Probably through active control, many hunter-gatherer populations stabilized about 30% below the maximum carrying capacity of their landscapes to provide food. (But some must not have, for the human population continued to grow.) This is near where wolves and weather keep populations of elk and may reflect a long term limit of carrying capacity.

Of course what a landscape will provide in food depends partly on how you use it. Agriculture let people fill the world more completely. (It supports up to 100 times more people per unit of land.) But agricultural productivity fell as the post glacial world warmed and dried, and as better soils were depleted and poorer soils exploited. Hierarchical agricultural societies and more limited diets meant shorter, sicker, more difficult lives for much of the population. Agricultural populations tend to oscillate around the maximum food supply, so starvation was common. New lands, new crops, new crop rotations, plant and animal breeding, new irrigation schemes kept agricultural productivity and the world population slowly rising, but at least in Europe, periodic starvation only ended with the Industrial Revolution and the exploitation of fossil fuels, the cheap transport of grain from new lands, new fertilizers from faraway deposits of seabirds or from inorganic minerals. Under solar powered agriculture, world population grew from 4 million people 12,000 years ago, in the twilight of the hunter-gatherers, to 750 million in 1750, the beginning of industrialization; and until then worlds like Europe, essentially filled, kept growing. The industrial and economic revolutions of our day continue to fill the world with more people. (Seven hundred fifty million to one billion people is a reasonable world population.) One might speculate that population growth (like economic growth) is an evolutionary imperative, but in fact women will control their fertility if it seems advantageous to them and if they are allowed to do so. If not for immigration, populations throughout the the industrial world would be falling—children are seen as too expensive. The economic emancipation of women from the control of men, universal education, better medical care (thus better childhood survival), and some sort of social safety net slows population growth to near zero even in relatively poor countries.

Control of economic growth however is another matter: growth in the economy is driven not just by growth in the population but, as the French textbook made clear, by our ideas of what life should be—that is, is the purpose of life to become rich? Most modern Americans consider the pursuit of riches, rather than happiness, the purpose of life in the American republic. Or rather, they think one implies the other. Early in the Industrial Revolution growth that was faster than that of population allowed Europeans to outrun the Malthusian dilemma (world output rose approximately 10 times world population from 1800-2000); and become prosperous. Now, as we approach other limits (another aspect of the dilemma), a writer defines sustainable growth as growth that does not cause the implosion (from environmental disaster) of the society from which it springs. But no growth is sustainable. And economists don't know how to run a capitalist economy that doesn't grow. In economic terms, efficiency, through competition over prices, drives down the need for labor (usually the greatest cost in producing a product); growth is then needed to maintain full employment. The constant production of new products (cell phones, tablet computers, android operating systems) maintains demand (and thus employment). So growth is necessary for economic stability. However people have ideas about what to do: for instance, taxing raw materials rather than capital and labor, by favoring the use of labor over the use of materials, should let a considerably lower level of material growth—perhaps a negative one—be profitable. Declining populations remove the need for growth. More equal societies tend to be less oriented towards growth. But growth is not yet over. In the coming world, new sources of energy will have to be developed, houses made more energy efficient, degraded lands, fisheries and rivers renovated, coastal and riverine settlements moved inland and uphill. The development of capitalist economics turned human greed (the anxiety and hunger for status) into a developmental force. The rational pursuit of profit has transformed our world more than technology, which is only a tool.

———————

"We're cooked," remarked Vaclav Smil, a Canadian expert on agriculture, energy and the environment at a conference on global warming with 10,000 participants in Montreal in 2006. He's right. The earth responds slowly to changes in its atmospheric gases and to the surface warming of its oceans. The extra carbon dioxide already in the atmosphere will take a century or more to exert its full warming potential. (So the ocean will continue to rise until 2100 even if carbon dioxide emissions were stopped tomorrow.) The carbon dioxide will eventually be absorbed by the deep ocean but because the ocean circulation takes so long (1000-10,000 years for one thermohaline revolution) surface waters become more and more saturated with carbon dioxide, and can absorb less and less. Feedback processes kick in. The forests in the western United States and across the boreal regions of Canada and Russia are collapsing from drought, insect damage and warmer temperatures. Drought stresses the trees, thawing permafrost uproots them, and insects are many times more abundant in the shorter, warmer winters and warmer summers. These dying forests will decay, or more likely, burn, putting hundreds of millions or billions, of tons of carbon dioxide into the atmosphere. As the permafrost below them thaws, it emits methane and carbon dioxide. So do warming boreal peat bogs. The tundra lakes, filled with water from the last ice age, expand as the ice beneath them melts, then drain away, exposing more bare ground to the sun. This soil also emits methane and carbon dioxide. As the sea ice melts in the Arctic and Antarctic, the ocean warms from the sun. The warmth melts the undersides of continental glaciers where they reach the sea, letting them slide seaward more quickly. Along the east Siberian coast, methane, produced by bacteria and locked in a water/ice lattice by cold and the weight of the seawater, bubbles up from the seabed. Such lattices store hundreds of billion tons of methane. As the sea warms (not much: 1°-2°C.) these hydrates become unstable. (A sudden release of methane from seafloor hydrates is thought to have been behind a dramatic warming 55 million years ago.) As they warm and dry further, the tropical peat swamp forests of Indonesia burn. (Clearing and burning tropical peat swamps to plant palm oil plantations has been a major contributor to global warming over the last 20 years.) The Amazon rain forest burns more frequently as it is logged, and as transpiration from the trees falls, rainfall falls, and the forest begins to collapse from west to east. The dead trees emit huge amounts of carbon dioxide. These are positive feedback processes put in place by human activity and a small amount of warming.

Climate change is a problem mostly for charismatic animals like us. Rapid climate change, like too much ultraviolet light, may make much of the green and blue world uninhabitable for many of its larger plants and animals, whose generation times are relatively long and whose ability to move is limited. Trees, for instance, will have trouble moving fast enough to accomodate a rapidly changing climate. For many species of amphibians, the current state of the world, awash in manmade chemicals that compromise their immune systems and in manmade ultraviolet light that scrambles the DNA in their eggs, in warm air from cleared lowlands swallowing up the clouds on wet mountaintops—which become sunny and dry, too dry for breeding frogs—and in diseases spread by amphibians from other continents, is already too much and many populations have collapsed. Fewer amphibians means fewer hawks, minks, raccoons, coatis, fewer nutrients brought from the margins of wetlands (where the frogs live) to the uplands, as urine, bones and dung. Higher seas, more powerful rains and windstorms, more severe droughts, rising or falling average temperatures will make much human infrastructure obsolete. Such infrastructure includes roofs, bridges, tunnels, seawalls, roads, farmland, water reservoirs, river works, riverbank housing, seaside lots, buildings in woodlands that burn or in regions with hurricanes or tornados. Places are occupied and houses built according to long term climatic averages; that is, human settlements are manifestations of climate. At the same time, chemical and nutrient pollution may make much of the landscape unstable or uninhabitable. Ecological collapse came for the Maya and the Tiahuanaco people in small changes of climate, for which their cultures, already stressed by overpopulation and habitat destruction, could not cope; for Native Americans in the 1500s and 1600s in microbes to which their immune systems were not adapted. When cultures and their support structures collapse, many people die. What keeps us from acting to alleviate the growing risk is fear: fear of change, fear of economic collapse.

At the least, since governments seem incapable of acting except in the face of calamity, we are in for a warming of 2°-3° C. (and then probably 4°-5° C.) and a sealevel rise of 1-4 meters (up to 25 meters by 2100 in one respectable scenario). This is catastrophic for low lying countries (those on river deltas or on islands) and for any settlements near the coast. Perhaps 10-20% of people worldwide will be displaced (about a billion people, perhaps 50 million in the United States), much low lying cropland destroyed, coastal aquifers turned brackish.

If warming can be kept to this level however, it may be less of a worry than the continuing contamination of the biosphere with bioaccumulating chemicals, with unforeseen effects on people and animals. Changes on this scale don't mean the end of people, but may mean the end of our civilization. Of course, people can plan for the coming changes and thus reduce their economic consequences and try (like our ancestral hunter-horticulturalists) to live within the green world. In 2020-2040, when the Arctic is mostly ice free in summer, there will be refugia for some arctic mammals (ringed seals, polar bears) about the northern tip of Greenland. (The several species of Arctic seals depend on different ice conditions and depths to the seabed and it is unlikely all species will survive. Polar bears evolved from grizzlies, with which they still can mate, and can evolve again.) Other refugia for cold adapted animals will exist about Antarctica; new animals will colonize that continent, perhaps from the other end of the globe, and evolve there into new animals. The connected wild and semi wild landscapes I have tried to describe, into which the manmade landscape of cities, roads, farms, and fiber farm forests fits, should let plants and animals move around in response to a changing climate. (Plants also move through landscape connections and connected landscapes have more species of plants.) Prudent homeowners may order seeds from forests 500 miles south, so as the trees around them die from winters that are too short or summers that are too hot and dry, others can take their place. Connected, functioning landscapes let small populations of organisms survive, which expand as conditions change. Audubon saw a chestnut-sided warbler once. A bird of the woodland edge, it is now one of the more abundant warblers in the settled parts of the northeastern forest.

The gift of fossil fuels was cheap, abundant energy. Our civilization runs on energy. (In an average year, the energy we derive from fossil fuels is 400 times the net potential energy fixed by plants from sunlight.) Without electricity and motor fuels our cities would start to collapse in 3 days (the length of their food supply), or a week (far too long to be without water). There are three reasons for hope. One of them is that the development of a low carbon, energy efficient economy would be immensely profitable. Carbon emissions become worth avoiding at $50-$100 per ton. The United States produces about 2 billion tons of anthropogenic carbon a year, or $100-$200 billion worth. This amounts to a rounding error in the current U.S. budget, but a potent economic incentive. Paying farmers $50-$100 a ton ($25-$100 an acre) for storing carbon would transform farming practices. Paying electricity producers $50-$100 per ton for avoided carbon (and guaranteeing the payments for 20 years, as the Germans do with their solar power premiums) would make photovoltaic power profitable. Under such conditions, photo-voltaic power generation, or generation from solar thermal power plants (which are more cost effective than photovoltaic panels but only work in sunny climates and on a large scale), makes much more sense than coal. Machines turn over every 20-50 years, appliances and cars every 3-10, commercial buildings every 30, so it wouldn't take long to transform our energy use. From 1800 to 1950 industrial energy in the United States changed from wood to coal and then to oil and natural gas, largely under economic incentives, but influenced by government policy. A reasonable estimate for shifting to a solar powered energy supply worldwide is 50 years, too long to prevent much climate change but better than doing nothing. A changeover in 20 years is probably possible. (A small Danish island did it in 10.) Buildings use 33-40% the total energy in the United States, 70% of the electricity. Efficient new (or retrofitted) buildings save 70-90% of this energy, and with solar collectors on their walls or roofs produce more daytime electricity than they use. Dispersed electric generation is much cheaper than the current centralized system. Efficient retrofitting of buildings means replacing windows with more energy efficient ones, super insulating walls and roofs, using daylight for lighting office spaces, and using more efficient lights, office equipment, heating and cooling equipment, perhaps in some climates replacing air handling equipment with building designs that exploit the natural bouyancy of air (screen doors, open windows, high ceilings). Replacing all the electric motors and lights in the United States with efficient ones would save half the United States' electricity production. If investment focused on reducing energy use (say, if utilities were allowed to profit from some of the energy saved), the United States after 20 years (the time between renovations) would use 10-25% of the energy it now uses to produce the same value of national income. Reducing energy use by 75% is not technically difficult, but would require public policies that more or less guaranteed the new investments (and a tax that kept the cost of using energy the same, so houses didn't get bigger and cars driven further and energy use overall rise). Reducing energy use also speeds up the time it takes to replace fossil fueled with solar energy, since so much less energy is required.

Using that much less energy makes renewable resources look good. Rough calculations from the 1980s indicate that covering roofs of houses in Britain with solar collectors would supply most of the daytime electricity needs of British households, and most of their hot water. (That is, Britain's needs at that time: not needs that were 75-90% less.) More generally, it is thought that surfaces of buildings in industrial countries could generate 15-50% of the countries' electricity needs; or all of them, with energy use reduced 90%. Putting collectors on the roofs of buildings in the United States, much of which is quite sunny, would supply more than the United States' daily electricity needs. If energy use fell by 75%, such collectors would provide 3 times the power we need, or our total energy needs, in the form of electricity (which is easily convertible to other forms). Some fossil fueled or nuclear power would be necessary to provide so called baseline power, but only about 5% of what we have now (which would provide 25-50% of total electricity needs). (Alternatively, a square 600 km by 600 km—approximately the size of Arizona—filled with concentrating solar collectors in the deserts of the U.S. Southwest would provide 500 million people with the current energy usage of Americans, which is—per capita—double that of the rest of the developed world. The current population of the U.S. is something over 300 million. As energy use falls, the square gets smaller.) Reducing energy use so much makes many things possible. About 9% of California's current electricity supply comes from the wind; that is somewhat more than 35% of what would be needed if electricity use fell by 75%. Reducing home heating and cooling through better insulation and design shifts part of the energy costs of heating and cooling to manufacturers. But the energy embodied in an efficient house equals 50 years of heating, rather than 3-5 years, as now. (This is better than the past: in 1850 a farmhouse in the colder parts of the United States burned the equivalent of the wood that went into it every year.) Using heat pumps (electrifying the heating or cooling system) reduces energy costs by another 80%, by eliminating the need for producing a source of heat or cold (the ground is the source: again, the actual reduction depends on how the electricity is produced).

Another reason for hope is that a nontoxic economy can also profitable. The European Union's classifying of carpet remnants as toxic waste (thus increasing disposal fees) made the Steelcase Company invent a nontoxic, recyclable carpet (no waste, period). Increasing dumping fees increased the recycling of materials from house demolitions in Denmark from 12% to 82% of the total (4% is the average in industrial countries). Developing non-toxic materials would mean transforming another aspect of the industrial system, and another immensely profitable enterprise, if government policy guaranteed the regulations for the lifetime of the investment. With forward looking policies (tweaking the market), the goal of capitalist effort would be zero net carbon emissions and a chemical industry based on nonbioaccumulating, recyclable, biodegradable materials. Taxes on labor, income and investment (high tax rates on investment require high rates of return) would slowly be replaced by taxes on climate change gases, non-renewable fuels, air traffic, pesticides, toxic chemicals (dioxins, benzene, bromine, chlorine), nutrients such as nitrogen and phosphorus, irrigation water, loss of topsoil, newly cut timber, aquifer depletion, landfill waste. As the cost of labor fell, businesses could hire more workers to remanufacture products, close the loops in material flows (so one company's waste becomes another's raw material), save energy, change raw materials and manufacturing processes.

The last reason for hope is a line on a graph. The problem of economic growth in capitalist societies is not easily solvable. Reducing the energy and materials needed for a modern lifestyle, making them nontoxic, and leaving room for the natural world to operate, helps. For the rest, capitalist societies tend to become more and more skewed in income over time and so redistributing income may be necessary for democratic institutions to survive. Perhaps then growth in wealth will become self-limiting. Growth in population is another matter. Here growth in wealth helps: no country in the world with a per capita income over $5000 (a tenth that of the U.S.) has a fertility rate much above the replacement level of 2.1 children per woman. Some writers think the current global fertility rate of 2.7 children per woman is an all time low. In our agricultural/industrial world, family size depends on infant mortality (that is, on health care and diet); the status of women (whether they have status as individuals or as childbearers, whether they can control their incomes, their fertility and their destinies); education; and (perhaps) on industrialization, which makes children a financial liability rather than an economic asset. If every woman had one child, world population would fall from 6.7 billion people (2008) to 1.6 billion by 2100. If she had two children, or slightly fewer than two, population would also fall, but more slowly. Such a slow transition to a world population of 1-2 billion should be manageable. That imaginary line on the graph is another reason for hope.

I end with notes for a web site:

There are 5 billion chickens in the United States at any one time. Feathers are a major waste product of the chicken industry. Circuit boards made from chicken feather keratin (a waxy substance in the feathers), coated with soybean resin, have a lower die-electric constant than standard boards, which are made from virgin plastics and coated with petroleum based hydrocarbons. Chicken feather boards can switch signals faster and support faster processing speeds.

Chickens turn calcium carbonate into eggshells at body temperature. Shellfish make calcium carbonate materials harder than cement at 40° Fahrenheit. Spiders spin silk stronger than kevlar at ambient temperatures. People manufacture portland cement by heating limestone to 2700° Fahrenheit and make Kevlar from petroleum based molecules boiled at several hundred degrees Centigrade in pressurized sulfuric acid. Eggshells, shellfish shells and spider webs are all biodegradable, and their manufacture, as far as is known, does not use hazardous chemicals. Can new manufacturing techniques mimic these processes?

Twenty-two million tons of food waste are buried in landfills in the United States each year. Bacteria have been used to turn it into a biodegradable plastic polymer. The yield of polymer is about 20% of the weight, so the process would produce about 4.5 million tons of polymer a year. To this feedstock could be added the many millions of tons of slaughterhouse waste currently reprocessed into animal feed. (Thirty percent of commercial cattle feed consists of reprocessed animal fats. Cows are no longer supposed to be fed cows, but they are fed pigs and chickens, which are fed cows. Mistakes are made, bags of feed get mixed up. Such hightech cannibalism gave us mad cow disease. Some European cattle feed in the 1980s may have contained scavenged human bodies, which are often only partly burned when put into the Ganges River, where their flesh is eaten by turtles.)

Food waste and slaughterhouse waste could also be composted, or processed into fuels. Steaming organic matter (food waste, landfill waste) under low pressure with a citric acid catalyst produces a fuel; heat, which can be recycled into the process; and biochar, a charcoal that greatly increases the productivity of agricultural soils.

Ethyl lactate made from fermented corn starch can replace many dangerous and polluting volatile organic compounds. Its use is competitive in cost with existing paints, paint thinners, glues, inks, dyes and circuit board cleaners (such as dichloroethylene and trichloroethylene, major pollutants of much U.S. ground water). Ethlyl lactate breaks down into carbon dioxide and water. It is a much better use for corn than distilling it into fuel.

Agricultural oils may find similar uses. The polycyclic aromatic hydrocarbons produced by tire wear (40,000 tons per year in the United States), that form a haze over Los Angeles freeways, derive from the petroleum based oils used to make the tires. Tires made from plant based oils (jojoba, soybean, cottonseed, hazelnut) might eliminate these hazardous chemicals, though this has not yet been shown. Jojoba is an oilseed shrub that grows in the desert without irrigation. (If irrigated, it can be irrigated with salty water.) Using less material lets us return to a chemical industry based on carbohydrates, rather than on petrochemicals.

Do we need the chemistry of chlorine? No, but changing the stream of chemical manufacturing processes of which it is a part would be a major headache. Approximately 1% of chlorine is used to disinfect water. Adding chlorine to water for the purpose of disinfection creates a whole family of chlorinated hydrocarbons, from the reaction of chlorine with organic molecules in the water. Many of these chemicals are cancer promoting or mutagenic; some mimic human hormones. Treating water with ultrasound, ozone, and ultraviolet light disinfects it without creating hazardous compounds.

Paper is commonly bleached with chlorine. Polyoxymetalate can also be used to bleach paper. It works as well as chlorine and the chemical is easily regenerated from the waste stream for reuse. Using a polyoxymetalate bleaching process lets paper mills increase the recycling of their process water and saves half their use of electricity. Modern paper mills can produce essentially no effluent and use very little new water (some evaporates in the paper-making process). Nontoxic, soybean based inks that float off waste paper in a warm water bath are collected and reused. Limiting their use of water and electricity and their production of pollutants lets paper mills locate in cities near their suppliers and markets, use a feedstock of urban waste paper, and save many dollars and millions of gallons of fuel in delivery costs. (The annual production of cellulose from waste paper in New York City amounts to 50-100% of that harvested from the forests of Brazil.)

Solvents based on citrus oils (say, from orange peels), and materials like ethyl lactate can adequately replace solvents based on chlorine. Cleaning with mild soap and water has been successfully used to replace dry cleaning fluids. Dry cleaning fluids like tetrachloroethylene, a neurotoxin and carcinogen, are ubiquitous in North American water and food supplies.

Placing 6 foot tall concrete bat boxes in clearings formerly occupied by tropical forest in Costa Rica attracts bats. The roosting bats drop seeds of forest plants in their manure (5-20 times as many seeds as in clearings without bat boxes). The plants that grow, especially the fast growing pioneer plants, provide cover for mammals, birds and insects that disperse more seeds. Replanting a forest this way is much cheaper and more efficient than replanting one using human labor.

Tuberculosis hospitals in Lima, Peru, with large windows and high ceilings had better air circulation than hospitals with mechanical ventilation systems. Especially in warm climates, passive airflow systems (which depend on the building's design), with roofed outdoor walkways and waiting rooms, may be preferable to mechanically ventilated closed systems.

The Steelcase Company recently invented a compostable upholstery fabric. Their reason for inventing it was the designation by the European Union of textile mill trimmings as hazardous waste. Disposal of the trimmings was going to cost the company considerably more than previously. The new fabric contains no mutagens, carcinogens or heavy metals. Its manufacture generates no toxic waste. Manufacturing costs for the material are less than for fabrics that use standard, hazardous materials. (An added benefit is that users of the fabric do not accumulate toxic chemicals through their skins.) The carpet is completely recyclable and can be remanufactured with little energy input indefinitely. The company leases its floor coverings. The manufacturing process allows it to renew the leased material at little cost to itself.

A tax on pollutants sends a clear, long term signal.

Melting mixed plastics produces a weak, unstable material. Pulvering mixed plastics in a ball mill under pressure in the presence of carbon dioxide breaks the plastics up sufficiently at the molecular level to let the molecules recombine. The hybrid polymer yields a homogeneous melt that can be formed into durable new plastics. Is this is a temporary solution for mixed plastics in the landfill, another side trip on plastics' long journey into oblivion; or a way to reconstitute plastics indefinitely?

Concrete production produces 5-8% of anthropogenic carbon dioxide globally. (Twenty percent of China's production: China is now the world's largest producer of greenhouse gases). Geopolymer concrete (E-crete) produces 10-20% of the greenhouse gases associated with the production of portland concrete. Adding alkali to silicates and aluminates derived from fly ash and slag (waste products of coal burning and steel production), plus gravel and sand, makes E-concrete. Unlike portland concrete, no carbon dioxide is produced during polymerization. No heating is required to produce the raw materials, which are waste products—though heat went into producing them originally. Geopolymer concrete is more porous than regular concrete, hardens faster, is more resistant to acid, fire and microbial attack.

An experimental new process makes cement from seawater and the carbon dioxide exhaust from coal burning power plants. Since this process aborbs much of the carbon dioxide emitted by the plant, it has the potential to make coal burning plants much less polluting. Mining and transporting coal are already so destructive that it would seem better to slowly phase coal out, but processes like these, by reducing the problems of coal ash and carbon dioxide, are a tremendous step forward. They could be adopted immediately (and would, if carbon were taxed).

Of course the production of all fossil fuels involves tremendous environmental destruction.

Beer bran, a byproduct of brewing beer from barley, adsorbs hazardous organic molecules, such as benzene and trichloroethylene. Beer bran is a waste product. The activated charcoal usually used as a filter requires heating coal to 900° Centigrade. (Another Brit idea: the English have an obsession with the uses of beer.)

Tomato sauce can be extracted from crushed tomatoes using a semipermeable membrane (so-called reverse osmosis technology). The process uses 30 times less energy than heat reduction. The sauce tastes better and has more nutritional value. (In general concentrating foods using direct osmosis rather than heat uses 95% less energy and produces foods of better nutritional value.)

Direct osmosis can also be used to desalinate water. Windmills could mechanically drive reverse osmosis desalination plants and pump the fresh water ashore. This would not require converting the wind power to electricity. The stream of salty waste water can be processed using rapid spray evaporation. Salt water is sprayed as a fine mist into a heating vessel. The salt falls to the floor, the vapor is condensed to pure water. The cost is of rapid spray evaporation is one-third that of regular desalination and virtually all the brine stream can be converted into fresh water and salt. (The brine stream, a polluting byproduct of desalination plants, is usually released into the sea, where it damages the flora and fauna of the seafloor.)

Titanium dioxide solar cells are currently 33% efficient, approximately double that of solid state silicon (the current standard). They are transparent. Their cost is not that much more than glass.

Methane digesters can be used to compost food waste. The extracted methane can be burned to generate electricity and the compost sold as fertilizer. This process does not decrease the global warming effect of the methane but burning it to produce electricity replaces the burning of other fuels. (The carbon dioxide and methane in landfills are recyclable: they come from renewable sources.) Food waste normally goes into landfills, which are currently the largest single source of methane in the United States. The contents of old landfills can also be put through methane digesters. In time, the one trillion aluminum cans old landfills contain (a year's supply of aluminum ingot, worth $21 billion) will make them worth mining.

Shallow landfills can be used as digesters in place. The landfill is capped with two impervious (low permeability) layers of clay, with a permeable layer of sand between. The methane trapped beneath the lower level of clay is extracted and used to generate electricity. The carbon dioxide extracted from the landfill gas is pumped back into the permeable layer at slightly above atmospheric pressure to keep oxygen from being drawn into the landfill as methane is drawn out. This keeps the production of methane up. (The methane production process is anaerobic.)

Processing the 200 million tons of manure confined animals produce annually in the United States in a methane digester would make electricity, a odorless liquid fertilizer, and processed solids for sale as a soil amendment. The anaerobic digestion process is sensitive to the properties of the feedstock (its temperature, liquidity, alkinality, pH, carbon-to-nitrogen ratio) and so must be properly overseen (a labor cost). Keeping animals in confinement is not something anyone except the owners of the facility should favor, but this a is a good way to dispose of their manure.

Texas Instruments built a green chip factory in Dallas, Texas, for 30% less per square foot than a conventional one. Water usage is 35% less, electricity usage 20% less. The lower costs make it competitive with plants in China or Singapore. (Do they clean their circuit boards with steam, ethyl lactate, or trichloroethylene?)

A writer estimates it would cost $23 billion a year to turn 10% of every region on earth into a national park. This figure includes land purchase and staffing costs. (It may also be too low.)

A petrochemical plant covering 300 acres, with some additional acreage in natural gas production facilities and pipelines, can produce the fiber grown on 600,000 acres of cotton. A tempting idea, since cotton grown conventionally is destructive of both soils and landscape. But the petrochemical industry has left us an overwhelming legacy of pollution. Extracting natural gas from coal and shale beds is not an environmentally friendly operation. Hemp, bamboo and sweet gum are much less demanding fiber crops than cotton. Perhaps there are nonpolluting ways to manufacture usable fibers, from environmentally friendly feedstocks.

With a 17-year cutting rotation, tree plantations for railway fuel in India would occupy 20 acres per mile of track. This constitutes a band about 200 feet deep along 80% of the track on one side. Approximately 5-10% of that acreage in photovoltaic panels would also work.

Organic vegetable soups have 6 times the salicylates of conventional vegetable soups. Pests feeding on the organically grown plants stimulates them to secrete salicylates. (In general, crop plants like potatoes can lose a third of their leaf area without reducing yields.) Salicylates (aspirin is one) are anti-inflammatories and anti-oxidents. They help prevent heart attacks, strokes, several cancers and may delay the onset of Alzheimer's disease. Similarly, organic tomato ketchup contains 2-3 times the cancer fighting lycopenes of non-organic ketchup. Temporary shortages of nutrients (as occur in organic agriculture) may also stimulate the production of anti-oxidants in plants.

A fleet average of 40 miles per gallon for cars and light trucks would save over a billion barrels of oil in the U.S. annually, more than we now import from the Persian Gulf. Cutting the speed limit to 55 miles per hour would let the present fleet save the same amount of oil.

Cars running on compressed air can go 200 miles at 30 miles per hour and refuel in three minutes at a cost of $2.50. Of course, faster speeds reduce the car's range.

Human urine makes up about 1% of the volume of waste water flow, but contains 80% of its nitrogen and 45% of its phosphorus. Urine could replace a quarter of the commercial fertilizer currently used for crop production. Removing 50-60% of the urine in waste water would turn sewage treatment plants into net producers of energy. With fewer nutrients in the water, microbes in the aeration tanks turn the remaining nitrogen and phosphorus into biomass much more quickly and efficiently. The biomass formed is richer and generates more methane (the source of the treatment plant's energy) in the anaerobic digesters. The transit time of the waste water through the plant is much reduced.

Urine separation toilets separate out urine and store it in tanks until it is collected and taken away to be converted to fertilizer. Such toilets require replumbing existing waste water systems. Separating the urine from the waste water flow itself would be simpler. Low flush toilets reduce the volume of waste water and toilets with separate flushes for urine and for solids reduce it further. This helps concentrate the urine. Once solids have been settled out, reverse osmosis membranes might further concentrate the urine. Then the urine/water mix is cooled until it is half frozen. Water freezes preferentially out of the mixed fluid. Water freezes as a pure substance, expelling other molecules from its crystal lattice, so most of the fertilizing elements remain in the liquid fraction, which is then decanted and reacted with magnesium oxide to produce struvite, an ammonium phosphate fertilizer.

One use for partially treated wastewater is for irrigation. One-tenth of the world's irrigated crops are grown with partially treated wastewater. The crops use the nutrients in the urine and the water is cleaned during its passage through the soil. Excess water moves as groundwater flow into streams or percolates into the soil, where it maintains groundwater levels. Such waters should not contain industrial waste (as they usually do), and unless further cleaned, should not be used on vegetable crops. Rivers receiving runoff from such fields are not swimmable or drinkable while farmers working in fields irrigated with sewage water suffer from rashes and infections from microbes in the sewage (both arguments for cleaning the water further).

Nanoparticles of iron seem to catalyze the breakdown of chlorinated solvents like trichloroethylene. If so, they could be used to decontaminate soil or ground water. (Bacteria recently isolated from sewage also break down the chloroethylenes. Chloroethylenes are neurotoxins and carcinagens now common in soils and drinking water.)

A cheap molecular sieve of zeolites will filter carbin dioxide from the exhause gases of power plants. The gas can be pumped into depleted oil wells, or reacted with magnesium oxide to create building blocks and the filter reused. But even at present prices, photovoltaic electricity is probably cheaper than that produced by coal burning plants that have to sequester their carbon dioxide. Solar panels may also be more efficient, as much of the energy in the coal is used up in its mining, transportation and burning (about half in modern plants). The further capture of toxic chemicals, heavy metals and carbon dioxide and the proper disposal of fly ash (perhaps in E-crete) would add to this energy use.

Raising cattle sustainably in Montana raised the profits of the ranchers over 20%. The cattle were raised on grass and given antibiotics only when they were sick.

There are 5 billion acres of currently degraded soils on the planet (more than 5 times the current cropland of the United States: degraded by human use). Revegetating them (perhaps farming them with perennial crops) could absorb most of the carbon dioxide now emitted by human activity. In the dry Sahel, for instance, acacia trees encouraged in the fields by local farmers (against professional agronomists' advice), form a virtuous circle. The trees add carbon and nitrogen to the soil. They provide shade and forage for cattle. More cattle mean more manure for the fields and better crop yields. More land can then be used for the trees (which remove carbon that would otherwise contribute to global warming and whose transpiration over large areas slightly raises rainfall). In California pasturelands, grass grows better under native blue oaks and cattle prefer to graze there.

Cattle also eat the oak seedlings, which must be protected if the trees are to regenerate. In general, modern farmers do not allow trees in their fields or pastures.

Converting crop wastes like corn stover or wheat straw to biochar in a small furnace at 180° C. with a citric acid catalyst makes a soil amendment—biochar charcoal—that greatly increases the productivity of agricultural soils and keeps the carbon out of the atmosphere for tens of thousands of years.

Injecting an appropriate mix of bacteria into sewers reduces odor and digests 50% of the solids before they reach the treatment plant. This clever idea uses the sewers as an extension of the plant.

Approximately 1.4 million pounds of human hair contained in mesh pillows (so the hair could be easily retrieved) would have soaked up the oil spilled by the Exxon Valdez in a week. (Barbershops in the City of London produce several times that in a year.) Exxon spent $2 billion on a high-tech cleanup which further damaged the environment and recovered 12% of the oil.

Selling 10% of the straw from wheatfields in Oregon as a feedstock for non-toxic paper pulp raised the earnings of the farmers by 25-50%. The rest of the straw is left as a stubble mulch. The effluent from the paper mill that processes the straw can be used as a fertilizer.

Land contaminated with cadmium and zinc can be cleaned up by cultivating a flowering brassica, a subspecies of *Thlassi caerulescens*, which accumulates the metal in its tissues (stems, leaves, flowers). After harvest, the plants are dried and burned and the metals recovered from the ash.

Radiation will kill food-borne pathogens like salmonella in chicken and ground beef; so will the anti-oxidants found in dried plums.

Miners removed 700 tons of gold from the California hills during the 1850s and 1860s, using 7000 tons of mercury to do this. (Mercury attracts finely divided gold. Woolen mats soaked with mercury were used to pick up particles of gold in crushed ore.) Much of the mercury ended up in San Francisco Bay, where 150 years later it makes the fish unsafe to eat. The level in the fish is falling and will probably reach a safe level in another 50 years. So contamination is not forever. (Size, sedimentation, bacterial action, dilution have helped heal the bay.) New Haven harbor on Long Island Sound however is still however too toxic for benthic organisms. New Haven's problem derives from the concentration of metalworking industries in Connecticut in the late nineteenth and early twentieth centuries and to the continuing transport of metals from the rivers to the harbor. The larvae of polychaete worms settle, but as soon as they bore into the mud and ingest the sediment they die.

Environmental damage may occur abruptly and be irreversible: processes are nonlinear, thresholds are crossed, buffering capacities exceeded, ecosystem resilience lost. Without warning, moose populations flip to a lower level, water plants disappear from estuaries, jellyfish and algal blooms replace striped bass and oysters. Since the new states are stable, such changes can be hard to reverse. Basic ecosystem services such as purifying the air, cleaning the water, maintaining the carbon dioxide balance of the atmosphere, decomposing wastes, filtering untraviolet light out of the solar spectrum, providing sources of new medicines or new knowledge, permitting recovery from natural disturbances, are not exchangeable for economic gain, at least not by an economy grounded in reality. There are few trade-offs in human interference with the natural environment: only losses. Mines leach heavy metals and other toxic compounds essentially forever, unless their drainage waters are filtered (forever) by men. The economic users of an environment need economic signals of the harm they are doing. Thus—one way or another—farmers and homeowners should pay if they pollute rivers or deplete groundwaters, trawlers pay for the damage they cause to the seabed, and chemical companies pay for the damage their products cause. Putting a realistic price on environmental damage would make much of it cease immediately. Taxes on pollutants like mercury provide a clear longterm signal of social intent.

The recovery of San Francisco Bay from massive mercury contamination, even after 200 years, is a hopeful sign. (A variety of *E. coli* has been engineered to take up mercury. It could be used to clean up waste water streams or polluted waterways.)

In the United States and Canada 10-15 million tons of salt are spread on roads annually. According to a 1987 study, each ton does $1400 of damage to roads and bridges. It also increases the saltiness of surface waters. This becomes a problem for people if the water is a source of drinking water. It is also a problem for aquatic organisms. Digesting whey (a dairy waste) with bacteria to produce acetate, and combining the acetate with limestone makes calcium magnesium acetate, a nontoxic, noncorrosive salt substitute that can be sprayed on roads in advance of storms. A new process makes calcium magnesium acetate more competitive in cost with salt, but it still costs considerably more. Looking at the whole costs of road maintenance would make the new material affordable. (It was probably affordable in 1987, at $1200 a ton, but states and localities paid for the salt, the federal government for road and bridge repairs.)

Electricity can be extracted from hot rocks 5 kilometers down at a rate of about 25 megawatts per cubic kilometer of granite for about 20 years. (Then the rocks must be let reheat.) Pumping water through the rocks uses about 20% of the power produced. There is sufficient geothermal heat within 10 kilometers (4 miles) of the earth's surface in the United States to provide all the 27 trillion kilowatt hours of electricity the United States used in 2005 for the next 2000 years. Such drilling distances are well within modern limits. Of course the energy is renewable. (The possibility of generating small earthquakes in some areas is a problem.)

Yam beans from the American tropics yield 35-70 tons of plant per acre. A crop fixes about 50 tons of nitrogen per acre per year. After rotenone (a natural pesticide) is extracted from the seeds, they can be pressed for oil, and the oil cake, similar to soybean meal, fed to pigs. All parts of the plant are edible. Its roots keep without refrigeration. It will grow in poor, dry soil. It is traditionally grown in Mexico along with corn and beans.

In the late 1990s the Danish island of Samsoe won a contest for Denmark's "renewable energy island." Winning was the result of a plan put together by an engineer who didn't live on the island but thought that its small size and steady winds would make it a good candidate. No prize monies or economic incentives came with the prize. Samsoe won through the efforts of one island resident who became interested in the idea and found funding for a position for himself to develop the project. About the same time, the Danish government passed a law requiring electric utilites to offer producers of wind power 10-year contracts at a fixed rate. Under such contracts the cost of the turbines was usually paid off in 8 years.

The changeover to renewable energy on Samsoe was created by this one man, who talked his neighbors into thinking renewable energy was a good idea. After the project started, thinking about how to use renewable energy (and make money at it) became a game. The island now produces more renewable energy than it uses, mostly from large wind turbines owned by the islanders or other investors, also from small backyard turbines, photovoltaic panels, and rooftop hot water heaters. Heat and hot water in several of the small villages (Samsoe has 4300 inhabitants) is provided by furnaces that burn straw and (in one case) wood chips. The residents of Samsoe still use fossil fuels in their cars, tractors and trucks but the island as a whole produces more renewable energy than it uses in renewable and fossil fueled energy combined. (A few farmers press diesel—or salad—oil from canola seeds, a major oilseed crop in Europe.) Samsoe did not make any attempt to save energy during the conversion and energy use is now the same, or perhaps slightly greater, than before.

Like the American Middle West, Samsoe is a radically simplified human habitat, lacking much of its original mammalian, avian, amphibian, insect and fishy life. Leaching of nutrients into groundwater, streams and estuaries is a major problem, as is overfishing in the surrounding seas. However nothing stops the island from going further and trying to become a sustainable landscape. The flat sandy fields of Holland and Denmark were long maintained with leaf mold, crop rotation, rock powders and manure. Land to protect streams and estuaries and for wildlife can be set aside, as can undersea habitat.

Like Samsoe, the United States could remake its energy environment. It would take guaranteed contracts for renewable electricity production for periods somewhat longer than the time needed to pay off the investment, new long distance power lines, much solar thermal investment in the West and Southwest, solar electric panels and water heaters on private walls and roofs. Besides that, buildings that use 50-75% less energy, electric motors and appliances that use 25-50% less energy, and cars that get 150-200 miles per gallon (all of this built of nontoxic materials) are not bad economic ideas. Whether we can stop global warming or construct a sustainable landscape isn't clear but we can quickly reduce our carbon footprint.

On the dry eastern slopes of the Casacade Mountains in Oregon in the Sisters Ranger District the United States Forest Service is trying to restore forests that have been altered by logging, replanting and fire suppression into dense forests that are susceptible to drought, root diseases, insects and catastrophic fire. In this area, prior to 1900, frequent low intensity fires had created open forests of sun loving trees; ponderosa pine dominated the high desert forests, along with some sugar pine, western white pine, western larch and coastal Douglas fir. (The last is intermediate in shade tolerance; it will grow in some shade.) Early logging removed most of the big trees, low intensity fires were suppressed, and the open woodlands were invaded by the shade tolerant white fir, which formed thick stands. Some parts of the forest were replanted to plantation conifers, also thickly. The dense mid-level fir canopy, with some remaining large pines above it, provided ideal habitat for spotted owls, which colonized the area. But the new forest was overcrowded and unstable. In the late 1980s and early 1990s, an epidemic of budworm, drought, root diseases (which spread easily in a crowded stand) and bark beetles killed most of the firs. Bark beetles also killed many of the ponderosa pines, which lose the vigor necessary to ward of insect attack in crowded stands. With so much fuel, a series of catastrophic wildfires burned large areas of the Sisters Ranger District (91,000 acres of 324, 000 in the district burned in 1991).

The Sisters Ranger District then came up with a plan to restore the original, fire resistant old growth forests. Most of the dead trees were removed (4-13 were left per acre for wildlife); all large, healthy pines and larches (the sun loving trees) were left; and as much fir as possible was removed. The amount of fir that could be removed was constrained by the need to provide habitat for the spotted owls, a protected species. Large firs were left along with the large pines and larches, whose numbers were too few to make a sufficiently dense stand. Some mid-level fir stands were left

amidst the taller sun loving trees for the owls. The thicker stands were surrounded by the more open historical forest of old growth pines, which is
now maintained by cutting and low intensity fires. High intensity fires that
invade the stands of fir die down when they reach the surrounding open
forest. The mid-level fir canopy regenerates quickly and will be rotated
through the whole area. This likely mimics the natural situation, since
ideal owl habitat consists of regenerating old growth woodland. This new
woodland, neither completely historical, nor what the forest developed into
with cutting and without fire, will produce sawtimber, poles, pulpwoods
and fuelwood. Some of the clearing does not produce a profit and has been
contracted out to prison labor; volunteer organizations also take part. A
major purpose of the management is to reduce the risk of catastrophic fire
and make the surrounding area safer for ongoing human settlement.

Further north, also on the dry side of the Cascades, in southern British Columbia, open pondersosa pine forests with an undergrowth of bunch
grasses and forbs slowly filled in with Douglas fir during the twentieth
century, as the low intensity fires that had maintained the stands were
suppressed. The fires had been set by the Lilloet First Nations people, a
Salish tribe that had historically burned the woods to regenerate collected
foods (huckleberry, raspberry, glacier lily, wild onion, spring beauty, buffalo berry, service berry) and to increase forage for mule deer. This dry,
warm woodland (July temperatures average 75° F.) had 5-40 ponderosa
pines per acre. The low intensity fires also allowed trees growing on moist
sites (red cedar and paper birch, along with many shrubs and herbs important to wildlife) to survive. (High intensity fires would have killed these
plants.) So the fires also maintained variety in the woodland. When the
fires were suppressed, the woods filled in with young conifers (up to 500
trees per acre) and in the late twentieth century large areas began burning
catastrophically, destroying houses and threatening nearby towns. Now an
attempt is being made to restore the historical old growth forest.

Most fires in western North America are started by lightning but
lightning is uncommon in this part of British Columbia. Studies of fire
scars on trees showed that fires burned at 5-10 year intervals for the last
400 years. These fires were almost certainly set by the Lilloets or their
predecessors. No one knows when the process started. The forest may have
been continuously manipulated by people from the time of its establish-

ment after the retreat of the glaciers. (One suspects this is how the scrub oak barrens, habitat of heath hens in New England, arose.) Burning of the herbaceous cover to renew it would have let only the fire-resistant pines survive. The goal of restoration is not the forest of a golden age, but a healthy, vigorous woodland adapted to today's world. (In north temperate landscapes, the golden age from the beginning included people, with their spears and firesticks.)

———————

I want to avoid proposing one of those totalitarian utopias in which people walk, take public transportation, drink tap water instead of soda, and borrow a communal car now and then for a trip to the country—those utopias so popular in the nineteenth and early twentieth centuries, and so terrible when put into operation. Of course I do end up proposing one. The engineering ideas of the very efficient hypercar, if applied to the vehicles of public transportation, would multiply their benefits several-fold (in some calculations, by 10 times). In other words, energy efficient public transportation, in cities where public transportation worked—that is, where zoning encouraged mixed development and dense human settlement along transportation corridors, and where bus lines, light rail lines and subsidized taxi services put everyone within reach of public transportation—would save even more energy and materials than a fleet of very efficient cars, one for every 1.5 persons. In terms of public monies, sudsidizing public transportation is much cheaper than subsidizing private cars. Depending on how you calculate it, the current public subsidy for cars in the United States is several hundred billion dollars annually (including things like road resurfacing, drilling subsidies for oil companies, military operations to protect oilfields). And in general what consumers pay directly for public transporation (their personal outlay: this is a political decision) is less than the cost of purchasing and maintaining a car. It is probably close to the cost of maintaining a car (gas, insurance, tires, oil changes, brake jobs). Of course those freeways and car repair shops represent jobs; many, many jobs if one traces the chain back to the manufacturers of steel, aluminum, plastics and concrete, and forward to the car dealers, insurance agencies, motor vehicle bureaus and banks. Public transportation is also a source of jobs, with its own construction, financing and maintenance streams. (Cars now represent less of a boon to the American economy. After World War II cars were made in the United States of materials mined and processed there, and fueled with oil pumped from Texas and Oklahoma. Their manufacture supported a middle class. Now half or more new cars are made abroad, of foreign materials—a loss of $15,000 each to the American economy—and fueled with foreign oil.)

People like me are run over by events. While writers like me speculate about what should be done, things take their course. It now seems most probable that world oil production will peak early in the third millennium (2010? 2020?). If demand is sufficient, production will be maintained, but at steadily rising cost. When the cost becomes too great, production of oil will fall. If the oil geologist Hubbert is right, the great bulk of oil production will have occupied the century between 1950 and 2050. A rise in the real price of oil will reverse the trend of the previous 50 years. (That steady fall in the price of energy is what made Paul Ehrlich lose his bet to Julian Simon—essentially that as the population rose, the economic world, not just the biological one, would become poorer and poorer. Ehrlich bet that as the better deposits of metal ores were used up and poorer ones exploited, the price of the refined metal would rise. But thanks to lower and lower energy costs and advances in the refining process (and the substitution of other products for metals, which put downward pressure on their price), that didn't happen. In many ways, their bet—one betting on human ingenuity, one on the limits of the world—encompass the arguments in this book. Ehrlich shouldn't have bet on prices. But on what? The state of native fish stocks ? The amount of carbon dioxide in the atmosphere? Not on the price of fish in the store!) As the real price of fuel rises, we may not need a tax to make cars more efficient, or public transportation more desirable. We may still need government fuel standards; Europeans now pay double or triple what Americans do for gasoline, have for the most part excellent public transportation systems, and still fill their roads with not-very-efficient cars. (But cars that are twice as efficient as ours.)

The end of oil doesn't necessarily mean the end of our dependence on fossil fuels, since at present rates of consumption considerable natural gas and several hundred years of coal are left (or fewer—everything depends on the rate of economic growth), but it gives one pause for thought. Perhaps, as some have argued, it would be better to use oil as a more or less recyclable lubricant, and coal and natural gas as feedstocks for a non-polluting, also recyclable chemical industry. This won't happen soon. Perhaps some coal and oil should be left in place—because it is the right thing to do, and as a resource for future generations. Renewable energies are the only permanent sources of supply on earth, which is to say they will last us for the next 500 million years, the time left out of the earth's 4 billion years of evolutionary time by the growing radiance of the sun. People not fond of the possibilities of renewable energy consider nuclear fusion, which, though not yet worked out technically, is currently the most long term solution to the problem of obtaining large amounts of high density energy. (Renewable sources like solar cells are so called low density sources because they occupy a lot a space.) While supplies of uranium on earth are limited, nuclear fission could also carry us for some time (less and less the faster we grow, maybe two or three generations of reactors), especially if we gave up our squeamishness over plutonium and used fast breeder reactors to process new fuel from old (and along the way help solve the problem of radioactive waste): fast breeder reactors are 60 times more efficient than once through ones. In the case of fusion power, estimates vary, but the deuterium in the oceans should last several thousand years at present rates of energy consumption. (Uranium in the oceans is also extremely abundant but diffuse and not now economic to extract.) After deuterium, there is helium-3 from the moon. Our civilization's dependence on energy is greatly underestimated. People compare the Internet Revolution to the Industrial Revolution: but the Internet Revolution sits squarely amidst the warm houses and cheap energy supplies of the Industrial Revolution: one is a subset of the other.

Bringing the underdeveloped world to the living standards of the developed one is not easy. An inhabitant of North America currently is responsible for the emission of the equivalent of 24 tons of carbon dioxide a year, a European 11, an average earth inhabitant 5 tons. Emissions must fall to one ton per person by 2050 (unless population grows, then emissions must be less) to keep the temperature rise below 2° C. So to maintain a livable world (about a third of the air pollution in Los Angeles now comes from China) energy use in the West has to fall by over 90%, and continue (with population) on a downward trend. Energy use and population in the developing world must also fall. The climate will still warm, but less. Continuing growth in energy use or population will obliterate any advantages of energy efficiency. (Internet use now absorbs 10% of the U.S. electricity supply.) One of the advantages of photovoltaic power is that energy supply can be integrated with human habitation—that is, put on the roofs of houses, in back yards, over parking lots. Separate structures and more land development aren't needed. But if we want this new solar powered world, we have to grasp it, and not wait for the invisible hand of the market to give it to us. Jevon's paradox states that improvements in energy efficiency lead to the purchase of more energy using goods (and to more profits for manufacturers), and thus to greater energy use. As energy (effectively) gets cheaper, we use more of it. One way out of this is to make up the difference in the real cost of energy with taxes—so as mileage per gallon goes up, so does the price of gas, thus the cost per mile remains the same. An increase of 2-4% a year in the cost of fuel would drive energy efficiency and keep energy use down, steering us toward an economy based less on consumption.

———————————

A "normal" level of carbon dioxide in the atmosphere is probably 280 parts per million (the level in 1750). The level of carbon dioxide is now 385 parts per million and rising at 2% per year (that is, it will double to 700 ppm by 2045-2050). A disastrous rise in temperature and sea level will supposedly occur after a global warming of 4°C., within the generally accepted range of temperature predicted for 2100 (4°-9° C. if we don't control carbon emissions). However, the most recent time carbon dioxide levels were at 350 ppm, sea level was 80 feet higher, so we may already be there, so to speak, the sea and atmosphere just haven't responded yet. Both land and sea have great thermal inertia. As the atmosphere warms, the sea warms, but more slowly. Sea level rises partly from the thermal expansion of water in the oceans, partly from more water flowing in from melting glaciers. The ocean has bulges and hollows because of changing currents and jet stream winds and is affected by shifting gravitational pulls from collapsing ice sheets, so sea level rise will vary considerably from place to place.

Melting large glaciers like the Greenland ice sheet or the Antarctic glaciers takes time (millennia or centuries, one century for the Greenland ice sheet under the most calamitous of recent scenarios), so sea level rise beyond 6-10 feet by 2100 is unlikely but 80 feet is possible. A sea level rise of 10 feet would displace tens of millions of people (in Long Island, Florida, the Gulf Coast, Bangladesh, Southeast Asia, China, about the Mediterranean, Baltic and North seas, in the Rhine Delta). Higher seas push river floods back upstream, into areas that didn't flood before, and makes the rice fields in the deltas of the great south Asian rivers (the Ganges, the Mekong, the Irrawaddy, the Red, the Pearl) unusable. The fields will become brackish estuaries and produce shrimp and fish. Barrier islands will move inland toward the coast and coastal aquifers (such as the Magothy under Long Island) will become too salty to drink. Mountain glaciers, smaller and fed by yearly snows, melt more quickly than continental ones. Those in the Andes that water the high terraces of Peru (most cultivable land in Peru is over 9000 feet) are almost gone. When they are gone and ground water levels fall, many crops will no longer grow. The Himalayan glaciers that feed the great rivers of India, Pakistan, China and Southeast Asia, are also melting. Without them, spring floods will be greater and summer flows lower. Much land now irrigated by these rivers will no longer be cultivable. Two billion people depend on its crops. Since groundwaters in India and China are already overpumped, the only way to maintain agricultural production will be with older water harvesting techniques, such as the systems of hand dug ponds ("tanks") that once caught the rains flowing through the valleys of monsoon India. Such systems work but are limited by rainfall.

Complete collapse of the continental ice sheets will raise sea level 250 feet. No civilization could withstand such changes and the changes in climate that went along with them but this is a problem for the future. In the meantime, other things will happen in the ocean. Its rising acidity from the dissolved carbon dioxide will cause its fisheries to collapse, as the shell forming algae at the center of food webs die. (All commercial fish stocks are predicted to collapse from overfishing by 2048, so we may have caught the last fish just in time.) Coral reefs will melt away and animals with calcium carbonate shells (clams, oysters, mussels) will go extinct. The Gulf Stream will slow greatly or shut down, ending the circulation of oxygenated water to the deep sea and suffocating the animals of the depths. As the sea stagnates, it will become perfused with hydrogen sulfide, a toxic gas. The change in ocean currents and surface temperatures will change weather patterns and make many parts of the earth (the east coast of North America, much of Mexico and Central America, South America south of the Amazon, parts of Southeast Africa, much of Southeast Asia) uninhabitable from constant storms, floods and drought.

The land warms more quickly than the sea. Much of the land on earth is between 30° North and 30° South (that is, about the equator). Some of this is now desert, some tropical forest and savannah. As the climate warms these forests and grasslands will be replaced by desert (though some pockets of vegetation in favored locations may remain). Desert conditions will spread south and north, encompassing most of the United States, southern Europe up to the latitude of Paris, northern South America, most of Africa, India, Southeast Asia and all of China: most of the inhabited world. The boreal forests and tundra of North America and Eurasia will be replaced by mixed deciduous woodland and grassland. Most flowering plants and large animals, unable to migrate quickly enough, or their way blocked by human settlements, will go extinct. The habitable parts of the world, where large animals can live and crops grow, will consist of the boreal regions (an immense landscape, its Siberian section unfortunately contaminated by radioactivity from the Soviet nuclear program), the west coast of Greenland, Iceland, New Zealand, Tasmania, southern Patagonia, western Antarctica. Some writers imagine highrise cites amidst intensively cultivated stony Arctic soils.

What will happen to people? Most, in both undeveloped and developed parts of the world, will die, probably not catastrophically, but slowly, from starvation and despair, as death rates climb by 15-20%. This happened in Russia recently (with a lesser rise in the death rate) after the collapse of the Soviet Union, and is still happening there today. (The collapse of the Soviet system explains why Russian troops stationed far from their home bases must return in spring to plant, and in fall to harvest, their potatoes.) It probably happened with the collapse of the Maya and the Aztec civilizations in Mexico and Central America, and the Sumerians in Mesopotamia. Industrial civilization can maintain itself in a desert, desalinating seawater, growing crops in greenhouses cooled by seawater and watered by its sweet condensation, mining copper, pumping oil out of the sand, fueling itself largely with solar panels, living underground where daytime temperatures average 150° Fahrenheit. Would it? As the economic blows worsen, and food, water and electricity become scarce, I doubt whether the retreat from the present will be orderly. Farmers will not plant with native grasses the fields they abandon. For one thing, they will have no money to do so. People imagine an orderly retreat to the Arctic coasts (forget about national boundaries) but this ignores the difficulties of feeding large populations, purifying polluted surface water, maintaining the infrastructure necessary to build roads, power stations, vehicles, cement plants in the north. As the seas rise, the water will flood the containment ponds of abandoned nuclear power stations, where the spent fuel rods are stored, oil refineries with their stored oil and chemicals, private houses with their toxic cleaners and pesticides. This material will spread to river deltas and inshore waters. Public zoos and private animal shelters will release their animals rather than let them starve: lions, tigers, elephants, camels, yaks may once again populate North America. Tropical plants will escape from botanical gardens into the new tropical habitat. Over a long time (20,000-100,000 years?), levels of carbon dioxide in the atmosphere will fall, and the ocean, cleansed of manmade and natural toxins, its circulation restored, will return to something like normal, and after another million years or more, new adaptive radiations will fill it with new creatures.

Perhaps people will watch some of this, as they wander the corners of the deserts with palms and springs, carrying their bows, and digging tools scavenged from former habitations (much of them now under water), and the great savannahs and woods of the Arctic and Antarctic coasts.

Against the backdrop of planetary evolution, our concerns seem petty. Mammalian extinctions peak every 2.5 million years, when the earth's orbital geometries combine to produce a more strongly seasonal climate (harsh winters, hot dry summers), which makes survival more difficult. Every 100,000-2,000,000 years a stony asteroid 1-2 kilometers in diameter strikes the earth. Dust from the collision and soot from widespread fires dim the sun, slowing or halting photosynthesis and cooling the climate worldwide. About once every 50 million years a nearby supernova explosion bathes the earth in sufficient X-ray radiation to kill most vertebrates (the microbial world remains more or less undisturbed). The evolution of life itself causes difficulties. The creation of an oxygenated atmosphere 2.5 billion years ago by newly evolved photosynthetic bacteria caused the first great extinctions: oxygen was poisonous to most existing organisms. About 2.3 billion years ago the photosynthetic microbes' removal of carbon from the atmosphere cooled the earth sufficiently to cause a glaciation that lasted 100 million years: the first snowball earth. Similar cooling events have occurred twice more, with the evolution of multicellular plants and of land plants. The deep roots of land plants greatly increased the rate of weathering, locking up carbon dioxide as calcium carbonate, and so the evolution of forests 360 million years ago was followed by another long period of glaciation. And in 500 million years the increasing brightness of the sun will make the earth too hot to inhabit. As the increasing radiance of the sun evaporates the oceans, fierce winds will sweep their water into the stratosphere, from where it will be lost to space. One billion years from now increasing temperature will have made the earth lifeless.

———

This is all far away. The climatic change and extinctions that are occuring now are our doing alone.

Bibliography

This book depends almost entirely on secondary sources. It is not a work of original scholarship or a reasoned argument so much as a pastiche of examples and data that support the central place of nature in the human world. The stories that scientists tell change (some have changed while I have written this book); even data change. I have tried to be accurate and up-to-date. But in a decade someone may write a book that reaches similar conclusions with entirely different stories.

While I think footnotes are superfluous in a work of this sort, some parts of it are more dependent on others' work than other parts. The description of rivers and salmon derive a good deal from David Montgomery's *King of Fish* and Alice Outwater's *Water* (both excellent books), the descriptions of western forests from Arno and Fiedler's *Mimicking Nature's Fire*, the logging rules for western forests from the work of Jerry Franklin. Other sections derive from many works: that on buffalo from the *Buffalo Book, Buffalo Management and Marketing, Lives of Game Animals, Mammals of the Northern Great Plains, Groundwater Exploitation in the High Plains, Prehistoric Hunters of the High Plains, Water, The World's Water 2000-2001, The Ecological Indian: Myth and History, Bring Back the Buffalo, The Destruction of the Bison, Ogallala: Water for a Dry Land*; that on forests from *The Work of Nature, Defining Sustainable Forestry, The Hidden Forest, Toward Forest Sustainability, Natural Capitalism, California Forests and Woodlands, Land Use and Watersheds, Deforesting the Earth, Americans and their Forests, Mimicking Nature's Fire*. Certain works I admired for their perfection as works of art (*The Work of Nature* by Yvonne Baskin) as well as for the information they contained. Many examples, facts, and factoids came from *New Scientist*, 1995-2008, others from *Ecology* and *Ecological Applications* (especially 1995-1997), some from *Natural History* (for instance, the story about the difficulties faced by the gray jay in the southern parts of its range), a few from *The New York Review of Books* and *The New York Times* (most of whose relevant articles were better reported in *New Scientist*). Fernand Braudel's *Capitalism and Material*

512

Life 1400-1800 opened a door in my mind and I.G. Simmons *Changing the Face of the Earth: Culture, Environment, History* pushed me through it. Some of the books listed contributed a single word to my essay, some a sentence, but all of them improved my understanding of the place of man in nature.

Journals (1995-2008)
Ecology
Ecological Applications
Natural History
New Scientist
The New York Review of Books
The New York Times

Books

Andersen, Tom. 2002. *This Fine Piece of Water: An Environmental History of Long Island Sound*. Yale University Press: New Haven and London.
Anderson, M. Kat. 2005. *Tending the Wild: Native American knowledge and the management of California's resources*. University of California Press: Berkeley.
Arno, S. and C. Fiedler. 2005. *Mimicking Nature's Fire: restoring fire-prone forests in the West*. Island Press: Washington, D.C.
Askins, R. 2000. *Restoring North American Birds: lessons from landscape ecology*. Yale University Press: New Haven, Conn.
Ayers, E. 1999. *God's Last Offer*. Four Walls, Eight Windows: New York and London.
Barnes, B., D. Zak, S. Denton and S. Spurr. 1998. *Forest Ecology, 4th Ed.* John Wiley: New York.
Baron, David. 2004. *The Beast in the Garden*. W. W. Norton and Company: New York and London.
Bartram, W. 1996. *Travels and Other Writings*. Edited by T. P. Slaughter. The Library of America: New York.
Baskin, Yvonne. 1997. *The Work of Nature: how the diversity of life sustains us*. Island Press: Washington, D.C. and Covelo, CA.
———. 2002. *A Plague of Rats and Rubbervines: the growing threat of species invasions*. Island Press/Shearwater Books: Washington, D.C.

Benrus, J. 1997. *Biomimicry: innovation inspired by nature*. William Morrow: New York.

Bradley, J. 1987. *Evolution of the Onondaga Iroquois: accomodating change, 1500-1665*. Syracuse University Press: Syracuse, NY.

Braudel, F. 1967. *Capitalism and Material Life 1400-1800*. Harper and Row: New York/ Evanston/ San Francisco/ London.

———. 1982. *The Wheels of Commerce: Civilization and Capitalism 15th-18th Century*. Harper and Row: New York/ Cambridge/ Philadelphia/ San Francisco/ London/ Mexico City/ Sao Paulo/ Sydney.

———. 1982. *The Perspective of the World: Civilization and Capitalism 15th-18th Century*. Harper and Row: New York/ Cambridge/ Philadelphia/ San Francisco/ London/ Mexico City/ Sao Paulo/ Sydney.

Brown, Nancy Marie. 2007. *The Far Traveller: Voyages of a Viking Woman*. Harcourt, Inc.: Orlando/Austin/New York/San Diego/London.

Bush, M. 1999. *Ecology of a Changing Planet, 2nd Ed.* Prentice Hall: New York.

Cairns, J., K. Dickson, and E. Herrick, ed. 1977. *Recovery and Restoration of Damaged Ecosystems: proceedings of the International Symposium on the Recovery of Damaged Ecosystems held at Virginia Polytechnic Institute and State University, Blacksburg, Virginia, on March 23-25, 1975*. University of Virginia Press: Charlottesville, VA.

Callenbach, E. 1996. *Bring Back the Buffalo: a sustainable future for America's Great Plains*. Island Press: Washington, D.C.

Calow, P. and G. Petts. 1992-1994. *The Rivers Handbook: hydrological and ecological principles: in two volumes*. Blackwell Scientific Publications: Oxford and Boston.

Carling, P. and G. Petts, ed. 1992. *Lowland Floodplain Rivers: geomorphological perspectives*. Wiley and Sons: Chichester and New York.

Carrol, C. 1973. *The Timber Economy of Puritan New England*. Brown University Press: Providence.

Castetter, E. and W. Bell. 1942. *Pima and Papago Indian Agriculture*. University of New Mexico Press: Albuquerque, New Mexico.

Cayton, A. 1999. *Frontier Indiana*. Indiana University Press: Bloomington and Indianopolis.

Changnon, S., ed. 1996. *The Great Flood of 1993*. Westview Press/Harper Collins: Boulder, CO and Oxford, England.

514

Cioc, M. 2002. *The Rhine: an eco-biography, 1815-2000.* University of Washington Press: Seattle, WA.

Clark, J. 1996. *Coastal Zone Management Handbook.* Lewis Publishers: Boca Raton/ New York/ London/ Tokyo.

Clover, Charles. 2006. *The End of the Line: how overfishing is changing the world and what we eat.* The New Press: New York and London.

Collman, James. 2001. *Naturally Dangerous: surprising facts about food, health and the environment.* University Science Books: Sausalito, CA.

Cooke, G. Dennis (et al.). 1993. *Restoration and Management of Lakes and Reservoirs, 2nd ed.* Lewis Publishers: Boca Raton, FL.

Coon, C. 1971. *The Hunting Peoples.* Little Brown: Boston.

Cronk, J. and M. Fennessy. 2001. *Wetland Plants: biology and ecology.* Lewis Publishers: Boca Raton, FL.

Cronin, W. 1983. *Changes in the Land: Indians, colonists, and the ecology of New England.* Hill and Wang/Farrar, Straus and Giroux: New York.

Crosby, A. 1986. *Ecological Imperialism: the biological expansion of Europe, 900-1900.* Cambridge University Press: Cambridge and New York.

Curtin, P., G. Brush, G. Fisher, ed. 2001. *Discovering the Chesapeake: the history of an ecosystem.* Johns Hopkins University Press: Baltimore and London.

Daily, G., and K. Ellison. 2002. *The New Economy of Nature: the quest to make conservation profitable.* Island Press/Shearwater Books: Washington/Covelo/London.

Dean, C . 1999. *Against the Tide: the battle for America's beaches.* Columbia University Press: New York.

Deffeyes, K. 2001. *Hubbert's Peak: the impending world oil shortage.* Princeton University Press: Princeton, New Jersey.

Dempsey, Dave. 2004. *On the Brink: the Great Lakes in the 21st century.* Michigan State University Press: East Lansing.

Diamond, J. 1997. *Guns, Germs, and Steel.* W. W. Norton: New York and London.

Diamond, A. and F. Filian, ed. 1987. *The Value of Birds: based on the proceedings of a symposium and workshop held at the XIX World Conference of the International Council for Bird Preservation, June 1986, Kingston, Ontario.* ICBP technical publication no. 6: Cambridge, England.

Dilsaver, L. and C. Colten, ed. 1992. *The American Environment: interpretation of past geographies.* Rowman and Littlefield: Lanham, MD.

Doolittle, W. 2000. *Cultivated Landscapes of Native North America*. Oxford University Press: New York.

Duncan, T. 1972. *Atlantic Islands: Madeira, the Azores and the Cape Verdes in seventeenth century commerce and navigation*. University of Chicago Press: Chicago and London.

Eden, M. 1990. *Ecology and Land Management in Amazonia*. Belhaven Press: London.

Eisenberg, E. 1998. *The Ecology of Eden*. Alfred A. Knoph: New York.

Erwin, Douglas H. 2006. *How Life on Earth Nearly Ended 250 Million Years Ago*. Princeton University Press: Princeton and Oxford.

Etherington, J.1982. *Environment and Plant Ecology, 2nd Edition*. Wiley: Chichester and New York.

Fagan, Brian. 2004. *The Long Summer: how climate changed civilization*. Basic Books: New York.

Flannery, T. 2001. *The Eternal Frontier*. Atlantic Monthly Press: New York.
———. 2005. *The Weather Makers*. Atlantic Monthly Press: New York.

Forman, R. 2003. *Road Ecology: science and solutions*. Island Press: Washington, D.C. and Covelo, CA.

Fradkin, M. 1981. *A River No More*. Alfred A. Knopf: New York.

Frison, G. 1991. *Prehistoric Hunters of the High Plains, 2nd Edition*. Academic Press: San Diego.

Galbraith, J. 1994. *A Journey Through Economic Time*. Houghton Mifflin: New York.

Geertz, C. 1963. *Agricultural Involution: the process of ecological change in Indonesia*. University of California Press: Berkeley.

George, J. 1998. *Everglades Wildguide: the natural history of Everglades National Park, Florida*. U.S. Department of the Interior: Washington, D.C.

Gimpel, J. 1976. *The Medieval Machine: the industrial revolution of the Middle Ages*. Holt, Rinehart and Winston: New York.

Gleick, P. 2000. *The World's Water 2000-2001: the biennial report on freshwater resources*. Island Press: Washington, D.C.

Le Goff, J. 2005. *The Birth of Europe*. Blackwell: Malden, MA.

Gosnell, Mariana. 2005. *Ice: the nature, the history and the uses of an astonishing substance*. Knopf: New York.

Graf, W. 1985. *The Colorado River: instability and basin management*. Association of American Geographers: Washington, D.C.

Gremillion, K., ed. 1997. *People, Plants and Landscapes: studies in paleoethnobotany.* University of Alabama Press: Tuscaloosa.

Grove, A. T. and Oliver Rackham. 2001. *The Nature of Mediterranean Europe: an ecological history.* Yale University Press: New Haven.

Grzimek, B. 1990. *Grzimek's Encyclopedia of Mammals.* McGraw-Hill: New York.

Guthrie, R. Dale. 1990. *Frozen Fauna of the Mammoth Steppe: the story of Blue Babe.* University of Chicago Press: Chicago.

Hammer, D., ed. 1989. *Constructed Wetlands for Wastewater Treatment: municipal, industrial and agricultural.* Lewis Publishers: Chelsea, Michigan.

Hardin, B. 1996. *A River Lost.* W. W. Norton: New York.

Hardin, G. 1993. *Living Within Limits: ecology, economics, and population taboos.* Oxford University Press: New York.

Herrick, J. W. and D. Snow. 1995. *Iroquois Medical Botany.* Syracuse University Press: Syracuse, NY.

Hershkowitz, A. 2002. *Bronx Ecology: blueprint for a new environmentalism.* Island Press: Washington, D.C.

Hinrichsen, D. 1998. *Coastal Waters of the World.* Island Press: Washington, D.C.

Hollander, Jack M. 2003. *The Real Environmental Crisis.* University of California Press: Berkeley and Los Angeles.

Hudson, J. 1994. *Making the Corn Belt.* Indiana University Press: Bloomington.

Huber, P. 1999. *Hard Green: saving the environment from environmentalists.* Basic Books: New York.

Hunt, R. L. 1993. *Trout Stream Therapy.* University of Wisconsin: Madison.

Isenberg, A. C. 2000. *The Destruction of the Bison: an environmental history, 1750-1920.* Cambridge University Press: New York.

Jackson, Dana and Laura Jackson, ed. 2002. *The Farm as Natural Habitat: reconnecting food systems with ecosystems.* Island Press: Washington, D.C.

Jackson, K. 1985. *Crabgrass Frontier: the suburbanization of the United States.* Oxford University Press: New York.

Jackson, Tim. 2009. *Prosperity Without Growth: economics for a finite planet.* Earthscan: London and Sterling, VA.

Jenish, D'Arcy. 2004. *Epic Wanderer: David Thompson and the mapping of the Canadian West.* University of Nebraska Press: Lincoln.

Jennings, D. and J. Helbring. 1983. *Buffalo Management and Marketing.* National Buffalo Association: Custer, SD.

Jones, J., D. Armstrong, R. Hoffman, and C. Jones. 1983. *Mammals of the Northern Great Plains.* University of Nebraska Press: Lincoln and London.

Keller, R. and M. Turek. 1998. *American Indians and National Parks.* University of Arizona Press: Tucson.

Kent, D., ed. 1994. *Applied Wetlands Science and Technology.* Lewis Publishers: Boca Raton/ Ann Arbor/ London/ Tokyo.

Kohm, K. and J. Franklin, ed. 1997. *Creating a Forestry for the 21st Century: the science of ecosystem management.* Island Press: Washington, D.C. and Covelo, CA.

Krech, Shepard III. 1999. *The Ecological Indian: myth and history.* W.W. Norton and Co.: New York and London.

Kromm, D. and S. White, ed. 1992. *Groundwater Exploitation in the High Plains.* University of Kansas Press: Lawrence, Kansas.

Kurlansky, Mark. 2006. *The Big Oyster.* Ballantine Books: New York.

Kusher, J. and M. Kentula. 1970. *Wetland Creation and Restoration: the status of the science.* Island Press: Washington, D.C.

Lamb. H.H. 1995. *Climate History and the Modern World, 2nd Edition.* Routledge: London and New York.

Leal, D. and R. Meiners. 2002. *Government vs. Environment.* Rowman and Littlefield: Lanham, MD.

Lenik, Edward J. 2002. *Picture Rocks: American Indian rock art in the Northeast woodlands.* University Press of New England: Hanover.

Lerner, S. 1997. *Eco-Pioneers.* MIT Press: Cambridge and London.

Lindenmayer, D. and J. Franklin. 2003. *Towards Forest Sustainability.* Island Press: Washington, D.C.

Loewen, J. 1999. *Lies Across America: what our historic sites get wrong.* New Press, distributed by W.W. Norton: New York.

Lopez, Barry. 1978. *Of Wolves and Men.* Charles Scribner's Sons: New York.

Luoma, J. 1999. *The Hidden Forest.* Henry Holt: New York.

MacKay, David. 2009. Sustainable Energy—without the hot air. UIT Cambridge Ltd.: Cambridge, England.

Madrick, Jeff. 2002. *Why Economies Grow: the forces that shape prosperity and how we can get them working again.* Basic Books: New York.

Manley, T. O. and P. L. Manley. 1999. *Lake Champlain in Transition: from research toward restoration.* American Geophysical Union: Washington, D.C.

Mann, Charles C. 2005. *1491.* Alfred A. Knopf: New York.

Manning, R. 1995. *Grassland: the history, biology, politics, and promise of the American prairie.* Viking: New York.

Marcus, J. and K. Flannery. 1996. *Zapotec Civilization: how urban society evolved in Mexico's Oaxaca Valley.* Thames and Hudson: London.

Marcus, R. and E. Fernald. 1975. *Florida: a geographical approach.* Kendall/ Hunt Publishing Company; Dubuque, IO.

McDaniel, Carl N. and John M. Gowdy. 2000. *Paradise for Sale: a parable of nature.* University of California Press: Berkeley and Los Angeles.

McEvoy, Thom. 2004. *Positive Impact Forestry: a sustainable approach to managing woodlands.* Island Press: Washington, D.C.

Melville, E. 1994. *A Plague of Sheep: environmental consequences of the conquest of Mexico.* Cambridge University Press: Cambridge/ New York/ Melbourne.

Mendelsohn, R . 2001. *Global Warming and the American Economy: a regional assessment of climate change impacts.* Edward Elgar: Cheltenham, U.K. and Northhampton, MA.

Minnis, P.E. 2000. *Ethnobotany: a reader.* University of Oklahoma Press: Norman.

Miyazaki, N., Z. Adeel, and K. Ohwada. 2005. *Mankind and the Oceans.* United Nations University Press: Tokyo.

Montgomery, David R. 2007. *Dirt: the erosion of civilization.* University of California Press: Berkeley and London.

———. 2003. *King of Fish: the thousand year run of salmon.* Westview Press/ Perseus Books Group: Cambridge, MA.

Nabhan, G. 1997. *The Culture of Habitat.* Counterpoint: Washington, D.C.

———. 1989. *Enduring Seeds: Native American agriculture and wild plant conservation.* North Point Press: San Francisco.

———.1982. *The Desert Smells Like Rain: a naturalist in Papago Indian country.* North Point Press/Farrar, Straus and Giroux: New York.

Naiman, R. J. and H. Decamps. 1990. *Ecology and Management of Aquatic-Terrestrial Ecotones.* UNESCO: Paris, France; and Parthenon Publishing Group: Camforte, UK and Park Ridge, NJ.

Oglesby, R., C. Carlson and J. McCann, editors. 1971. *River Ecology and Man.* Academic Press: New York and London.

de Onis, J. 1992. *The Green Cathedral: sustainable development of Amazonia.* Oxford University Press: New York and Oxford.

Opie, J. 1993. *Ogallala: water for a dry land.* University of Nebraska Press: Lincoln.

Outwater, A. 1996. *Water: a natural history.* Basic Books: New York.

Pace, R. 1998. *The Struggle for Amazon Town: Gurupa revisited.* Lynne Rienner: Boulder and London.

Patrick, R., E. Ford and J. Quarles. 1987. *Groundwater Contamination in the United States, 2nd Edition.* University of Pennsylvania Press: Philadelphia.

Pauly, D. and J. L. Maclean. 2003. *In a Perfect Ocean: the state of fisheries and ecosystems in the North Atlantic Ocean.* Island Press: Washington, D.C.

Pearce, Fred. 2006. *When Rivers Run Dry.* Beacon Press: Boston.

Pereira, H. 1973. *Land Use and Water Resources in Temperate and Tropical Climates.* Cambridge University Press: Cambridge.

Pielou, E. C. 1998. *Freshwater.* University of Chicago Press: Chicago and London.

Pimm, Stuart L. 1991. *The Balance of Nature?.* University of Chicago Press: Chicago and London.

———. 2001. *The World According to Pimm: a scientist audits the earth.* McGraw-Hill: New York, Chicago, *et al.*

Platt, R. H., R. Rowntree, and P. Muick. 1994. *The Ecological City: preserving and restoring urban biodiversity.* University of Massachusetts Press: Amherst, MA.

Pollan, Michael. 2006. *The Omnivore's Dilemma: a natural history of four weeks.* Penguin Press: New York.

Ponting, C. 1992. *A Green History of the World.* St. Martin's Press, New York.

Powell, R. 1982. *The Fisher: life history, ecology, and behavior.* University of Minnesota Press: Minneapolis.

Pretty, J. 1995. *Regenerating Agriculture: Policies and Practice for Sustainability and Self-Reliance.* London International Institute for Environment and Development/A Joseph Henry Press Book: London.

Prince, H. 1997. *Wetlands of the American Midwest.* University of Chicago Press: Chicago and London.

Pruter, A. T. and D. L. Alverson. 1972. *The Columbia River Estuary and Adjacent Ocean Waters.* University of Washington Press: Seattle.

Reader, J. 1999. *Africa: a biography of the continent.* Vintage Books/Random House: New York.

Reisner, M. 1993. *Cadillac Desert.* Penguin Books: New York.

———. 1991. *Game Wars: the undercover pursuit of wildlife poachers.* Penguin Books: New York.

Richardson, B. 1994. *People of Terra Nullius: betrayal and rebirth in aboriginal Canada.* Douglas and McIntyre: Vancouver and Toronto; and University of Washington Press: Seattle.

Robbins, R. 1999. *Global Problems and the Culture of Capitalism.* Allyn and Bacon: Needham Heights, MA.

Roberts, Callum. 2007. *The Unnatural History of the Sea.* Island Press/Shearwater Books: Washington/Covelo/London.

Rose, Gene. 2000. *San Joaquin: a river betrayed.* Word Dancer Press: Clovis, CA.

Rosenzweig, M. 2003. *Win-Win Ecology: how the earth's species can survive in the midst of human enterprise.* Oxford University Press: Oxford and New York.

Royte, Elizabeth. 2005. *Garbage Land.* Little Brown: New York.

Ruddiman, William. 2005. *Plows, Plagues and Petroleum: how humans took control of climate.* Princeton University Press: Princeton, NJ.

Russell, E. 1997. *People and the Land through Time.* Yale University Press: New Haven and London.

Sacks, Oliver and Jesse Cohen, ed. 2003. *The Best American Science Writing 2003.* Ecco Press: New York.

Satterlind, D. 1972. *Wildland Watershed Management.* The Ronald Press: New York.

Sawin, J. 2004. *Mainstreaming Renewable Energy in the 21st Century.* Worldwatch Paper 169. Worldwatch Institute: Washington, D.C.

Scarpino, P. 1985. *Great River: an environmental history of the upper Mississippi.* University of Missouri Press: Columbia.

Schlesinger, W. 1997. *Biogeochemistry: an analysis of global change, 2nd Edition.* Academic Press: San Diego/ London/ Boston/ New York/ Sidney/ Tokyo/ Toronto.

Schneider, Paul. 2006. *Brutal Journey: the epic story of the first crossing of North America.* Henry Holt: New York.

Schumm, S. A. 2003. *The Fluvial System.* Wiley: New York.

Schwartz, M. 1997. *A History of Dogs in the Early Americas*. Yale University Press: New Haven.

Scott, W. B. and E. J. Crossman. 1973. *Freshwater Fishes of Canada*. Fisheries Research Board of Canada: Ottawa.

Seton, E. 1953. *Lives of Game Animals*. C. T. Branford: Boston.

Shea, W. R. 1994. *Energy Needs in the Year 2000: ethical and environmental perspectives*. Watson Publications International: Canton, MA.

Sides, H. 2006. *Blood and Thunder*. Doubleday: New York/ London/ Sydney/ Auckland.

Simmons, I. G. 2008. *Global Environmental History*. University of Chicago Press: Chicago and London.

———. 1989. *Changing the Face of the Earth*. Blackwell: Oxford, UK, and Cambridge, MA.

———. *The Ecology of Natural Resources*. 1974. Halsted Press: New York.

Smil, V. 2008. *Global Catastrophes and Trends: the next fifty years*. MIT Press: Cambridge, MA, and London, England.

———. 2003. *Energy at the Crossroads: global perspectives and uncertainties*. MIT Press: Cambridge, MA.

———. 2002. *The Earth's Biosphere*. MIT Press: Cambridge, MA, and London, England.

———. 2000. *Feeding the World: a challenge for the twenty-first century*. MIT Press: Cambridge, MA.

———. 1997. *Cycles of Life: civilization and the biosphere*. Scientific American Library: New York.

Smith, N. 1999. *The Amazon River Forest: a natural history of plants, animals, and people*. Oxford University Press: New York.

Smith, L., R. Pederson, and R. Kaminski. 1989. *Habitat Management for Migrating and Wintering Waterfowl in North America*. Texas Tech University Press: Lubbock, TX.

Souder, W. 2000. *A Plague of Frogs: the horrifying true story*. Hyperion: New York.

Stager, Curt. 1999. *Field Notes from the Northern Forest*. Syracuse University Press: Syracuse, NY.

Steingraber, S. 1997. *Living Downstream: an ecologist looks at cancer and the environment*. A Merloyd Lawrence Book/ Addison-Wesley: Reading, MA.

Stilgoe, J. 1982. *Common Landscape of America, 1580 to 1845.* Yale University Press: New Haven.

Stuart, David E. 2000. *Anasazi America.* University of New Mexico Press: Albuquerque.

Taber, Richard and Neil Payne. 2003. *Wildlife Conservation and Human Welfare.* Krieger Publishing Company: Malabar, Florida.

Tennant, Alan. 2004. *On the Wing: to the edge of the earth with the peregrine falcon.* Alfred A. Knopf: New York.

Terborgh, J. 1989. *Where Have all the Birds Gone?* Princeton University Press: Princeton, NJ.

Thirsk, J. 2000. *The English Rural Landscape.* Oxford University Press: Oxford and New York.

Thomas, J. and D. Toweill. 1982. *The Elk of North America.* Stackpole: Harrisburg, PA.

Vaillant, John. 2005. *The Golden Spruce: a true story of myth, madness, and greed.* A. A. Knopf Canada: Toronto.

Van Driesche, J. 2000. *Nature out of Place: biological invasions in the global age.* Island Press: Washington, DC.

Volk, T. 1998. *Gaia's Body: toward a physiology of earth.* Copernicus/Springer-Verlag: New York

Weeks, W. 1997. *Beyond the Ark.* Island Press: Washington, D.C. and Covelo, CA.

Weidensaul, Scott. 1999. *Living on the Wind: across the Hemisphere with migratory birds.* North Point Press/Farrar, Straus, and Giroux: New York.

Weisman, Alan. 2007. *The World Without Us.* A Thomas Dunne Book/ St. Martin's Press: New York

Weller, M. 1987. *Freshwater Marshes: ecology and wildlife management.* University of Minnesota Press: Minneapolis.

Whitney, G. 1994. *From Coastal Wilderness to Fruited Plain.* Cambridge University Press, Cambridge.

Whyte, I. 1995. *Climate Change and Human Society.* Arnold: London and New York.

Wigmosta, M. and S. Burges. 2001. *Land Use and Watersheds: human influence on hydrology and geomorphology in urban and forest areas.* American Geophysical Union: Washington, D.C.

Wild, Anthony. 2004. *Coffee: a dark history.* W. W. Norton: New York.

www.ingramcontent.com/pod-product-compliance
Lightning Source LLC
Chambersburg PA
CBHW031810170526
45157CB00001B/27